Nonlinear Structural Engineering

Demeter G. Fertis

Nonlinear Structural Engineering

With Unique Theories and Methods to Solve Effectively Complex Nonlinear Problems

With 119 Figures and 61 Tables

Springer

Demeter G. Fertis
186 Court Drive,
Suite 301,
Fairlawn,
Ohio 44333, USA
e-mail: dgfertis@msn.com or dgfertis@hotmail.com

Library of Congress Control Number: 2006923692

ISBN-10 3-540-32975-7 Springer Berlin Heidelberg New York
ISBN-13 978-3-540-32975-6 Springer Berlin Heidelberg New York

Springer is a part of Springer Science+Business Media.

springer.com

Printed in Germany

Typesetting by SPi using a Springer LaTeX macro package
Cover design: Estudino Calamar, Viladasens

Printed on acid-free paper SPIN 11672029 62/3100/SPI 5 4 3 2 1 0

About the author

Demeter G. Fertis is professor of civil engineering at the University of Akron. He was previously an associate professor at the University of Iowa and at Wayne State University, and a research engineer in the Michigan State Department of Transportation. Dr. Fertis was also visiting professor at the National Technical University in Athens, Greece. During his teaching career, he taught more than twenty graduate courses on different subjects, in the Civil, Mechanical, and Engineering Mechanics Departments. He has received a BS degree (1952) in civil engineering and urban planning and an MS degree (1955) in civil engineering from Michigan State University, East Lansing, and a Doctor of Engineering degree (1964) from the National Technical University of Athens, Greece. Dr. Fertis has consulted for NASA, Ford Motor Company, the Atomic Power Development Associates, General Motors, Boeing Aircraft Company, Lockheed California Company, Goodyear Aerospace, and the Department of the Navy. He has developed patents, which received international attention and used by professional engineers. He is the author of many books, published by major publishers, and numerous articles in professional journals and proceedings. A member of ASCE, ASME, the American Academy of Mechanics, Who's Who in America, Who's Who in the World, and many other organizations, Professor Fertis has served as a professional journal editor and as a member of many national and international technical committees.

Dr. Fertis is now an Emeritus Professor at the University of Akron and devotes his time doing scholarly research, writing scholarly books, giving lectures, and advancing technology by developing new methodologies for the solution of complicated engineering problems.

For more information about the author see his website
http://www.demetergfertis.com

No problem in nonlinear engineering is simple no matter how small or how unimportant it may appear to be. Its nonlinear behavior is the one that makes this problem complex and you need to have a complete understanding of this nonlinear behavior in order to provide a reasonable solution; and **this is the purpose of this book.**

The Author

Preface

The practicing design engineer, who deals with the design of structures and structural and mechanical components in general, is often confronted with nonlinear problems and he/she needs to develop a design procedure that deals effectively with such types of problems. Flexible members, structures subjected to blast and earthquake, suspension bridges, aircraft structural elements, and so on, are only a few examples where understanding of their nonlinear behavior is extremely important for an adequate and safe design.

Many of our nonlinear structures are composed of beam elements that can be taken apart from the structure, and their behavior can be studied by satisfying appropriate boundary conditions. Once we are in a position to understand completely the behavior of the nonlinear beam problem, we can then expand our knowledge effectively so that it includes a complete understanding of the nonlinear behavior of two-dimensional and three-dimensional structures and structural components.

In part, the purpose of this book is to concentrate its efforts on the nonlinear static and dynamic analysis of structural beam components that are widely used in everyday engineering applications. The analysis and design of the beam component can become very complicated when it is subjected to a large deformation, or when its material is permitted to be stressed well beyond its elastic limit and all the way to failure. The problem becomes even more complicated when the cross-sectional geometry of the member changes along its length, or when the modulus of elasticity of its material varies along its length. Therefore, such beam problems deserve special consideration. The book also includes a reasonable treatment regarding the nonlinear analysis of inelastic plates, suspension bridges and their failures, multistory buildings subjected to strong earthquakes, as well as many other interesting nonlinear problems, such as thick cylinders, inelastic torsion, inelastic vibrations, inelastic analysis of flexible members, and so on.

The future of engineering is becoming increasingly nonlinear, and both the engineering student and the practicing engineer should be prepared for it. The material included in this book is carefully selected in order to provide a good start in understanding and comprehending important aspects of nonlinear analysis. It is written in such a way that it can be conveniently used as a textbook in courses on nonlinear structural analysis, nonlinear analysis of mechanical components, and so on, so that the engineering student can start preparing himself/herself for the very challenging nonlinear problems of the present and future. Students in their senior year, or first-year graduates, should start to acquire and fully comprehend such knowledge in order to prepare themselves for the rugged road lying ahead.

Both the engineering student and the practicing engineer can also use this book as a self-study book, because each subject in the book is explained in detail with many examples and illustrations. The content of the book reflects the many years of experience and research of the author and his collaborators, and unique methods and theories based on exact mathematical derivations and models, have been developed to deal conveniently and effectively with many challenging problems in nonlinear structural engineering and mechanics.

These theories and methods are general, and they can be used successfully in many areas of engineering, such as civil, mechanical, aeronautical, structural design, computer science, space technology, mechanics, automobile engineering, manufacturing, processing and production engineering, as well as in other related engineering, physics, and applied mathematics fields.

The first chapter of the book deals with basic theories and principles associated with the nonlinear deformations of flexible members. It discusses the state of the art of such problems, the associated difficulties, the elastica theory, the Euler–Bernoulli nonlinear differential equation, the dependence of the moment and stiffness of the flexible member on the geometry of the deformation, and many other related subjects. It also discusses the general theory of equivalent systems for both linear and nonlinear deformations. The development of this theory and method is based on exact mathematical derivations in the form of equivalent linear and pseudolinear systems that simplify many of the complex problems in the nonlinear analysis of structural components. The method and theory are general, and the structural or mechanical component can have any variation in moment of inertia and modulus of elasticity along its entire length, as well as complex loading conditions, and still be convenient to solve.

In this manner, the initially complex nonlinear problem is converted into a much simpler pseudolinear one, which permits us to solve it by using well-known linear methods of analysis, including the finite element method. In the finite element method, the initially nonlinear element is converted into a mathematically derived pseudolinear element, which makes the application of the method to nonlinear problems more effective and accurate. The method and theory are general, and the structural and mechanical components are permitted to have any variation in moment of inertia and modulus of elasticity along their entire length, and be subjected to complicated loading conditions. Such variations are easily incorporated in the solution of the nonlinear problem.

The second and third chapters provide solution methodologies for various types of uniform and variable stiffness flexible beam problems with a large variety of loading conditions. The solution methodologies of such complex problems are made convenient, and they are based on the exact mathematical solution of the problem. The mathematical simplifications that are introduced are very convenient and very accurate, so that the student, or the practicing engineer, will be able to develop complete understanding of the given problem and draw reliable and useful engineering design conclusions. Each examined case is thoroughly explained and analyzed with sufficient number of examples to illustrate both theory and application.

The fourth chapter deals with the inelastic analysis of structural or mechanical components. This is an extremely important topic because many structures, such as the ones subjected to blast and earthquake, for example, or structures where weight control is important, are often permitted to be stressed well beyond the elastic limit of their material. Under such loading conditions, it becomes extremely important for the design engineer to know how a structure reacts when its material is stressed beyond the elastic limit and all the way to failure, so that he/she will be able to provide a safe design.

When a structural component enters the inelastic range, the modulus of elasticity along its length becomes variable and it must be taken into consideration in the analysis. The method of the equivalent systems, in combination with Timoshenko's reduced modulus concept, has been used effectively and reliably in this chapter to deal with this problem. Reliable methodologies have been developed that examine the deformation and stress characteristics of such loading conditions and also provide reliable methodologies for the computation of reliable ultimate loads.

The fifth chapter deals with the vibration analysis of flexible structural or mechanical components. The purpose of this chapter is to derive the general nonlinear differential equations of motion regarding the free vibration of flexible members of uniform and/or variable cross section along their length. Unique solution methodologies are developed that simplify the solution of very complex problems. Although the derived nonlinear differential equation applies to both large and small vibrational amplitudes, the detailed analysis, however, is concentrated on small oscillation vibration superimposed on the large nonlinear static displacements, which define the static equilibrium position of the flexible member.

The mathematically derived dynamically equivalent system is exact, linearized, and known methods of linear vibration analysis can be used to solve it. Galerkin's finite element method and the energy method of Lord Rayleigh are used in this chapter for this purpose.

The sixth and final chapter deals, in part, with fundamental nonlinear aspects of major suspension bridges, such as the Tacoma Narrows suspension bridge, and the Rion-Antirion, the longest cable-stayed suspension bridge in the world. Important suspension bridge catastrophic failures caused by nonlinear phenomena, as well as other catastrophic engineering failures, are included in the discussion, in order to illustrate the importance of thoroughly understanding the structural and mechanical nonlinear behavior. Many engineers in the past have learned the hard way from their structural failures, and we hope that from now such failures can be prevented.

The sixth chapter includes also other important subjects such as the inelastic analysis of thin plates, the inelastic earthquake response of multistory buildings, eccentrically loaded columns, the inelastic analysis of members with axial restraints, elastic and inelastic analysis of thick cylinders, inelastic torsion, inelastic analysis of flexible members, inelastic vibration, and other topics.

I wish to express my thanks and appreciation to my collaborators and my students for their kind support over the past decades and for the endless discussions on the various subjects. My special thanks and gratitude go to my wife, Anna, for her valuable suggestions during the writing of this text, her constant encouragement, and for the electronic form of the figures and the manuscript. I also wish to thank Springer for making my work available to the academic and professional audiences. In particular I wish to thank Dr. Dieter Merkle, Editor Director (Europe), Engineering, Dr. Christoph Baumann, Engineering Editor, and the editorial and production teams for their fine handling regarding the preparation of the manuscript.

Akron, Ohio Demeter G. Fertis

Contents

1 **Basic Theories and Principles of Nonlinear Beam Deformations** 1
1.1 Introduction 1
1.2 Brief Historical Developments Regarding the Static and the Dynamic Analysis of Flexible members 1
1.3 The Euler–Bernoulli Law of Linear and Nonlinear Deformations for Structural Members 8
1.4 Integration of the Euler–Bernoulli Nonlinear Differential Equation 10
1.5 Simpson's One-Third Rule 14
1.6 The Elastica Theory 16
1.7 Moment and Stiffness Dependence on the Geometry of the Deformation of Flexible Members 22
1.8 General Theory of the Equivalent Systems for Linear and Nonlinear Deformations 29
1.8.1 Nonlinear Theory of the Equivalent Systems: Derivation of Pseudolinear Equivalent Systems 30
1.8.2 Nonlinear Theory of the Equivalent Systems: Derivation of Simplified Nonlinear Equivalent Systems 40
1.8.3 Linear Theory of the Equivalent Systems 44

2 **Solution Methodologies for Uniform Flexible Beams** 63
2.1 Introduction 63
2.2 Pseudolinear Analysis for Uniform Flexible Cantilever Beams Loaded with Uniformly Distributed Loading Throughout their Length 64
2.3 Pseudolinear Analysis for Uniform Simply Supported Beams Loaded with a Uniformly Distributed Loading Throughout their Length 71
2.4 Flexible Uniform Simply Supported Beam Loaded with a Vertical Concentrated Load 76

2.5 Uniform Statically indeterminate Single Span Flexible Beam Loaded with a Uniformly Distributed Load w_o on its Entire Span ... 82
2.6 Uniform Statically Indeterminate Single Span Flexible Beam Subjected to a Vertical Concentrated Load ... 86
2.7 Flexible Uniform Cantilever Beam Under Combined Loading Conditions ... 90
2.8 Flexible Uniform Cantilever Beam Under Complex Loading Conditions ... 96
2.8.1 Application of Equivalent Pseudolinear Systems ... 96
2.8.2 Deriving Simpler Nonlinear Equivalent Systems ... 101

3 Solution Methodologies for Variable Stiffness Flexible Beams ... 105
3.1 Introduction ... 105
3.2 Flexible Tapered Cantilever Beam with a Concentrated Vertical Load at its Free End ... 105
3.3 Doubly Tapered Flexible Cantilever Beam Subjected to a Uniformly Distributed Loading ... 113
3.4 Solution of the Problem in the Preceding Section by Using a Simplified Nonlinear Equivalent System ... 119
3.5 Flexible Tapered Simply Supported Beam with Uniform Load ... 121
3.6 Flexible Tapered Simply Supported Beam Carrying a Trapezoidal Load ... 125
3.7 Using an Alternate Approach to Derive a Simpler Equivalent Nonlinear System of Constant Stiffness ... 128
3.7.1 Application to Cantilever Flexible Beam Problems ... 129
3.7.2 Application to Flexible Simply Supported Beam Problems ... 133

4 Inelastic Analysis of Structural Components ... 143
4.1 Introduction ... 143
4.2 Theoretical Aspects of Inelastic Analysis ... 144
4.2.1 The Theory and Concept of the Reduced Modulus E_r ... 144
4.2.2 Application of the Method of the Equivalent Systems for Inelastic Analysis ... 155
4.3 Inelastic Analysis of Simply Supported Beams ... 165
4.4 Ultimate Design Loads Using Inelastic Analysis ... 172

5 Vibration Analysis of Flexible Structural Components ... 185
5.1 Introduction ... 185
5.2 Nonlinear Differential Equations of Motion ... 186
5.2.1 The general Nonlinear Differential Equation of Motion ... 186
5.2.2 Small Amplitude Vibrations of Flexible Members ... 189

5.3 Application of the Theory and Method . . . 193
5.3.1 Free Vibration of Uniform Flexible Cantilever Beams . . . 193
5.3.2 Free Vibration of Flexible Simply supported Beams . . . 204
5.4 The Effect of Mass Position Change During the Vibration of Flexible Members . . . 211
5.5 Galerkin's Finite Element Method (GFEM) . . . 213
5.6 Vibration of Tapered Flexible Simply Supported Beams Using Galerkin's FEM . . . 221
5.7 Concluding Remarks . . . 224

6 Suspension Bridges, Failures, Plates, and Other Types of Nonlinear Structural Problems . . . 229
6.1 Introduction . . . 229
6.2 Brief Discussion on Fundamental Aspects of Suspension Bridges . . . 229
6.3 The Collapse of the Tacoma Narrows Suspension Bridge . . . 232
6.4 Other Failures and What We Learn from Them . . . 235
6.5 Eccentrically Loaded Columns . . . 237
6.6 Inelastic Analysis of Members with Axial Restraints Using Equivalent Systems . . . 240
6.7 The Longest Cable-Stayed Suspension Bridge in the World . . . 253
6.8 Inelastic Analysis of Thin Rectangular Plates . . . 259
6.9 Inelastic Earthquake Response of Multistory Buildings . . . 269
6.9.1 Resistant R of a Structure . . . 270
6.9.2 Multistory Buildings Subjected to Strong Earthquakes . . . 276
6.10 Elastic and Inelastic Analysis of Thick-Walled Cylinders Subjected to Uniform External and Internal Pressures . . . 285
6.10.1 Elastic Analysis of Thick Cylinders . . . 285
6.10.2 Inelastic Analysis of Thick Cylinders . . . 290
6.11 Inelastic Analysis of Members of Non-Rectangular Cross Sections . . . 294
6.12 Torsion Beyond the Elastic Limit of the Material . . . 297
6.13 Vibration Analysis of Inelastic Structural Members . . . 299
6.14 Inelastic Analysis of Flexible Members . . . 307

Appendix A Acceleration Impulse Extrapolation Method (AIEM) . . . 323

Appendix B Computer Program Using the AIEM for the Elastoplastic Analysis in Example 6.5 . . . 327

References . . . 329

1

Basic Theories and Principles of Nonlinear Beam Deformations

1.1 Introduction

The minimum weight criteria in the design of aircraft and aerospace vehicles, coupled with the ever growing use of light polymer materials that can undergo large displacements without exceeding their specified elastic limit, prompted a renewed interest in the analysis of flexible structures that are subjected to static and dynamic loads. Due to the geometry of their deformation, the behavior of such structures is highly nonlinear and the solution of such problems becomes very complex. The solution complexity becomes immense when flexible structural components have variable cross-sectional dimensions along their length. Such members are often used to improve strength, weight and deformation requirements, and in some cases, architects and planners are using variable cross-section members to improve the architectural aesthetics and design of the structure.

In this chapter, the well known theory of *elastica* is discussed, as well as the methods that are used for the solution of the elastica. In addition, the solution of flexible members of uniform and variable cross-section is developed in detail. This solution utilizes equivalent pseudolinear systems of constant cross-section, as well as equivalent simplified nonlinear systems of constant cross-section. This approach simplifies a great deal the solution of such complex problems. See, for example, Fertis [2, 3, 5, 6], Fertis and Afonta [1], and Fertis and Lee [4].

This chapter also includes, in a brief manner, important historical developments on the subject and the most commonly used methods for the static and the dynamic analysis of flexible members.

1.2 Brief Historical Developments Regarding the Static and the Dynamic Analysis of Flexible members

By looking into past developments on the subject, we observe that the static analysis of flexible members was basically concentrated in the solution of simple elastica problems. By the term *elastica*, we mean the determination of the exact shape of the deflection curve of a flexible member. This task was carried out by using various types of analytical (closed-form) methods and techniques, as well as various kinds of numerical methods of analysis, such as the finite element method. Numerical procedures were also extensively used to carry out the complicated mathematics when analytical methods were used.

The dynamic analysis of flexible members was primarily concentrated in the computation of their free frequencies of vibration and their corresponding mode shapes. The mode shapes were, one way or another, associated with large amplitudes. In other words, since the free vibration of a flexible member is taking place with respect to its static equilibrium position, we may have large static amplitudes associated with the static equilibrium position and small vibration amplitudes that take place about the static equilibrium position of the flexible member. We may also have large vibration amplitudes that are nonlinearly connected to the static equilibrium position of the member. This gives some fair idea about the complexity of both static and dynamic problems which are related with flexible member.

A brief history of the research work associated with the static and the dynamic analysis of flexible members is discussed in this section. Since the member, in general, can be subjected to both elastic and inelastic behavior, both aspects of this problem are considered.

The deformed configuration for a uniform flexible cantilever beam loaded with a concentrated load P at its free end is shown in Fig. 1.1a. The free-body diagram of a segment of the beam of length x_o is shown in Fig. 1.1b. Note the difference in length between the projected length x in Fig. 1.1a, or Fig. 1.1b, and the length x_o along the length of the member. The importance of such lengths, as well as the other items in the figure, are explained in detail later in this chapter and in following chapters of the book.

The basic equation that relates curvature and bending moment in its general sense was first derived by the brothers, Jacob and Johann Bernoulli, of the well-known Bernoulli family of mathematicians. In their derivation, however, the constant of proportionality was not correctly evaluated. Later on, by following a suggestion that was made by Daniel Bernoulli, L. Euler (1707–1783) rederived the differential equation of the deflection curve and proceeded with the solution of various problems of the elastica [7–10]. J.L. Lagrange (1736–1813) was the next person to investigate the elastica by considering a uniform cantilever strip with a vertical concentrated load at its free end [8, 10–12]. G.A.A. Plana (1781–1864), a nephew of Lagrange, also worked on the elastica problem [13] by correcting an error that was made in Langrange's investigation of the elastica. Max Born also investigated the elastica by using variational methods [14].

Since Bernoulli, many mathematicians, scientists, and engineers researched this subject, and many publications may be found in the literature. The methodologies used may be crudely categorized as either *analytical* (closed-form), or based on *finite element* techniques. The analytical approaches are

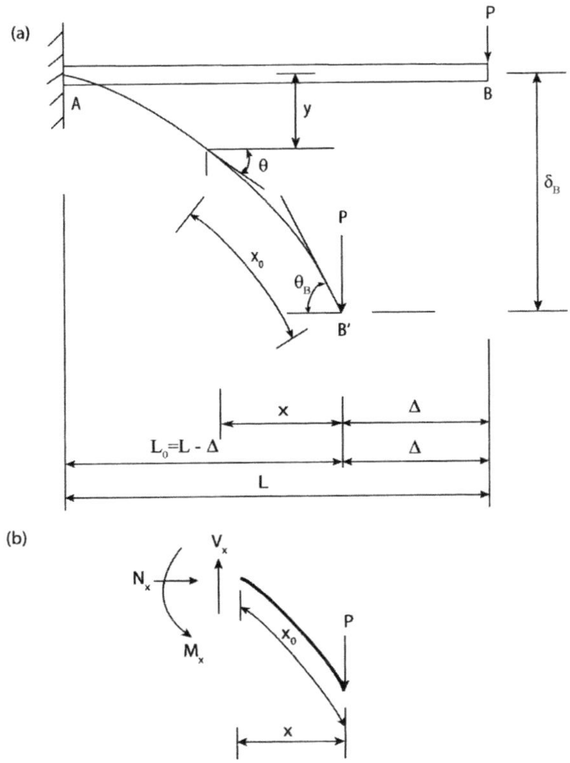

Fig. 1.1. (**a**) Large deformation of a cantilever beam of uniform cross section. (**b**) Free-body diagram of a beam element

based on the Euler–Bernoulli law, while in the finite element method the purpose is to develop a procedure that permits the solution of complex problems in a straightforward manner.

The more widely used analytical methods include *power series, complete and incomplete elliptic integrals*, numerical procedures using the fourth order *Runge–Kutta method*, and the author's method of the *equivalent systems* which utilizes *equivalent pseudolinear systems* and *simplified nonlinear equivalent systems.*

In the *power series* method, the basic differential equation is expressed with respect to x_o, i.e.

$$\frac{d\theta}{dx_0} = \frac{M}{E_1 I_1 f(x_0) g(x_0)} \tag{1.1}$$

where $f(x_o)$ and $g(x_o)$ represent the variation of the moment of inertia $I(x_o)$ and the modulus of elasticity $E(x_o)$, respectively, with respective reference values I_1 and E_1, respectively. Note that for uniform members and linearly elastic materials we have $f(x_o) = g(x_o) = 1.00$. Also note that θ is the angular rotation along the deformed length of the member as shown in Fig. 1.1a.

For constant E, Eq. (1.1) is usually expressed in terms of the shear force V_{x_0} as follows:

$$EI_1 \frac{d}{dx_0}\left\{f(x_0)\frac{d\theta}{dx_0}\right\} = -V_{x_0}\cos\theta \tag{1.2}$$

or, for members of uniform I,

$$EI\frac{d^2\theta}{dx_0^2} = -V_{x_0}\cos\theta \tag{1.3}$$

In order to apply the power series method, we express θ in Eqs. (1.2) and (1.3) as a function of x_o by using the following Maclaurin's series:

$$\theta(x_0) = \theta(c) + (x_0 - c)\theta'(c) + \frac{(x_0 - c)^2}{2!}\theta''(c) + \frac{(x_0 - c)^3}{3!}\theta'''(c) + \cdots \tag{1.4}$$

where c is any arbitrary point along the arc length of the flexible member.

The difficulties associated with the utilization of power series is that for variable stiffness members subjected to multistate loadings, θ depends on both x and x_o. The coordinates x and x_o are defined as shown in Fig. 1.1.

The *method of elliptic integrals* so far is used for simple beams of uniform E and I that are loaded only with concentrated loads. For a uniform beam that is loaded with either a concentrated axial, or a concentrated lateral load, the governing differential equation is of the form

$$\frac{d^2\theta}{dx_0^2} = K\Gamma(\theta) \tag{1.5}$$

where K is an arbitrary constant, and Γ(θ) is a linear combination of cos θ and sin θ. The nonlinear differential equation given by Eq. (1.5) may be integrated by using the elliptic integral method, which requires some certain knowledge of elliptic integrals. The difficulty associated with this method is that it cannot be applied to flexible members with distributed loads, or to flexible members with variable stiffness.

In the *fourth order Runge–Kutta method* the nonlinear differential equations are given in terms of the rotation θ, as shown by Eqs. (1.2) and (1.3). The difficulty associated with this method is that for multistate loadings the expressions for the bending moment involve integral equations which are functions of the large deformation. In such cases, the application of the Runge–Kutta method becomes extremely difficult. However, if θ is only a function of x_o, then the method can be easily applied.

The *method of the equivalent systems*, which was developed initially by the author and his collaborators in order to simplify the solution of complicated linear statics and dynamics problems [5,6,15–30], was extended by the author and his students during the past fifteen years for the solution of nonlinear statics and dynamics problems [1–3, 5, 6, 31–51]. Both elastic and inelastic ranges are considered, as well as the effects of axial compressive forces in both ranges. The solution of the nonlinear problem is given in the form of *equivalent pseudolinear systems*, or *simplified equivalent nonlinear systems*, which permit very accurately and rather conveniently the solution of extremely complicated nonlinear problems. A great deal of this work is included in this text. Once the pseudolinear system is derived, linear analysis may be used to solve it because its static or dynamic response is identical, or very closely identical, to that of the original complex nonlinear problem. For very complex nonlinear problems, it was found convenient to derive first a simplified nonlinear equivalent system, and then proceed with pseudolinear analysis. Much of this work is included in this text in detail and with application to practical engineering problems.

We continue the discussion with the research by K.E. Bisshoppe and D.C. Drucker [52]. These two researchers used the power series method to obtain a solution for a uniform cantilever beam, which was loaded (1) by a concentrated load at its free end, and (2) by a combined load consisting of a uniformly distributed load in combination with a concentrated load at the free end of the member. J.H. Lau [53] also investigated the flexible uniform cantilever beam loaded with the combined loading, consisting of a uniformly distributed load along its span and a concentrated load at its free end, by using the power series method. He proved that superposition does not apply to large deflection theory, and he plotted some load–deflection curves for engineering applications. P. Seide [54] investigated the large deformation of an extensional simply supported beam loaded by a bending moment at its end, and he found that reasonable results are obtained by the linear theory for relatively large rotations of the loaded end.

Y. Goto et al. [55] used elliptic integrals to derive a solution for plane elastica with axial and shear deformations. H.H. Denman and R. Schmidt [56] solved the problem of large deflection of thin elastica rods subjected to concentrated loads by using a Chebyshev approximation method. The finite difference method was used by T.M. Wang, S.L. Lee, and O.C. Zienkiewicz [57] to investigate a uniform simply supported beam subjected (1) to a nonsymmetrical concentrated load and (2) to a uniformly distributed load over a portion of its span.

The Runge–Kutta and Gill method, in combination with Legendre Jacobi forms of elliptic integrals of the first and second kind, was used by A. Ohtsuki [58] to analyze a thin elastic simply supported beam under a symmetrical three-point bending. The Runge–Kutta method was also used by B.N. Rao and G.V. Rao [59] to investigate the large deflection of a cantilever beam loaded by a tip rotational load. K.T. Sundara Raja Iyengar [60] used the power series method to investigate the large deformation of a simply supported beam under the action of a combined loading consisting of a uniformly distributed load and a concentrated load at its center. At the supports he considered (1) the reactions to be vertical, and (2) the reactions to be normal to the deformed beam by including frictional forces. He did not obtain numerical results. He only developed the equations.

G. Lewis and F. Monasa [61] investigated the large deflection of a thin cantilever beam made out of nonlinear materials of the Ludwick type, and C. Truesdall [62] investigated a uniform cantilever beam loaded with a uniformly distributed vertical load. R. Frisch-Fay in his book *Flexible Bars* [63] solved several elastica problems dealing with uniform cantilever beams, uniform bars on two supports and initially curved bars of uniform cross section, under point loads. He used elliptic integrals, power series, the principle of elastic similarity, as well as Kirchhoff's dynamical analogy to solve such problems.

Researchers such as J.E. Boyd [64], H.J. Barton [65], F.H. Hammel, and W.B. Morton [66], A.E. Seames and H.D. Conway [67], R. Leibold [68], R. Parnes [69], and others also worked on such problems. In all the studies described above, with the exception of the research performed by the author and his collaborators, analytical approaches which include arbitrary stiffness variations and arbitrary loadings, were not treated. This is attributed to the difficulties involved in solving the nonlinear differential equations involved. These subjects, by including elastic, inelastic, and vibration analysis, as well as cyclic loadings, are treated in detail by the author, as stated earlier in this section, and much of this work is included in this text and the references at the end of the text.

Because of the difficulties involved in solving the nonlinear differential equations, most of the early investigators turned their efforts to the utilization of the finite element method to obtain solutions. However, in the utilization of the finite element method, difficulties were developed, as stated earlier, regarding the representation of rigid body motions of oriented bodies subjected to large deformations.

K.M. Hsiao and F.T. Hou [70] used the small deflection beam theory, by including the axial force, to solve for the large rotation of frame problems by assuming that the strains are small. The total stiffness matrix was formulated by superimposing the bending, geometric, and linear beam stiffness matrices. An incremental iterative method based on the Newton–Raphson method, combined with a constant arc length control method, was used for the solution of the nonlinear equilibrium equations.

Y. Tada and G. Lee [71] adopted nodal coordinates and direction cosines of a tangent vector regarding the deformed configuration of elastic flexible beams. The stiffness matrices were obtained by using the equations of equilibrium and Galerkin's method. Their method was applied to a flexible cantilever beam loaded at the free end. T.Y. Yang [72] proposed a matrix displacement formulation for the analysis of elastica problems related to beams and frames. A. Chajes [73] applied the linear and nonlinear incremental methods, as well as the direct method, to investigate the geometrically nonlinear behavior of elastic structures. The governing equations were derived for each method, and a procedure outline was provided regarding the plotting of the load–deflection curves. R.D. Wood and O.C. Zienkiewicz [74] used a continuum mechanics approach with a Lagrangian coordinate system and isoparametric element for beams, frames, arches, and axisymmetric shells. The Newton–Raphson method was used to solve the nonlinear equilibrium equations.

Some considerable research work was performed on nonlinear vibration of beams. D.G. Fertis [2,3,5] and D.G. Fertis and A. Afonta [39,40] applied the method of the equivalent systems to determine the free vibration of variable stiffness flexible members. D.G. Fertis [2, 3], and D.G. Fertis and C.T. Lee [38, 41, 48] developed a method to be used for the nonlinear vibration and instabilities of elastically supported beams with axial restraints. They have also provided solutions for the inelastic response of variable stiffness members subjected to cyclic loadings. D.G. Fertis [49, 51] used equivalent systems to determine the inelastic vibrations of prismatic and nonprismatic members as well as the free vibration of flexible members.

S. Wionowsky-Krieger [75] was the first one to analyze the nonlinear free vibration of hinged beams with an axial force. G. Prathap [76] worked on the nonlinear vibration of beams with variable axial restraints, and G. Prathap and T.K. Varadan [77] worked on the large amplitude vibration of tapered clamped beams. They used the actual nonlinear equilibrium equations and the exact nonlinear expression for the curvature. C. Mei and K. Decha-Umphai [78] developed a finite element approach in order to evaluate the geometric nonlinearities of large amplitude free- and forced-beam vibrations. C. Mei [79], D.A. Evensen [80], and other researchers worked on nonlinear vibrations of beams.

Analytical research work regarding the inelastic behavior of flexible structures is very limited. D.G. Fertis [2, 3, 49] and D.G. Fertis and C.T. Lee [2–4, 47, 49] did considerable research work on the inelastic analysis of flexible bars using simplified nonlinear equivalent systems, and they have studied the general inelastic behavior of both prismatic and nonprismatic members. G. Prathap and T.K. Varadan [81] examined the inelastic large deformation of a uniform cantilever beam of rectangular cross section with a concentrated load at its free end. The material of the beam was assumed to obey the stress–strain law of the Ramberg–Osgood type. C.C. Lo and S.D. Gupta [82] also worked on the same problem, but they used a logarithmic function of strains for the regions where the material was stressed beyond its elastic limit.

F. Monasa [83] considered the effect of material nonlinearity on the response of a thin cantilever bar with its material represented by a logarithmic stress–strain function. Also J.G. Lewis and F. Monasa investigated the large deflection of thin uniform cantilever beams of inelastic material loaded with a concentrated load at the free end. Again the stress–strain law of the material was represented by Ludwick relation.

In the space age we are living today, much more research and development is needed on these subjects in order to meet the needs of our present and future high technology developments. The need to solve practical nonlinear problems is rapidly growing. Our structural needs are becoming more and more nonlinear. I hope that the work in this text would be of help.

The problem of inelastic vibration received considerable attention by many researchers and practicing engineers. Bleich [86], and Bleich and Salvadory [87], proposed an approach based on normal modes for the inelastic analysis of beams under transient and impulsive loads. This approach is theoretically sound, but it can be applied only to situations where the number of possible plastic hinges is determined beforehand, and where the number of load reversals is negligible. Baron et al. [88], and Berge and da Deppo [89], solved the required equation of motion by using methods that are based on numerical integration. This, however, involved concentrated kink angles which are used to correct for the amount by which the deflection of the member surpasses the actual elastic–plastic point. The methodology is simple, but the actual problem may become very complicated because multiple correction angles and several hinges may appear simultaneously. Lee and Symonds [90], have proposed the method of rigid plastic approximation for the deflection of beams, which is valid only for a single possible yield with no reversals. Toridis and Wen [91], used lumped mass and flexibility models to determine the response of beams.

In all the models developed in the above references, the precise location of the point of reversal of loading is very essential. A hysteretic model where the location of the loading reversal point is not required and where the reversal is automatically accounted for, was first suggested by Bonc [92] for a spring-mass system, and it was later extended by Wen [93] and by Iyender and Dash [94]. In recent years Sues et al. [95] have provided a solution for a single degree of freedom model for degrading inelastic model. This work was later extended by Fertis [2, 3] and Fertis and Lee [38], and they developed a model that adequately describes the dynamic structural response of variable and uniform stiffness members subjected to dynamic cyclic loadings. In their work, the material of the member can be stressed well beyond its elastic limit, thus causing the modulus E to vary along the length of the member. The derived differential equations take into consideration the restoring force behavior of such members by using appropriate hysteretic restoring force models.

The above discussion, is not intended to provide a complete historical treatment of the subject, and the author wishes to apologize for any unintentional omission of the work of other investigators. It provides, however, some insight regarding the state of the art and how the ideas regarding these very important subjects have been initiated.

1.3 The Euler–Bernoulli Law of Linear and Nonlinear Deformations for Structural Members

From what we know today, the first public work regarding the large deformation of flexible members was given by L. Euler (1707–1783) in 1744, and it was continued in the appendix of his well known book *Des Curvis Elastics* [7]. According to Euler, when a member is subjected to bending, we cannot neglect the slope of the deflection curve in the expression of the curvature unless the deflections are small. Euler has published about 75 substantial volumes, he was a dominant figure during the 18th century, and his contributions to both pure and applied mathematics made him worthy of inclusion in the short list of giants of mathematics – Archimedes (287–212 BC), I. Newton (1642–1727), and C. Gauss (1777–1855).

We should point out, however, that the development of this theory took place in the 18th century, and the credits for this work should be given to Jacob Bernoulli (1654–1705), his younger brother Johann Bernoulli (1667–1748), and Leonhard Euler (1707–1783). Both Bernoulli brothers have contributed heavily in the mathematical sciences and related areas. They also worked on the mathematical treatment of the Greek problems of *isochrone, brahistochrone, isoperimetric figures*, and *geodesies*, which led to the development of the new calculus known as the *calculus of variations.* Jacob also introduced the word *integral* in suggesting the name *calculus integrals.* G.W. Leibniz (1646–1716) used the name *calculus summatorius* for the inverse of the *calculus differentialis.*

The Euler–Bernoulli law states that the bending moment M is proportional to the change in the curvature produced by the action of the load. This law may be written mathematically as follows:

$$\frac{1}{r} = \frac{d\theta}{dx_0} = \frac{M}{EI} \tag{1.6}$$

where r is the radius of curvature, θ is the slope at any point x_o, where x_o is measured along the arc length of the member as shown in Fig. 1.1a, E is the modulus of elasticity, and I is the cross-sectional moment of inertia.

Figure 1.1a depicts the large deformation configuration of a uniform flexible cantilever beam, and Fig. 1.1b illustrates the free-body diagram of a segment of the beam of length x_o. Note the difference in length size between x and x_o in Fig. 1.1b. For small deformations we usually assume that $x = x_o$. For small deformations we can also assume that $L = L_o$ in Fig. 1.1a, because under this condition the horizontal displacement Δ of the free end B of the cantilever beam would be small.

In rectangular x, y coordinates, Eq. (1.6) may be written as

$$\frac{1}{r} = \frac{y''}{\left[1 + (y')^2\right]^{3/2}} = -\frac{M}{EI} \tag{1.7}$$

where

$$y' = \frac{dy}{dx} \tag{1.8}$$

$$y'' = \frac{d^2y}{dx^2} \tag{1.9}$$

and y is the vertical deflection at any x. We also know that

$$y' = \tan\theta \quad \text{or} \quad \theta = \tan^{-1} y' \tag{1.10}$$

Equation (1.7) is a second order nonlinear differential equation, and its exact solution is very difficult to obtain. This equation shows that the deflections are no longer a linear function of the bending moment, or of the load, which means that the principle of superposition does not apply. The consequence is that every case that involves large deformations must be solved separately, since combinations of load types already solved cannot be superimposed. The consequences become more immense when the stiffness EI of the flexible member varies along the length of the member. We discuss this point of view in greater detail, with examples, later in this chapter.

When the deformation of the member is considered to be small, y' in Eq. (1.7) is small compared to 1, and it is usually neglected. On this basis, Eq. (1.7) is transformed into a second order linear differential equation of the form

$$\frac{1}{r} = y'' = -\frac{M}{EI} \tag{1.11}$$

The great majority of practical applications are associated with small deformations and, consequently, reasonable results may be obtained by using Eq. (1.11). For example, if $y' = 0.1$ in Eq. (1.7), then the denominator of this equation becomes

$$\left[1 + (0.1)^2\right]^{3/2} = 0.985 \tag{1.12}$$

which suggests that we have an error of only 1.52% if Eq. (1.11) is used.

1.4 Integration of the Euler–Bernoulli Nonlinear Differential Equation

Figure 1.2 depicts the large deformation configuration of a tapered flexible cantilever beam loaded with a concentrated vertical load P at its free end. In this figure, y is the vertical deflection of the member at any x, and θ is its rotation at any x. We also have the relations

$$y' = \frac{dy}{dx} \tag{1.13}$$

$$y'' = \frac{d^2y}{dx} \tag{1.14}$$

and

$$y' = \tan\theta \quad \text{or} \quad \theta = \tan^{-1} y' \tag{1.15}$$

In rectangular x, y coordinates, the Euler–Bernoulli law for large deformation produced by bending may be written as [2, 3] (see also Eq. (1.7):

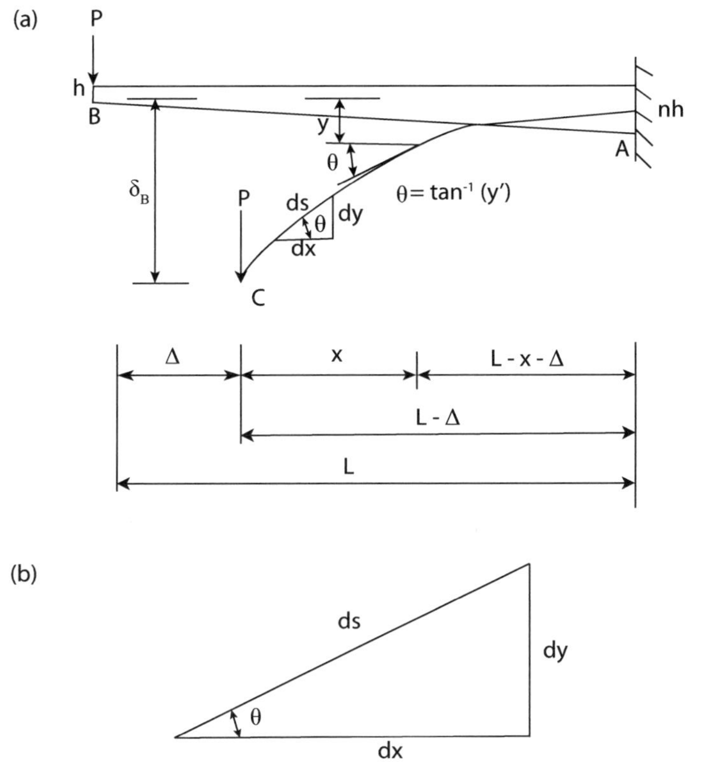

Fig. 1.2. (**a**) Tapered flexible cantilever beam loaded with a vertical concentrated load P at the free end. (**b**) Infinitesimal beam element

$$\frac{y''}{\left[1+(y')^2\right]^{3/2}} = -\frac{M_x}{E_x I_x} \tag{1.16}$$

where M_x is the bending moment produced by the loading on the beam. E_x is the modulus of elasticity of its material, and I_x is its cross-sectional moment of inertia.

Since the loading on the beam can be arbitrary and E_x and I_x can be variable, we may rewrite Eq. (1.16) in a more general form as follows:

$$\frac{y''}{\left[1+(y')^2\right]^{3/2}} = -\frac{M_x}{E_1 I_1 f(x)\, g(x)} \tag{1.17}$$

where f(x) is the moment of inertia function representing the variation of I_x with I_1 as a reference value, and g(x) is the modulus function representing the variation of E_x with E_1 as a reference value. If E and I are constant, then $g(x) = f(x) = 1.00$.

We integrate Eq. (1.17) by making changes in the variables. We let $y' = p$ and then $y'' = p'$. Thus, from Eq. (1.16), we obtain

$$\frac{p'}{[1+p^2]^{3/2}} = \lambda(x) \tag{1.18}$$

where

$$\lambda(x) = \frac{M_x}{E_x I_x} \tag{1.19}$$

Now we rewrite Eq. (1.18) as follows:

$$\frac{dp/dx}{[1+p^2]^{3/2}} = \lambda(x) \tag{1.20}$$

By multiplying both sides of Eq. (1.20) by dx and integrating once, we find

$$\int \frac{dp}{[1+p^2]^{3/2}} = \int \lambda(x)\,dx \tag{1.21}$$

We can integrate Eq. (1.21) by making the following substitutions:

$$p = \tan\theta \tag{1.22}$$

$$dp = \sec^2\theta\,d\theta \tag{1.23}$$

By using the beam element shown in Fig. 1.2b and applying the Pythagorean theorem, we find

$$(ds)^2 = (dx)^2 + (dy)^2 \quad \text{or} \quad ds = \left[(dx)^2 + (dy)^2\right]^{1/2} \tag{1.24}$$

$$\begin{aligned}\frac{ds}{dx} = \left[1 + \left(\frac{dy}{dx}\right)^2\right]^{1/2} &= \left[1 + (\tan\theta)^2\right]^{1/2} \\ &= \left[1+p^2\right]^{1/2}\end{aligned} \tag{1.25}$$

Thus,

$$\cos\theta = \frac{dx}{ds} = \frac{1}{[1+p^2]^{1/2}} \tag{1.26}$$

and from Eq. (1.22), we find

$$\sin\theta = p\cos\theta = \frac{p}{[1+p^2]^{1/2}} \tag{1.27}$$

By substituting Eqs. (1.22) and (1.23) into Eq. (1.21) and also making use of Eqs. (1.26) and (1.27), we find

$$\int \frac{\sec^2 \theta \; d\theta}{\left[1 + \frac{\sin^2 \theta}{\cos^2 \theta}\right]^{3/2}} = \int \lambda\,(x)\,dx \tag{1.28}$$

or, by performing trigonometric manipulations, Eq. (1.28) reduces to the following equation:

$$\int \cos\theta \; d\theta = \int \lambda\,(x)\,dx \tag{1.29}$$

Integration of Eq. (1.29), yields

$$\sin\theta = \varphi\,(x) + C \tag{1.30}$$

where the function $\varphi(x)$ represents the integration of $\lambda(x)$.

Equation (1.30) may be rewritten in terms of p and y' by using Eq. (1.27). Thus,

$$\frac{p}{[1+p^2]^{1/2}} = \varphi\,(x) + C \tag{1.31}$$

$$\frac{y'}{\left[1+(y')^2\right]^{1/2}} = \varphi\,(x) + C \tag{1.32}$$

where C is the constant of integration which can be determined from the boundary conditions of the given problem. If we will solve Eq. (1.32) for $y'(x)$, we obtain the following equation:

$$y'\,(x) = \frac{\varphi\,(x) + C}{\sqrt{1 - [\varphi\,(x) + C]^2}} \tag{1.33}$$

Integration of Eq. (1.33) yields the large deflection $y(x)$ of the member. Thus,

$$y\,(x) = \int_0^x \frac{\varphi\,(\eta) + C}{\sqrt{1 - [\varphi\,(\eta) + C]^2}} d\eta \tag{1.34}$$

This shows that when M_x/E_xI_x is known and it is integrable, then the Euler–Bernoulli equation may be solved directly for $y'(x)$ as illustrated in the solution of many flexible beam problems in [2, 3]. In the same references, utilization of pseudolinear equivalent systems is made, which simplify a great deal the solution of such problems. A numerical integration may be also used for Eq. (1.34), or Eq. (1.16), by using the Simpson's rule discussed in the following section of this text.

1.5 Simpson's One-Third Rule

Simpson's one-third rule is one of the most commonly used numerical method to approximate integration. It is used primarily for cases where exact integration is very difficult or impossible to obtain. Consider, for example, the integral

$$\delta = \int_a^b f(x)\, dx \tag{1.35}$$

between the limits a and b. If we divide the integral between the limits x=a and x=b into n equal parts, where n is an even number, and if $y_0, y_1, y_2, \ldots, y_{n-1}, y_n$ are the ordinates of the curve $y = f(x)$, as shown in Fig. 1.3, then, according to Simpson's one-third rule we have

$$\int_a^b f(x)\, dx = \frac{\lambda}{3}\left(y_0 + 4y_1 + 2y_2 + 4y_3 + \cdots + 2y_{n-2} + 4y_{n-1} + y_n\right) \tag{1.36}$$

where

$$\lambda = \frac{b-a}{n} \tag{1.37}$$

Simpson's rule provides reasonably accurate results for practical applications.

Let it be assumed that it is required to determine the value δ of the integral

$$\delta = \int_0^L x^2 dx \tag{1.38}$$

We divide the length L into 10 equal segments, yielding $\lambda = 0.1L$. By applying Simpson's rule given by Eq. (1.36), and noting that $y = f(x) = x^2$, we find

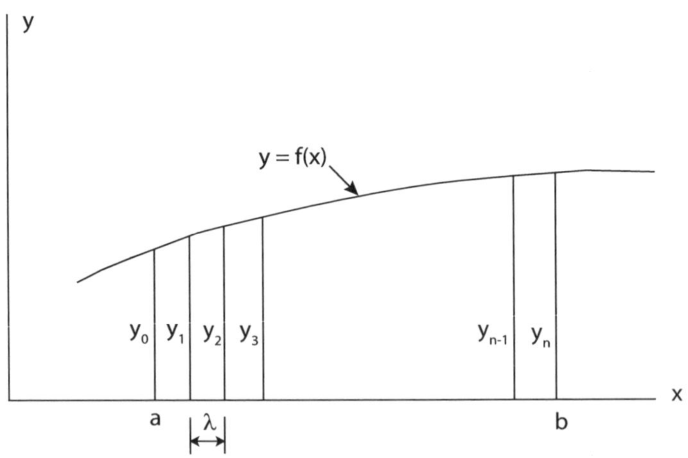

Fig. 1.3. Plot of a function y = f(x)

$$\delta = \frac{0.1L}{3}\Big[(1)(0)^2 + (4)(0.1)^2 + (2)(0.2)^2 + (4)(0.3)^2 + (2)(0.4)^2$$
$$+ (4)(0.5)^2 + (2)(0.6)^2 + (4)(0.7)^2 + (2)(0.8)^2 + (4)(0.9)^2 + (1)(1)^2\Big]L^2$$
$$= \frac{L^3}{3}$$

Note that for $\lambda = 0.1L$, the values of f(x) are $y_o = (0)^2$, $y_1 = (0.1L)^2$, $y_2 = (0.2L)^2$, and so on. In this case, the exact value of the integral is obtained.

As a second example, let it be assumed that it is required to find the value δ of the integral

$$\delta = \int_0^L x^3 dx \tag{1.39}$$

Again, we subdivide the length L into 10 equal segments, yielding $\lambda = 0.1L$. In this case, the Simpson's one-third rule yields

$$\delta = \frac{0.1L}{3}\Big[(1)(0)^3 + (4)(0.1)^3 + (2)(0.2)^3 + (4)(0.3)^2 + (2)(0.4)^3$$
$$+ (4)(0.5)^3 + (2)(0.6)^3 + (4)(0.7)^3 + (2)(0.8)^3 + (4)(0.9)^3 + (1)(1)^3\Big]L^3$$
$$= \frac{0.75L^4}{3} = \frac{L^4}{4}$$

The exact value of the integral is obtained again in this case.

More complicated integrals may be also evaluated in a similar manner, as shown later in this text. For example, let it be assumed that it is required to determine the length L of a flexible bar given by the integral

$$L = \int_0^{840} \left[1 + (y')^2\right]^{1/2} dx \tag{1.40}$$

where

$$y'(x) = \frac{G(x)}{\left\{1 - [G(x)]^2\right\}^{1/2}} \tag{1.41}$$

and

$$G(x) = 1.111\,(10)^{-6}\,x^2 - 0.783922 \tag{1.42}$$

Equation (1.40) is an extremely important equation in nonlinear mechanics for the analysis of flexible bars subjected to large deformations [2,3]. It relates the length L of the bar with the slope y' at points along its deformed shape.

For illustration purposes, we use here n=10, and from Eq. (1.37) we obtain

$$\lambda = \frac{840 - 0}{10} = 84$$

From Eq. (1.40), we note that

$$\mathrm{f}(\mathrm{x}) = \left[1 + (\mathrm{y}')^2\right]^{1/2} \tag{1.43}$$

The values of f(x) at x = 0, 84, 168, ..., 840 are designated as $y_o, y_1, y_2, \ldots, y_{10}$, respectively, and they are obtained by using Eq. (1.43) in conjunction with Eqs. (1.42) and (1.41). For example, for x=0, we have

$$\mathrm{G}(0) = -0.783922$$

$$\mathrm{y}'(0) = \frac{-0.783922}{\sqrt{1-(-0.783922)^2}} = -1.262641$$

$$\mathrm{y}_0 = \mathrm{f}(0) = \sqrt{1+(-1.262641)^2} = 1.610671$$

For x=84 in., we have

$$\mathrm{G}(84) = 1.111\,(10)^6\,(84)^2 - 0.783922 = 0.776083$$

$$\mathrm{y}'(84) = \frac{-0.776083}{\sqrt{1-(-0.776083)^2}} = -1.230645$$

$$\mathrm{y}_1 = \mathrm{f}(84) = \sqrt{1+(-1.230645)^2} = 1.585713$$

In a similar manner, the remaining points $y_2, y_3, \ldots, y_{10}$, can be determined. On this basis, Eq. (1.36) yields

$$\begin{aligned} \mathrm{L} = \frac{84}{3}\,&[1.610671 + (4)\,(1.585713) + (2)\,(1.518561) + (4)\,(1.426963) \\ &+ (2)\,(1.328753) + (4)\,(1.236242) + (2)\,(1.156021) + (4)\,(1.090986) \\ &+ (2)\,(1.042370) + (4)\,(1.011280) + 1] \\ = \frac{84}{3}\,&(38.106817) = 1,067\,\mathrm{in.} \end{aligned}$$

It should be realized that the value obtained for L is an approximate one, but better accuracy can be obtained by using larger values for the parameter n in Eq. (1.37). For practical applications, however, the design engineer usually has a fair idea about their accuracy requirements, and satisfactory and safe designs can be obtained by using approximate solutions.

1.6 The Elastica Theory

The exact shape of the deflection curve of a flexible member is called the elastica. The most popular elastica problem is the solution of the flexible uniform cantilever beam loaded with a concentrated load P at the free end, as shown in Fig. 1.1a.

The large deformation configuration of this cantilever beam caused by the vertical load P is shown in Fig. 1.2a. Note that the end point B moved to point C during the large displacement of point B. The beam is assumed to be inextensible and, consequently, the arc length AC of the deflection curve is equal to the initial length AB. We also assumed that the vertical load P remained vertical during the deformation of the member.

The expression for the bending moment M_x at any $0 \leq x \leq L_o$ may be obtained by using the free-body diagram in Fig. 1.1b and applying statics, i.e.,

$$M_x = -Px \tag{1.44}$$

In rectangular coordinates, the Euler–Bernoulli equation is given by Eq. (1.16), That is,

$$\frac{y''}{\left[1+(y')^2\right]^{3/2}} = -\frac{M_x}{E_x I_x} \tag{1.45}$$

where E_x is the modulus of elasticity along the length of the member, and I_x is the moment of inertia at cross sections along its length. By substituting Eq. (1.44) into Eq. (1.45) and assuming that E and I are uniform, we obtain

$$\frac{y''}{\left[1+(y')^2\right]^{3/2}} = \frac{Px}{EI} \tag{1.46}$$

Equation (1.45) may be also expressed in terms of the arc length x_o by using Eq. (1.6). That is

$$E_{x_0} I_{x_0} \frac{d\theta}{dx_0} = -M_x \tag{1.47}$$

By using Eq. (1.44) and assuming that E and I are constant along the length of the member, we find

$$\frac{d\theta}{dx_0} = \frac{Px}{EI} \tag{1.48}$$

By differentiating Eq. (1.48) once with respect to x_o, we obtain

$$\frac{d^2\theta}{dx_0^2} = \frac{P}{EI}\cos\theta \tag{1.49}$$

By assuming that

$$E_x I_x = E_1 I_1 g(x_0) f(x_0) \tag{1.50}$$

where $g(x_o)$ represents the variation of E_x with respect to a reference value E_1, and $f(x_o)$ represents the variation of I_x, with respect to a reference value I_1, we can differentiate Eq. (1.47) once to obtain

$$\frac{d}{dx_0}\left\{E_1 I_1 g(x_0) f(x_0) \frac{d\theta}{dx_0}\right\} = -V_{x_0}\cos\theta \tag{1.51}$$

For members of uniform cross section and of linearly elastic material, we have $g(x_o) = f(x_o) = 1.0$.

Equations (1.46) and (1.49) are nonlinear second order differential equations and exact solutions of these two equations are not presently available. Elliptic integral solutions are often used by investigators (see, e.g., Frisch-Fay [63]), but they are very complicated. This problem, as well as many other flexible beam problems, is discussed in detail in later sections of this chapter and other chapters of the book, where convenient methods of analysis are developed by the author and his collaborators to simplify the solution of such very complicated problems.

The integration of Eq. (1.46) may be carried out as discussed in Section 1.4. By using Eqs. (1.19) and (1.44), we write $\lambda(x)$ as follows:

$$\lambda(x) = -\frac{M_x}{EI} = \frac{Px}{EI} \tag{1.52}$$

Thus,

$$\varphi(x) = \int \lambda(x)\,dx = \frac{Px^2}{2EI} \tag{1.53}$$

On this basis, Eq. (1.32) yields

$$\frac{y'}{\left[1+(y')^2\right]^{1/2}} = \varphi(x) + C$$

or, by substituting for $\varphi(x)$ using Eq. (1.53), we obtain

$$\frac{y'}{\left[1+(y')^2\right]^{1/2}} = \frac{Px^2}{2EI} + C \tag{1.54}$$

where C is the constant of integration which can be determined by applying the boundary condition of zero y' at $x = L_o = (L-\Delta)$. By using this boundary condition in Eq. (1.54), we find

$$C = -\frac{P(L-\Delta)^2}{2EI} \tag{1.55}$$

By substituting Eq. (1.55) into Eq. (1.54), we obtain

$$\frac{y'}{\left[1+(y')^2\right]^{1/2}} = G(x) \tag{1.56}$$

where

$$G(x) = \frac{P}{2EI}\left[x^2 - (L-\Delta)^2\right] \tag{1.57}$$

Thus, by solving Eq. (1.56) for y', we obtain

$$y'(x) = \frac{G(x)}{\left\{1 - [G(x)]^2\right\}^{1/2}} \tag{1.58}$$

The same expression could be obtained directly from Eq. (1.33) by substituting for $\varphi(x)$ and C, which are the expressions given by Eqs. (1.53) and (1.55), respectively.

The large deflection y at any $0 \leq x \leq L_o$ may now be obtained by integrating once Eq. (1.58) and satisfying the boundary condition of zero deflection at $x = L_o$ for the evaluation of the constant of integration. It should be noted, however, that G(x) in Eq. (1.57) is a function of the unknown horizontal displacement Δ of the free end of the beam. The value of Δ may be determined from the equation

$$L = \int_0^{L_0} \left[1 + (y')^2\right]^{1/2} dx \tag{1.59}$$

by using a trial-and-error procedure. That is, we assume a value of Δ in Eq. (1.58) and then carry out the integration in Eq. (1.59) to determine the length L of the member. The procedure may be repeated for various values of Δ until the correct length L is obtained. This procedure is explained in detail in the numerical examples at the end of this section, as well as in many other sections of this text.

The integration in Eq. (1.59) becomes in many cases more convenient if we introduce the variable

$$\xi = \frac{x}{L - \Delta} \tag{1.60}$$

$$d\xi = \frac{dx}{L - \Delta} \tag{1.61}$$

On this basis, Eq. (1.59) may be written as

$$L = (L - \Delta) \int_0^1 \left\{1 + [y'(\xi)]^2\right\}^{1/2} d\xi \tag{1.62}$$

or, by using Eqs. (1.57) and (1.58) and the variable ξ, we can write Eq. (1.62) as follows:

$$L = (L - \Delta) \int_0^1 \frac{1}{\left\{1 - [G(\xi)]^2\right\}^{1/2}} d\xi \tag{1.63}$$

where

$$G(\xi) = \frac{P(L - \Delta)^2}{2EI}\left(\xi^2 - 1\right) \tag{1.64}$$

It should be pointed out again here that Eqs. (1.45) and (1.47) are second order nonlinear differential equations which describe the exact shape of the deflection curve of the flexible beam. In conventional applications these equations are linearized by neglecting the square of the slope $(y')^2$ in Eq. (1.45) as being small compared to unity. This assumption is permissible, as stated earlier, provided that the deflections are very small when they are compared with the length of the beam. For flexible bars, where deflections are large when they are compared with the length of the member, this assumption is not permissible, and Eq. (1.45), or Eq. (1.47), must be used in its entirety. This means that the deflections are no longer a linear function of the bending moment, or of the load, and consequently the principle of superposition does not apply. Therefore, every case involving large deformations has to be solved independently, since combinations of load types already solved cannot be superimposed. The situation yields much greater consequences when the stiffness EI of the member is also variable.

Example 1.1 For the uniform flexible cantilever beam in Fig. 1.1a, determine the rotation θ_B and the horizontal displacement Δ of its free end B. Assume that $P = 0.4$ kip (1.78 kN), $L = 1{,}000$ in. (25.4 m), and $EI = 180 \times 10^3\ \text{kip in.}^2$ ($516.54 \times 10^3\ \text{N m}^2$).

Solution: In order to obtain a solution to this problem we can use Eqs. (1.57), (1.58), and (1.54). From Eq. (1.57), by substituting the appropriate values for the stiffness EI of the member and its length L, we find

$$\begin{aligned} G(x) &= \frac{0.4}{(2)(180)(10)^3}\left[x^2 - (1{,}000 - \Delta)^2\right] \\ &= 1.111\,(10)^{-6}\left[x^2 - (1{,}000 - \Delta)^2\right] \end{aligned} \tag{1.65}$$

The horizontal displacement Δ of the free end of the member may be evaluated by applying a trial-and-error procedure using Eq. (1.59).

The trial-and-error procedure may be initiated by assuming values of Δ until we find the one that satisfies Eq. (1.59). For example, if we assume $\Delta = 160$ in. (4.064 m), Eq. (1.65) yields

$$G(x) = 1.111\,(10)^{-6}x^2 - 0.783922 \tag{1.66}$$

By using the assumed value for Δ, we find $L_0 = 1{,}000 - 160 = 840$ in. (21.336 m). Therefore,

$$y'(x) = \frac{G(x)}{\left\{1 - [G(x)]^2\right\}^{1/2}} \tag{1.67}$$

$$L = \int_0^{840}\left[1 + (y')^2\right]^{1/2} dx \tag{1.68}$$

where G(x) is as shown by Eq. (1.66).

The integration of Eq. (1.68) may be carried out numerically with sufficient accuracy by using Simpson's rule, as shown in Section 1.5, or by using other known numerical procedures. The Simpson's One-Third rule will be used here.

According to this rule, the integral of Eq. (1.68) may be evaluated from the general equation given by Eqs. (1.36) and (1.37). We rewrite these equation as shown below

$$\int_a^b f(x)\,dx = \frac{\lambda}{3}\left(y_0 + 4y_1 + 2y_2 + 4y_3 + \cdots + 2y_{n-2} + 4y_{n-1} + y_n\right) \quad (1.69)$$

where

$$\lambda = \frac{b - a}{n} \quad (1.70)$$

$y_0, y_1, y_2, \ldots, y_n$ are the ordinates of the curve $y = f(x)$, and n is the number of equal parts we use between the limits $x = a$ and $x = b$.

For illustrative purposes, we use $n = 10$, and from Eq. (1.70) we obtain

$$\lambda = \frac{840 - 0}{10} = 84$$

From Eq. (1.68), we note that

$$f(x) = \left[1 + (y')^2\right]^{1/2} \quad (1.71)$$

The values of f(x) at $x = 0, 84, 168, \ldots, 840$ are designated as $y_0, y_1, y_2, \ldots, y_{10}$, respectively, and they are obtained by using Eq. (1.71) in conjunction with Eqs. (1.66) and (1.67). For example, for $x = 0$, we have

$$G(0) = -0.783922$$
$$y'(0) = \frac{-0.783922}{\sqrt{1 - (-0.783922)^2}} = -1.262641$$
$$y_0 = f(0) = \sqrt{1 + (-1.262641)^2} = 1.610671$$

For $x = 84$

$$G(84) = 1.111\,(10)^{-6}\,(84)^2 - 0.783922 = -0.776083$$
$$y'(84) = \frac{-0.776083}{\sqrt{1 - (-0.776083)^2}} = -1.230645$$
$$y_1 = f(84) = \sqrt{1 + (-1.230645)^2} = 1.585713$$

In a similar manner the remaining points $y_2, y_3, \ldots, y_{10}$ can be determined. On this basis, Eq. (1.69) yields

$$\begin{aligned} L = \frac{84}{3}\,[&1.610671 + (4)(1.585713) + (2)(1.518561) + (4)(1.426963) \\ &+ (2)(1.328753) + (4)(1.236242) + (2)(1.156021) + (4)(1.090986) \\ &+ (2)(1.042370) + (4)(1.011280) + 1] \\ = \frac{84}{3}&(38.106817) = 1,067\,\text{in.}\ (27.10\,\text{m}) \end{aligned}$$

Since the actual length $L = 1{,}000$ in. (25.4 m), the procedure may be repeated with a new value for Δ. With computer assistance, the correct value of Δ was found to be 183.10 in. (4.65 m).

With known Δ, the value of y' and rotation θ_B at the free end B of the member may be obtained by using Eqs. (1.65), (1.67), and (1.10). From Eq. (1.65), we obtain

$$G(0) = 1.111\,(10)^{-6}\left[0 - (1{,}000 - 183.10)^2\right] = -0.741399$$

and from Eq. (1.67), we find

$$y'(0) = \frac{-0.741399}{\sqrt{1 - (-0.741399)^2}} = \frac{-0.741399}{0.671065} = -1.104810$$

Therefore, from Eq. (1.10), we find

$$\theta_B = \tan^{-1} y'(0) = 47.85^\circ$$

Also, Eq. (1.67), in conjunction with Eq. (1.65), can be used to determine the values of $y'(x)$, and consequently those of θ, at other points x. This procedure is explained in detail in later parts of this text.

The values of θ_B and Δ were also determined by using elliptic integrals. The results obtained are $\theta_B = 47.44^\circ$, and $\Delta = 181.67$ in. (4.61 m).

1.7 Moment and Stiffness Dependence on the Geometry of the Deformation of Flexible Members

To comprehend the various methods and methodologies developed in this text, as well as their application to practical engineering problems, we should realize first that the expressions for the bending moment M_x and the moment of inertia I_x of the flexible member are generally nonlinear functions of the large deformation of the member. These two quantities may be expressed as a function of x and x_o as follows:

$$M_x = M(x, x_0) \tag{1.72}$$

$$I_x = I_1 f(x, x_0) \tag{1.73}$$

where x is the abscissa of center line points of the deformed configuration of the member, x_o is the arc length of the deformed segment, I_1 is the reference moment of inertia, and $f(x, x_o)$ is a function representing the variation of I_x. For visual observation see, for example, Fig. 1.1. On this basis, the Euler–Bernoulli law given by Eq. (1.16) becomes a *nonlinear integral* differential equation that is extremely difficult to solve.

To reduce the complexity of such types of problems, we express the arc length $x_o(x)$ of the flexible member in terms of its horizontal displacement $\Delta(x)$, where $0 \leq x \leq (L - \Delta)$. See Fig. 1.1a or Fig. 1.2a. This is accomplished as follows:

$$x_0(x) = x + \Delta(x) \tag{1.74}$$

We also know, as shown from the discussion of Sect. 1.4, that the expression $x_o(x)$ is an integral function of the deformation and it can be expressed as

$$x_0(x) = \int_0^x \left\{1 + [y'(x)]^2\right\}^{1/2} dx \tag{1.75}$$

The derivation of Eq. (1.74) can be initiated by considering a segment dx_o before and after deformation, as shown in Fig. 1.4. By using the Pythagorean theorem we write

$$[dx_0]^2 = [dx]^2 + [dy]^2 \tag{1.76}$$

If we assume that

$$dx_0 = dx + d\Delta(x) \tag{1.77}$$

and then substitute into Eq. (1.76), we obtain

$$[dx + d\Delta(x)]^2 = [dx]^2 + [dy]^2 \tag{1.78}$$

or

$$dx + d\Delta(x) = \left\{1 + [y'(x)]^2\right\}^{1/2} dx \tag{1.79}$$

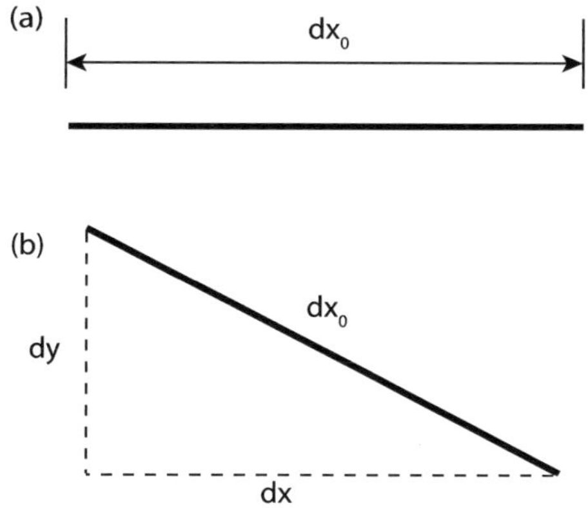

Fig. 1.4. (**a**) Undeformed configuration of an arc length segment dx_o. (**b**) Deformed configuration of dx_o

Integration of Eq. (1.79) with respect to x, yields the expression

$$x + \Delta(x) = \int_0^x \left\{1 + [y'(x)]^2\right\}^{1/2} dx \tag{1.80}$$

This expression provides the same results as Eqs. (1.74) and (1.75).

If we consider flexible members where one of their end supports is permitted to move in the horizontal direction, such as cantilever beams, simply supported beams, etc., approximate expressions for the variation of $\Delta(x)$ may be written and used to facilitate the solution of the nonlinear flexible beam problems. The cases of $\Delta(x)$ that have been investigated by the author and his collaborators and are proven to provide accurate results, are as follows [2, 3]

$$\Delta(x) = \text{constant} = \Delta \tag{1.81}$$

$$\Delta(x) = \Delta \frac{x}{L_0} \tag{1.82}$$

$$\Delta(x) = \Delta \sqrt{\frac{x}{L_0}} \tag{1.83}$$

$$\Delta(x) = \Delta \sin \frac{\pi x}{2L_0} \tag{1.84}$$

where Δ is the horizontal displacement of the movable end, and $L_o = (L - \Delta)$. The plots of the variations of $\Delta(x)$ given by Eqs. (1.81)–(1.84) are shown in Fig. 1.5. We can see from this figure that Δ is an upper limit. Even using Eq. (1.81) which is an upper limit, we obtain reasonably accurate results with error less than 3%. This shows that the variation of the bending moment M_x, and, consequently, the deformation of the member, are largely dependent

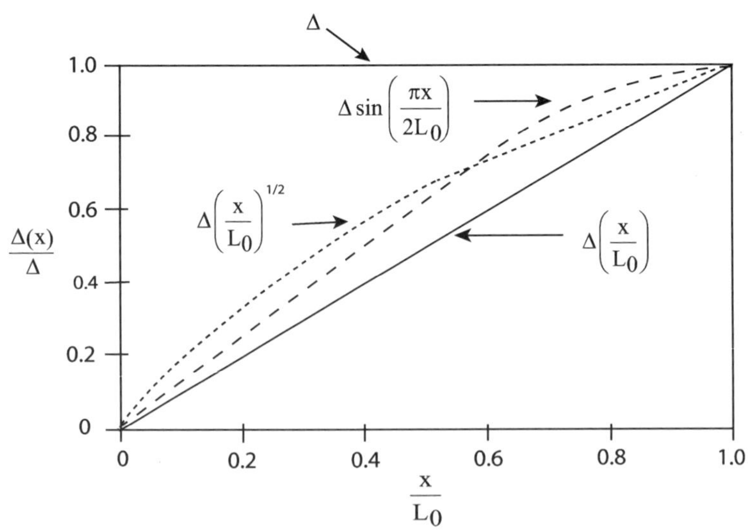

Fig. 1.5. Graphs of various cases of $\Delta(x)$

upon the boundary condition of $\Delta(x)$ at the moving end of the member, and it is rather insensitive to the variation of $\Delta(x)$ between the ends of the member. This is particularly true when the deformations are very large.

The variable moment of inertia I_x of a flexible member, as stated earlier, is a nonlinear function of the deformation. For tapered members that are loaded with concentrated loads only, the variation of the depth h(x) of the member may be approximated by the expression

$$h(x) = (n-1)\left[\frac{1}{n-1} + \frac{x}{L-\Delta}\right] h \tag{1.85}$$

where x is the abscissa of points of the centroidal axis of the member in its deformed configuration, n represents the taper, h is a reference height, and L is the undeformed length of he member. The error of 3% or less associated with the use of Eq. (1.85) is considered small for practical applications. Under this assumption, the solution of flexible members loaded with concentrated loads only, will not require the utilization of integral equations or the use of Eqs. (1.81)–(1.84). This point of view is amply illustrated later.

The following examples illustrate the application of the preceding theory.

Example 1.2 For the tapered flexible cantilever beam shown in Fig. (1.6), derive its exact integral nonlinear differential equation. Also suggest reasonable approximate ways to simplify the complexity of the problem. Assume that the modulus of elasticity E_x is constant and equal to E.

Solution: The Euler–Bernoulli nonlinear differential equation is

$$\frac{y''}{\left[1+(y')^2\right]^{3/2}} = -\frac{M_x}{E_1 I_1 f(x)\, g(x)} \tag{1.86}$$

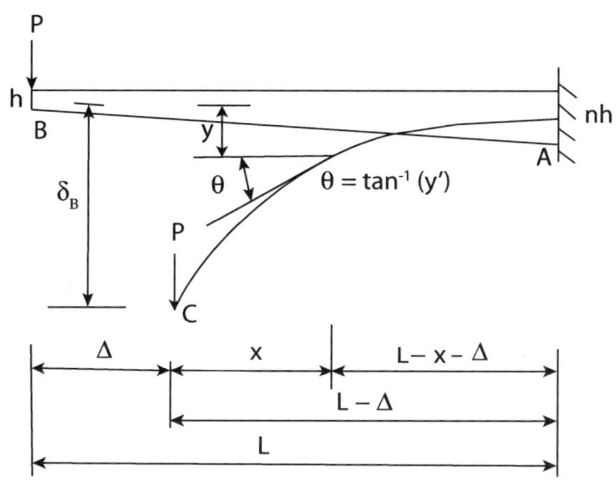

Fig. 1.6. Tapered flexible cantilever beam loaded with a vertical concentrated load P at the free end

where f(x) is the moment of inertia function with moment of inertia I_1 as a reference value, and g(x) represents the variation of E_x with respect to a reference value E_1. In this case we assume that the modulus of elasticity is constant and equal to E, thus making g(x) = 1.

By selecting the moment of inertia I_B at the free end B of the tapered beam as the reference value, the variation of the moment of inertia I_x at any $0 \leq x \leq (L - \Delta)$, where Δ is the horizontal displacement of the free end B, see Fig. 1.6, is written as follows:

$$\begin{aligned} I_x &= \frac{bh^3}{12}\left[f\left(x_0\right)\right] \\ &= I_B\left[1 + \frac{(n-1)}{L}x_0\right]^3 = I_B f\left(x\right) \end{aligned} \tag{1.87}$$

where

$$I_B = \frac{bh^3}{12} \tag{1.88}$$

$$x_0 = \int_0^x \left\{1 + \left[y'\left(x\right)\right]^2\right\}^{1/2} dx \tag{1.89}$$

$$f\left(x\right) = \left[1 + \frac{(n-1)}{L}x_0\right]^3 \tag{1.90}$$

and b is the constant width of the tapered member.

From Fig. 1.6 we observe that the bending moment M_x at any x from the free end C, is given by the expression

$$M_x = -P_x \tag{1.91}$$

By substituting Eqs. (1.87) and (1.91) into Eq. (1.86) and noting that g(x)=1, we find

$$\frac{y''}{\left[1 + (y')^2\right]^{3/2}} = \frac{P}{EI_B}\frac{x}{\left\{1 + \frac{(n-1)}{L}\int_0^x\left[1 + \left(y'\left(x\right)\right)^2\right]^{1/2} dx\right\}^3} \tag{1.92}$$

Equation (1.92) is the exact integral nonlinear differential equation representing the given problem and its solution in general, is very complicated. Therefore, reasonable approximation must be used to simplify the problem. A reasonable simplification would be to use the approximate expression for h(x) given by Eq. (1.85). On this basis, by using Eq. (1.85) and applying the well known expression $bh^3(x)/12$ for beam with rectangular cross section, we find

$$I_x = \frac{bh^3}{12}\left[1 + (n-1)\frac{x}{L - \Delta}\right]^3 = I_1 f\left(x\right) \tag{1.93}$$

where

$$I_1 = I_B = \frac{bh^3}{12} \tag{1.94}$$

and

$$f(x) = \left[1 + (n-1)\frac{x}{L-\Delta}\right]^3 \tag{1.95}$$

On this basis, by substituting Eqs. (1.91), (1.94) and (1.95) into Eq. (1.86), we obtain

$$\frac{y''}{\left[1 + (y')^2\right]^{3/2}} = \frac{P(L-\Delta)^3}{EI_B} \frac{x}{[(n-1)x + (L-\Delta)]^3} \tag{1.96}$$

Equation (1.96) is the simplified nonlinear differential equation representing the tapered cantilever beam in Fig. 1.6. This equation, as shown later in the text, is far easier to solve compared to Eq. (1.92), and provides accurate results.

Example 1.3 For the uniform flexible cantilever beam loaded with a uniformly distributed load w_o as shown in Fig. 1.7, determine its exact integral nonlinear differential equation. Also suggest reasonable approximate ways to simplify the complexity of the problem. Assume that the modulus of elasticity E_x is constant and equal to E.

Solution: The large deformation configuration of the member is depicted in Fig. 1.7. The bending moment M_x at any distance $0 \le x \le L_o$, where $L_o = L - \Delta$, is

$$M_x = -w_0 x_0 \frac{x}{2} \tag{1.97}$$

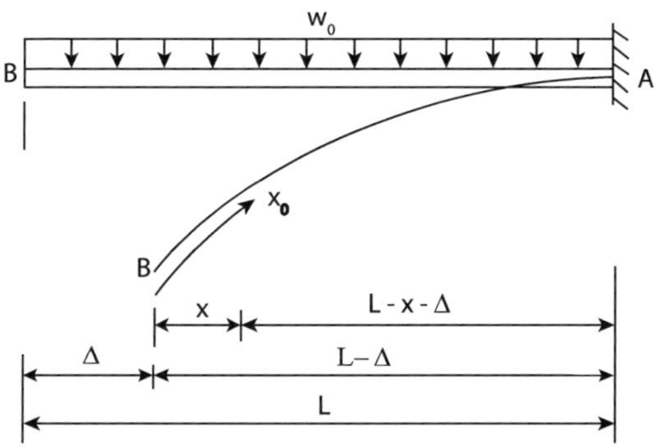

Fig. 1.7. Uniform cantilever beam loaded with a uniformly distributed load w_o over its entire span

From Eq. (1.75) we have

$$x_0 = \int_0^x \left\{1 + [y'(x)]^2\right\}^{1/2} dx \tag{1.98}$$

By substituting Eq. (1.98) into Eq. (1.97), we have

$$M_x = -\frac{w_0 x}{2} \int_0^x \left\{1 + [y'(x)]^2\right\}^{1/2} dx \tag{1.99}$$

By substituting Eq. (1.99) into the Euler–Bernoulli equation given by Eq. (1.86) and realizing that the stiffness EI of the flexible member is constant, we find

$$\frac{y''}{\left[1 + (y')^2\right]^{3/2}} = \frac{w_0 x}{2EI} \int_0^x \left\{1 + [y'(x)]^2\right\}^{1/2} dx \tag{1.100}$$

Equation (1.100) is the exact nonlinear integral differential equation for the cantilever beam in Fig. 1.7, and its solution is again very difficult.

We can simplify, however, the solution of Eq. (1.100) by assuming that x_o in Eq. (1.97) is given by Eq. (1.74). We rewrite this equation for convenience:

$$x_0(x) = x + \Delta(x) \tag{1.101}$$

Reasonable expressions to use for the horizontal displacement $\Delta(x)$ in Eq. (1.101), are given by Eqs. (1.81)–(1.84). If we decide to use Eq. (1.81), where the horizontal displacement Δ of the free end B of the flexible member is assumed to remain constant, then Eq. (1.101) yields

$$x_0(x) = x + \Delta \tag{1.102}$$

By substituting Eq. (1.102) into Eq. (1.97), we find

$$M_x = -\frac{w_0 x}{2}(x + \Delta) \tag{1.103}$$

and the Euler–Bernoulli nonlinear differential equation yields,

$$\frac{y''}{\left[1 + (y')^2\right]^{3/2}} = \frac{w_0 x}{2EI}(L - \Delta) \tag{1.104}$$

The solution of Eq. (1.104) is by far the most convenient to use when it is compared to Eq. (1.100)

If we make the assumption that $\Delta(x)$ is given by Eq. (1.82), then Eq. (1.101) yields

$$x_0(x) = x + \frac{\Delta x}{L_0} \tag{1.105}$$

and the expression for the bending moment given by Eq. (1.97), becomes

$$M_x = -\frac{w_0 x}{2}\left(x + \frac{\Delta x}{L_0}\right) \tag{1.106}$$

Therefore, by substitution, the Euler–Bernoulli equation yields

$$\frac{y''}{\left[1 + (y')^2\right]^{3/2}} = \frac{w_0 x}{2EI}\left(x + \frac{\Delta x}{L_0}\right) \tag{1.107}$$

Again, the nonlinear differential equation given by Eq. (1.107) is much simpler to solve, compared to the integral nonlinear differential equation given by Eq. (1.100). Similar simplifications are obtained by using Eq. (1.83) or Eq. (1.84) for the function $\Delta(x)$.

1.8 General Theory of the Equivalent Systems for Linear and Nonlinear Deformations

In the preceding sections it was shown that the solution of the Euler–Bernoulli differential equation becomes extremely difficult when the deformation of the member under consideration is large, and when the stiffness variation and loading conditions along its length vary arbitrarily. For flexible members, that is members which are subjected to large deformation, even simple cases of loading and constant stiffness will complicate a great deal the solution of the flexible problem.

In this section *the theory of the equivalent systems*, as it was developed by the author and his collaborators [2, 3, 5, 6, 15, 18], will be developed here for both linear and nonlinear problems. The emphasis, however, is concentrated in the solution of nonlinear problems, because this is the purpose of this text. For more information on linear systems the reader may consult the work of the author in [6, 84], as well as in other references at the end of this book.

The theory of the equivalent systems is general, and it applies to many structural problems which incorporate arbitrary variations of loading and moment of inertia along the length of a member, as well as variations of the modulus of elasticity of its material. The modulus of elasticity variations that are extensively examined in later sections of this book are the ones produced by large loadings that cause the material of the member to be stressed beyond its elastic limit and all the way to failure. In such cases, the modulus E will vary along the length of the member.

The purpose of the theory of the equivalent systems is to provide a much simpler mathematical model, in terms of *pseudolinear equivalent systems* and simplified *nonlinear equivalent systems*, that can be used to solve the extremely complicated nonlinear problem with well known simple methods of linear analysis. The pseudolinear equivalent system will always have a uniform stiffness EI throughout its equivalent length, and its loading will be different from the one acting on the original system. In other words, the length and loading of the pseudolinear equivalent system will be different, but its response will be identical to that of the original system.

1.8.1 Nonlinear Theory of the Equivalent Systems: Derivation of Pseudolinear Equivalent Systems

The derivation of pseudolinear equivalent systems of constant stiffness EI may be initiated by employing the Euler–Bernoulli law of deformations given by Eqs. (1.7) and (1.16). We rewrite Eq. (1.16) as follows:

$$\frac{y''}{\left[1+(y')^2\right]^{3/2}} = -\frac{M_x}{E_x I_x} \tag{1.108}$$

where the bending moment M_x, the modulus of elasticity E_x and the moment of inertia I_x are assumed to vary in any arbitrary manner.

The curvature of the member represented by the left-hand side of Eq. (1.108) is geometrical in nature, and it requires that the parameters M_x, E, and I on the right-hand side of the same equation to be also associated with the deformed configuration of the member. When the loading on the member is distributed and/or the cross-sectional moment of inertia is variable, the expressions for these parameters are in general nonlinear integral equations of the deformation and contain functions of horizontal displacement. That is, the bending moment M_x, depth h_x of the member, and moment of inertia I_x are all functions of both x and x_o. This is easily observed by examining the deformed configuration of the doubly tapered cantilever beam in Fig. 1.8. Therefore, the bending moment M_x has to be defined with respect to the deformed segment. On the other hand, the total load acting on an undeformed segment of a member does not change after the segment is deformed.

The variable stiffness $E_x I_x$, as discussed earlier, may be expressed as

$$E_x I_x = E_1 I_1 g(x) f(x) \tag{1.109}$$

where g(x) represents the variation of E_x with respect to a reference value E_1, and f(x) represents the variation of I_x with respect to a reference value I_1. If the member has a constant modulus of elasticity E and a constant moment of inertia I throughout its length, then g(x) = f(x) =1.00, and $E_x I_x$ = EI. In this case, the constant stiffness EI becomes the reference stiffness value $E_1 I_1$

By substituting Eq. (1.109) into Eq. (1.108), we have

$$\frac{y''}{\left[1+(y')^2\right]^{3/2}} = -\frac{1}{E_1 I_1}\,\frac{M_x}{g(x) f(x)} \tag{1.110}$$

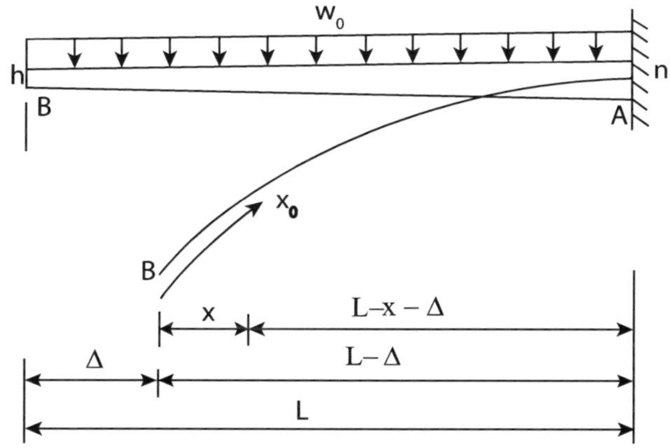

Fig. 1.8. Doubly tapered cantilever beam loaded with a uniformly distributed load w_o

If we integrate, hypothetically, Eq. (1.110) twice, the expression for the large transverse displacement y may be written schematically as

$$y(x) = \frac{1}{E_1 I_1} \int \left\{ - \int \left[1 + (y')^2 \right]^{3/2} \frac{M_x dx}{g(x) f(x)} \right\} dx + C_1 \int dx + C_2 \quad (1.111)$$

where C_1 and C_2 are the constants of integration which can be determined by using the boundary conditions of the member.

If we consider a member that has a constant stiffness $E_1 I_1$ and with an identical length and reference system of axes with the one used for Eq. (1.111), then we can write the expression for its large deflection y_e as follows:

$$y_e = \frac{1}{E_1 I_1} \int \left\{ - \int \left[1 + (y'_e)^2 \right]^{3/2} M_e dx \right\} dx + C'_1 \int dx + C'_2 \quad (1.112)$$

In Eq. (1.112), M_e is the bending moment at any cross section x, and C'_1 and C'_2 are the constants of integration.

On this basis, the deflection curves expressed by y and y_e in Eqs. (1.111) and (1.112), respectively, will be identical if

$$C_1 = C'_1 \quad \text{and} \quad C_2 = C'_2 \quad (1.113)$$

$$\int \left\{ - \int \left[1 + (y')^2 \right]^{3/2} \frac{M_e dx}{f(x) g(x)} \right\} dx = \int \left\{ - \int \left[1 + (y'_e)^2 \right]^{3/2} M_e dx \right\} dx \quad (1.114)$$

The conditions imposed by Eq. (1.113) are easily satisfied if the two members have the same arc length and boundary conditions. Equation (1.114) will be satisfied if $y_e{}' = y'$ and

$$M_e = \frac{M_x}{f(x) g(x)} \quad (1.115)$$

By following this way of thinking and examining Eq. (1.114), we conclude that the identity given by Eq. (1.114) will be satisfied when

$$\left[1 + (y_e')^2\right]^{3/2} M_e = \left[1 + (y')^2\right]^{3/2} \frac{M_x}{f(x)\, g(x)} \tag{1.116}$$

When the deformation of the members is small, then the angular rotations $(y')^2$ and $(y_e')^2$ may be neglected as being small compared to 1, and Eq. (1.116) reduces to Eq. (1.115). Thus, for small deflections, the moment diagram M_e of the equivalent system of constant stiffness E_1I_1 can be obtained from Eq. (1.115). Its equivalent shear force V_e and equivalent loading w_e can be obtained from Eq. (1.115) by differentiation. That is,

$$V_e = \frac{d}{dx}(M_e) = \frac{d}{dx}\left[\frac{M_x}{f(x)\, g(x)}\right] \tag{1.117}$$

$$w_e = -\frac{d}{dx}(V_e)\cos\theta = -\frac{d^2}{dx^2}\left[\frac{M_x}{f(x)\, g(x)}\right]\cos\theta \tag{1.118}$$

where $\cos\theta \approx 1$ when the rotations θ of the member are small. The equivalent constant stiffness system in this case is linear, and linear small deflection theory can be used for its solution.

When the deflections and rotations are large, $(y')^2$ and $(y_e')^2$ in Eqs. (1.110) and (1.116) cannot be neglected. By examining these two equations we observe that the moment M_e' of the equivalent pseudolinear system of constant stiffness E_1I_1 should be obtained from the equation

$$M_e' = \left[1 + (y')^2\right]^{3/2} M_e = \left[1 + (y')^2\right]^{3/2} \frac{M_x}{f(x)\, g(x)} = \frac{z_e}{f(x)\, g(x)} M_x \tag{1.119}$$

where

$$z_e = \left[1 + (y')^2\right]^{3/2} \tag{1.120}$$

Also we note that $\theta = \tan^{-1}(y')$ represents the slope of the initial nonlinear system.

By solving Eq. (1.110) for y'', we obtain

$$y'' = -\frac{1}{E_1I_1}\left[1 + (y')^2\right]^{3/2} \frac{M_x}{f(x)\, g(x)} \tag{1.121}$$

By substituting Eqs. (1.115) and (1.120) in Eq. (1.121) we find

$$y'' = \frac{M_e'}{E_1I_1} \tag{1.122}$$

Equation (1.122) is a pseudolinear differential equation and represents the pseudolinear equivalent system of constant stiffness E_1I_1. Therefore, it can be treated and solved as a linear differential equation once the moment diagram M_e' of the pseudolinear equivalent system is known.

The shear force V_e' and loading w_e' of the equivalent constant stiffness pseudolinear system may be determined from the expressions

$$V_e' = \frac{d}{dx}\left(M_e'\right) = \frac{d}{dx}\left[1 + (y')^2\right]^{3/2} M_e = \frac{d}{dx}\left[\frac{z_e}{f(x)\,g(x)}\right] M_x \tag{1.123}$$

$$\begin{aligned} w_e' &= -\frac{d}{dx}\left(V_e'\right)\cos\theta = \frac{d^2}{dx^2}\left[1 + (y')^2\right]^{3/2} M_e \cos\theta \\ &= -\frac{d^2}{dx^2}\left[\frac{z_e}{f(x)\,g(x)}\right] M_x \cos\theta \end{aligned} \tag{1.124}$$

When the equivalent constant stiffness pseudolinear system is obtained, elementary linear deflection theory and methods can be used to solve it. This is appropriate, because the deflections and rotations obtained by solving the pseudolinear system are identical to those of the original nonlinear system.

It should be also noted at this point that the equivalent moment diagram M_e given by Eq. (1.115), represents the moment M_e at any point x of a nonlinear equivalent system of length L equal to that of the original nonlinear system and of constant stiffness E_1I_1. This proves that we can also obtain equivalent simplified nonlinear systems by using Eq. (1.115). Consequently, the linearization of the initial variable stiffness flexible member, is obtained by multiplying M_e by the expression $\left[1 + (y')^2\right]^{3/2}$, as shown in Eq. (1.119). This is an extremely important observation, because, as it is shown later in this text, many complicated nonlinear problems with complex loadings and stiffness variations can be solved conveniently by obtaining first a simpler nonlinear equivalent system of constant stiffness E_1I_1 and then use it to proceed with pseudolinear analysis to obtain a pseudolinear equivalent system. This approach is amply illustrated in later parts of the text. Note also that V_e and w_e in Eqs. (1.117) and (1.118), respectively, give the shear force and loading, respectively, of the nonlinear equivalent system of constant stiffness E_1I_1.

We can simplify a great deal the mathematics regarding the computation of V_e' and w_e', or V_e and w_e, by approximating the shape of the moment diagram represented by Eq. (1.119), or the one represented by Eq. (1.115), with a few straight lines judiciously selected. This approximation simplifies to a large extent the derivation of pseudolinear and simplified nonlinear equivalent systems of constant stiffness. On this basis, the loading on the pseudolinear, or the equivalent linear system, will always consist of concentrated loads. This approach is amply illustrated in the following example.

Example 1.4 Determine a pseudolinear equivalent system for the uniform flexible cantilever beam of Example 1.1. By using the pseudolinear system determine the deflection and rotation at the free end. Show also how deflections and rotations can be determined at other points along the length of the member.

Solution: In Example 1.1, it was found that the horizontal displacement Δ of the flexible cantilever beam in Fig. 1.1a is 183.10 in. (4.65 m). The function G(x), in terms of Δ, is given by Eq. (1.65). By substituting the value of Δ into this equation, we find

$$\begin{aligned} G(x) &= 1.111\,(10)^{-6}\left[x^2 - (1{,}000 - 183.10)^2\right] \\ &= 1.111\,(10)^{-6}\left[x^2 - 667{,}325.61\right] \end{aligned} \tag{1.125}$$

From Eq. (1.67), the expression for $y'(x)$ is

$$y'(x) = \frac{G(x)}{\left\{1 - [G(x)]^2\right\}^{1/2}} \tag{1.126}$$

and from Eq. (1.119), the moment diagram M'_e of the pseudolinear equivalent system can be determined from the expression

$$M'_e = z_e M_x = \left[1 + (y')^2\right]^{3/2} M_x \tag{1.127}$$

where M_x, is the bending moment at any location x of the original system. The values of y' can be determined by using Eqs. (1.125) and (1.126). Note that $g(x) = f(x) = 1, EI = 180 \times 10^3$ kip in.$^2 (516.54 \times 10^3$ N m$^2), P = 0.4$ kip (1.78 kN) and L = 1,000 in. (25.4 m).

The bending moment M_x at any x between zero and $L_o = L - \Delta = 816.9$ in. (20.75 m), may be obtained from the equation

$$M_x = -Px = 0.4x \tag{1.128}$$

Table 1.1 provides the calculated values of $G(x)$, $y'(x)$, z_e, M_x and M'_e at various positions x between zero and 816.9 in. (20.75 m). By using the values of $M_e{}'$ in the last column of Table 1.1, we plot the moment diagram M'_e of the pseudolinear system as shown in Fig. 1.9a. We approximate the *shape* of $M_e{}'$ by four straight lines as shown in the same figure. The juncture points of these four straight lines may be located on above, or below the $M_e{}'$ curve, represented by the solid line in Fig. 1.9a. The purpose here is to approximate the shape of the $M_e{}'$ curve so that the areas added to this diagram and the areas subtracted from this diagram approximately balance each other. This can be done judiciously without any computations, because the accuracy of the calculated rotations and deflections depends mostly upon retaining the general shape of the $M_e{}'$ curve during its approximation with straight lines, but they are not very sensitive as to how accurately this approximation is performed. Even with moderately large errors in the approximation of $M_e{}'$, we obtain, for practical purposes, reasonable values for deflections and rotations. The reason for this is that, mathematically you get from moment to rotation and then to deflection by integration. When an appreciable error is introduced in the moment diagram curve, it reduces substantially by the time it gets to

Table 1.1. Values of G(x), y′(x), z_e, M_x and $M_e'(x)$ at various locations $0 \leq x \leq 816.9$ in.(1 in. = 0.0254 m, 1 kip in. = 113 N m, 1 kip = 4.448 kN)

(1) x (in.)	(2) G(x) Eq. (1.125)	(3) y′(x) Eq. (1.126)	(4) z_e Eq. (1.120)	(5) M_x (kip in.) Eq. (1.128)	(6) M_e' (kip in.) Eq. (1.127)
0	−0.7414	−1.1049	3.3095	0	0
100.0	−0.7303	−1.0689	3.1361	−40.00	−125.44
200.0	−0.6970	−0.9720	2.7121	−80.00	−216.97
300.0	−0.6414	−0.8360	2.2144	−120.00	−265.73
400.0	−0.5636	−0.6822	1.7739	−160.00	−283.82
500.0	−0.4636	−0.5232	1.4375	−200.00	−287.50
600.0	−0.3414	−0.3632	1.2043	−240.00	−289.03
700.0	−0.1970	−0.2009	1.0611	−280.00	−297.12
800.0	−0.0304	−0.0304	1.0014	−320.00	−320.45
816.9	0	0	1.000	−326.76	−326.76

rotation and deflection. This point was extensively investigated by the author and his students.

By applying statics, the shear force diagram is plotted as shown in Fig. 1.9b. For example,

$$V_1 = \frac{-170 - 0}{136.9} = -1.2418 \text{ kip } (5.5235 \text{ kN})$$
$$V_2 = \frac{-280 - (-170)}{200} = -0.55 \text{ kip } (2.4464 \text{ kN})$$

and so on. The *equivalent pseudolinear system* has length $L_o = L - \Delta = 816.9$ in. (20.75 m), and the loading is as shown in Fig. 1.9c. The loading is obtained by using Fig. 1.9b and applying statics. For example,

$$P_1 = 1.2418 \text{ kip } (5.5235 \text{ kN}) \quad \text{(downward)}$$
$$P_2 = 1.2418 - 0.55 = 0.6918 \text{ kip } (3.0771 \text{ kN}) \quad \text{(upward)}$$

and so on.

The pseudolinear system in Fig. 1.9c is, for practical purposes, an excellent approximation of the original nonlinear system. Linear methods of analysis, or available formulas from handbooks, may be used to solve the pseudolinear system for rotations and deflections. For example, the deflection and rotation at the free end B of the original system in Fig. 1.1a, may be determined by using the pseudolinear system in Fig. 1.9c and calculating its rotation and deflection at B′ by using handbook formulas, or known methods of basic mechanics, such as the moment–area method. Superposition is permissible, because the pseudolinear system is linear.

In order to use the moment–area method, we divide the M_e' diagram in Fig. 1.9a by the constant stiffness $EI = 180 \times 10^3$ kip in.^{2}(516.54×10^3 N m^2)

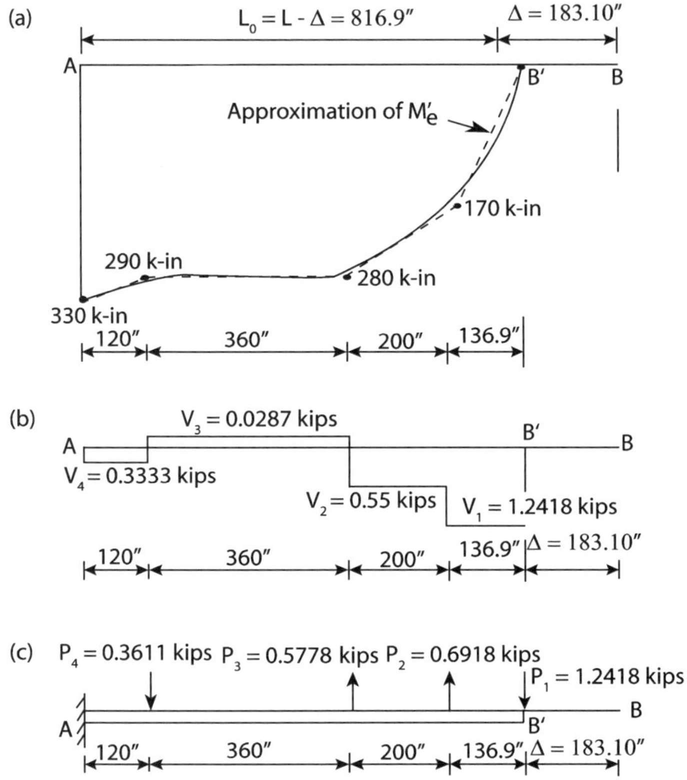

Fig. 1.9. (**a**) Moment diagram M_e' of the pseudolinear system with its shape approximated with four straight lines. (**b**) Shear force diagram. (**c**) Equivalent pseudolinear system of length $L_o = L - \Delta = 816.9$ in. (1 in. = 0.0254 m, 1 kip in. = 113 N m, 1 kip = 4.448 kN)

to obtain the M_e'/EI diagram. The deflection $\delta_{B'}$ at the free end B' of the pseudolinear system is equal to the first moment of the M_e'/EI diagram between A and B', taken about B'. By using the straight-line approximation of M_e' shown by the dashed lines in Fig. 1.9a, dividing it by EI and then taking its first moment about B', we find

$$
\begin{aligned}
\delta_{B'} &= \frac{1}{EI}\left[\frac{1}{2}(170)(136.9)\left(\frac{2}{3}\right)(136.9) + (170)(200)(236.9)\right. \\
&\quad + \frac{1}{2}(110)(200)(270.2333) + (280)(360)(516.9) + \frac{1}{2}(10)(360)(576.9) \\
&\quad \left. + (290)(120)(756.9) + \frac{1}{2}(40)(120)(776.9)\right] \\
&= \frac{1}{EI}[98,170,244]
\end{aligned}
$$

or

$$\delta_{B'} = \frac{1}{180(10)^3}[98,170,244] = 545.39 \text{ in. } (13.85 \text{ m})$$

Also, by using the moment–area method, $y_{B'}{}'$ at the free end of the pseudolinear system would be equal to the total $M_e{}'/EI$ area between points A and B′. Thus, by using again Fig. 1.9a, we obtain

$$\begin{aligned} y_{B'}{}' &= \frac{1}{EI}\left[\frac{1}{2}(170)(136.9) + (170)(200) + \frac{1}{2}(110)(200) + (280)(360)\right. \\ &\qquad \left. + \frac{1}{2}(10)(360) + (290)(120) + \frac{1}{2}(40)(120)\right] \\ &= \frac{1}{EI}[196,436.5] \end{aligned}$$

or

$$y_{B'}{}' = \frac{1}{180(10)^3}[196,436.5] = 1.0913$$

Thus,

$$\theta_{B'} = \tan^{-1}(y_B{}') = 47.50°$$

The calculated values $\delta_{B'}$ and $\theta_{B'}$ that were obtained by using the pseudolinear system in Fig. 1.9c are very accurate, at least for practical applications, and they are closely identical to the analogous exact values at the end B that can be obtained by solving directly the original member in Fig. 1.1a. In fact, the direct solution was obtained by integrating directly the Euler–Bernoulli equation given by Eq. (1.46). On this basis we found that the deflection and rotation at the end B of the nonlinear member in Fig. 1.1a, are $\delta_B = 523.27$ in. (13.29 m) and $\theta_B = 47.86°$. If we can assume that these last two values are the exact values, then the error by using the pseudolinear system would be 4.22% for the deflection and 0.75% for the rotation. However, if it is required, the accuracy can be improved if we will approximate the moment diagram of the pseudolinear system, shown in Fig. (1.9a), with more straight lines. The purpose in using the pseudolinear system is to simplify the solution of complicated nonlinear problems with complicated loading conditions and moment of inertia variations. This point of view is clearly demonstrated throughout this text.

By using direct integration, Table 1.2 has been prepared which shows the variation of δ_B, Δ_B, and θ_B at the free end B of the flexible beam for the indicated values of the concentrated vertical load P at the free end B. The second column of the table gives the values of δ_B that were obtained by neglecting y' and proceeding with linear analysis. If we compare these values with the analogous ones shown in the third column of the table, we note that the error by using linear analysis becomes unreasonably huge as the load P increases, and such linear procedures should not be used for flexible members.

By using nonlinear analysis, the large deflection configurations of the flexible member for values of the load P = 1, 3, and 10 kip, are shown plotted

Table 1.2. Values of δ_B, Δ_B, and θ_B for various values of P and comparisons with linear theory (1 in. = 0.0254 m, 1 kip = 4.448 kN)

load P	linear analysis	nonlinear analysis		
(kips)	δ_B (in.)	δ_B (in.)	Δ_B (in.)	θ_B (deg.)
0.2	370.37	328.61	67.36	28.90
0.4	740.74	523.27	183.10	47.86
0.6	1,111.11	629.00	281.29	59.42
0.8	1,481.48	691.58	356.71	66.87
1.0	–	732.14	414.95	71.95
2.0	–	821.37	577.22	83.23
2.5	–	841.64	621.12	85.48
3.0	–	856.20	653.84	86.92
5.0	–	888.86	731.69	89.00
10.0	–	921.40	810.27	89.61

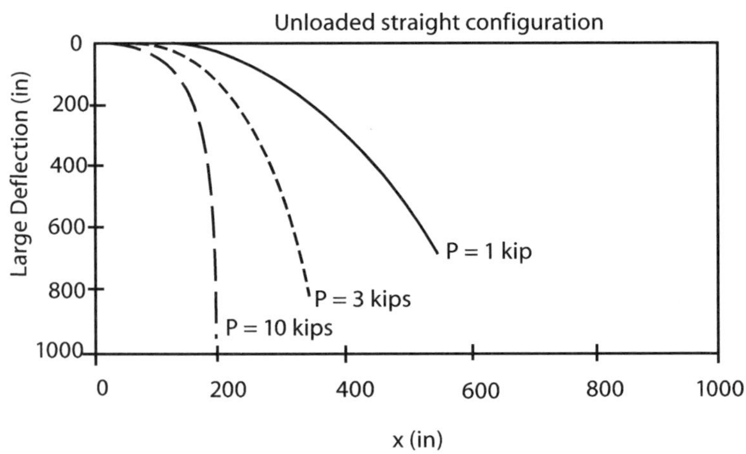

Fig. 1.10. Large deflection curves for P = 1, 3, and 10 kip (1 kip = 4.448 kN, 1 in. = 0.0254 m)

in Fig. 1.10. Note that for P = 10 kip (44.48 kN), the member is practically hanging in the vertical direction.

Deflections and Rotations at Any x Between Zero and L_o

We can also use the pseudolinear system in Fig. 1.9c to determine the vertical deflection y and rotation y′ at any $0 \leq x \leq L_o$. Since this system is now linear, we can use any linear method of analysis, or existing handbook formulas to do it. For this problem, the moment–area method and handbook formulas are convenient to use. Since the pseudolinear system is linear, superposition can be used. That is, if you prefer, you can solve the pseudolinear system in Fig. 1.9c for each concentrated vertical load separately and superimpose the results.

It should be pointed out, however, that the length L_o of the pseudolinear system is not equal to the length L of the original system, and the same applies for the values of x and x_o. The pseudolinear system provides the values y and y' at any $0 \leq x \leq L_o$, but for each x there corresponds a value of $0 \leq x_o \leq L$, where L is the length of the original member, which can be determined from Eq. (1.75). We rewrite this equation below:

$$x_0(x) = \int_0^x \left\{1 + [y']^2\right\}^{1/2} dx \tag{1.129}$$

The Simpson's rule given by Eq. (1.36) in Sect. 1.5 can be used for this purpose.

Let it be assumed, for example, that we used the pseudolinear system in Fig. 1.9c, we then applied the moment–area method, and we determined the values of y and y' at x = 100 in. (2.54 m), and we want to know the value of x_o of the original system in Fig. 1.1a that corresponds to x = 100 in. (2.54 m). We can do this by using Simpson's One-Third rule.

If we use this rule and assume n = 10, then Eq. (1.70) yields

$$\lambda = \frac{100 - 0}{10} = 10 \tag{1.130}$$

By examining Eq. (1.29), we note that the function f(x) to be used in Eq. (1.69), is

$$f(x) = \left\{1 + [y'(x)]^2\right\}^{1/2} dx \tag{1.131}$$

The values of f(x) at $x = 0, 10, 20, \ldots, 100$, are designed as $y_o, y_1, \ldots, y_{10}$, respectively, and they are determined by using Eq. (1.131) in conjunction with Eqs. (1.65) and (1.67), which are as follows:

$$G(x) = 1.111(10)^{-6}[x^2 - 667,325.61] \tag{1.132}$$

$$y'(x) = \frac{G(x)}{\{1 - [G(x)]^2\}^{1/2}} \tag{1.133}$$

For example, at x = 0, we have

$$G_0 = -0.783922$$

$$y'(0) = \frac{-0.783922}{\sqrt{1 - (-0.783922)^2}} = -1.262641$$

$$y_0 = f(0) = \sqrt{1 + (-1.262641)^2} = 1.610671$$

At x = 10 in. (0.254 m), we have

$$G(10) = 1.111(10)^{-6}[10^2 - 667,325.61] = -0.741288$$

$$y'(10) = \frac{-0.741288}{\sqrt{1 - (-0.741288)^2}} = \frac{-0.741288}{0.671188} = -1.104442$$

$$y_1 = f(10) = \sqrt{1 + (-1.104442)^2} = 1.489896$$

In a similar manner, the remaining values of $y_2, y_3, \ldots, y_{10}$, are determined.
By substituting into Eq. (1.69), we find

$$\begin{aligned} x_0 = \frac{10}{3} & [1.610671 + (4)(1.489896) + (2)(1.489079) + (4)(1.487724) \\ & +(2)(1.485832) + (4)(1.483413) + (2)(1.480479) \\ & +(4)(1.477039) + (2)(1.473105) + (4)(1.468699) + 1.463833] \\ = \frac{10}{3} & [44.558578] = 148.53 \text{ in. } (3.77 \text{ m}) \end{aligned}$$

Therefore, we can conclude that when we calculate the deflection or the rotation of the equivalent pseudolinear system at $x = 100$ in. (2.54 m), the corresponding position x_o on the original nonlinear system is 148.53 in. (3.77 m). In other words, there is a complete mathematical correspondence between x and x_o which defines the nature of the pseudolinear equivalent system. It should be also pointed out that the deflection curve of the equivalent pseudolinear system is identical to the one of the original nonlinear system.

1.8.2 Nonlinear Theory of the Equivalent Systems: Derivation of Simplified Nonlinear Equivalent Systems

It was stated earlier that the large deformations of flexible members are no longer a linear function of the bending moment or of the applied load and, consequently, the principle of superposition is not applicable. This restriction creates enormous difficulties when we try to solve beam problems with more elaborate loading conditions. The solution becomes even more complicated when the moment of inertia of a flexible member varies arbitrarily along its length. The flexible cantilever beam in Fig. 1.11, loaded as shown, illustrates a mild case of an elaborate combined loading condition coupled with a variable depth along the length of the member. A reasonable solution to such types of problems can be obtained if a simpler equivalent mathematical model is first obtained that accurately (or exactly) represents the initial complicated mathematical problem. This may be accomplished by reducing the initial nonlinear problem into a simpler equivalent nonlinear problem that can be solved more

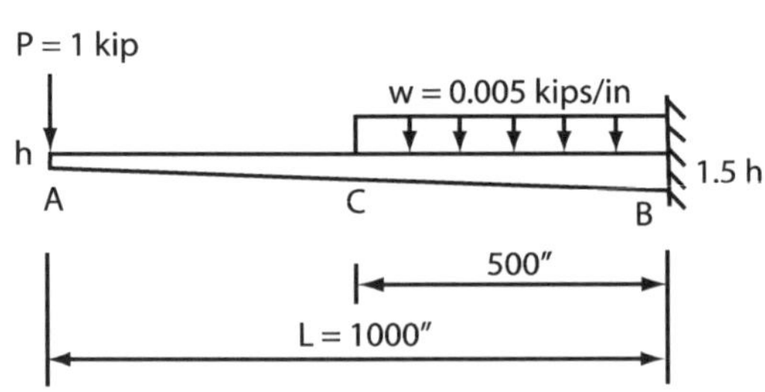

Fig. 1.11. Tapered flexible cantilever beam loaded as shown

conveniently by using either pseudolinear analysis as discussed earlier, or by utilizing available solutions (or solution methodologies) of nonlinear analysis.

The derivation of constant stiffness nonlinear equivalent systems, can be carried out by using Eq. (1.108) and substituting for the variable stiffness E_xI_x the expression given by Eq. (1.109), yielding

$$\frac{y''}{[1+(y')^2]^{3/2}} = -\frac{1}{E_1I_1}\frac{M_x}{g(x)f(x)} \tag{1.134}$$

or

$$\frac{y''}{[1+(y')^2]^{3/2}} = -\frac{M_e}{E_1I_1} \tag{1.135}$$

where

$$M_e = \frac{M_x}{g(x)f(x)} \tag{1.136}$$

Equation (1.135) is the nonlinear differential equation of an equivalent system of constant stiffness E_1I_1, whose bending moment M_e at any cross section is given by Eq. (1.136). Therefore, the variable stiffness nonlinear system represented by Eq. (1.134), and the one of constant stiffness represented by Eq. (1.135), will have identical deflection curves. Therefore, we can conclude that Eq. (1.135) may be used to solve the variable stiffness problem by applying nonlinear analysis. In order to make the solution easier, the shape of the M_e diagram represented by Eq. (1.136) may be approximated with a few straight lines. This procedure will produce a simpler constant stiffness equivalent nonlinear system that is always loaded with a few concentrated loads as it was accomplished earlier for the pseudolinear system.

The following example illustrates the application of the theory and methodology.

Example 1.5 The tapered flexible cantilever beam in Fig. 1.12a is loaded with a distributed triangular load $w_o = 0.01$ kip in.$^{-1}$(1,751.27 N m^{-1}), and a concentrated vertical load $P = 1$ kip (4.448 kN) at its free end. At the free end B the stiffness $EI_B = 180 \times 10^3$ kip in.^{2}(516,551 N m^2). Determine a simplified nonlinear equivalent system of constant stiffness EI_B. The width of the member is constant and equal to b, and the modulus of elasticity E is constant.

Solution: The depth h_x at any distance x from the free end B of the member is given by the equation

$$h_x = h\left(1+\frac{x}{2L}\right) \tag{1.137}$$

Therefore, for rectangular cross sections, the moment of inertia I_x at any distance x from the end B of the member is

$$I_x = \frac{bh_x^3}{12} = \frac{bh^3}{12}\left[1+\frac{x}{2L}\right]^3 = I_Bf(x) \tag{1.138}$$

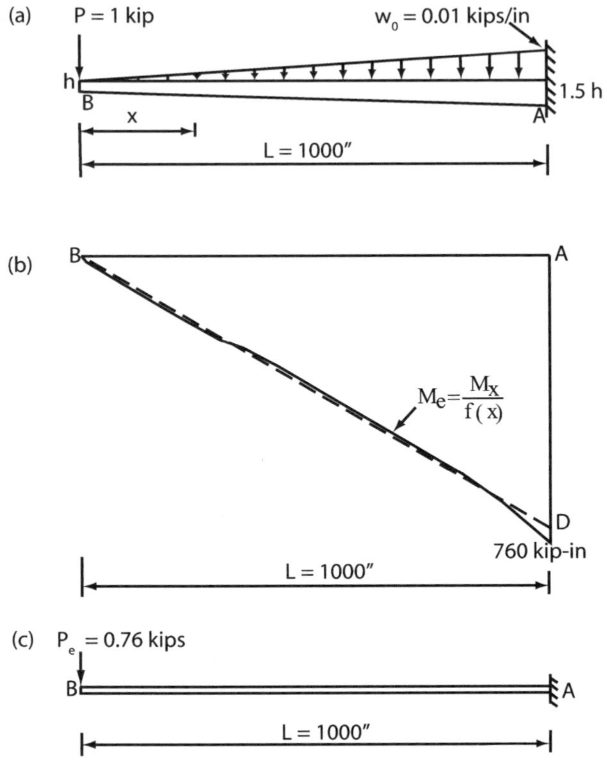

Fig. 1.12. (**a**) Original tapered cantilever beam loaded as shown. (**b**) Moment diagram M_e of the simplified nonlinear equivalent system of constant stiffness EI_B. (**c**) Simplified nonlinear equivalent system (1 in. = 0.0254 m, 1 kip = 4,448 N, 1kip in. = 113 N m)

where h_x is given by Eq. (1.137) and

$$I_B = \frac{bh^3}{12} \tag{1.139}$$

$$f(x) = \left[1 + \frac{x}{2L}\right]^3 \tag{1.140}$$

The moment diagram M_e of the simplified nonlinear equivalent system of constant stiffness EI_B, can be obtained by using Eq. (1.136). In this equation the modulus function g(x) is constant and equal to one. In the same equation, the moment M_x at any distance x from the free end B of the member, can be determined by using the original member in Fig. 1.12a and applying statics. For convenience, Table (1.3) has been prepared, where the first column of the table shows the selected values of x, the second column gives the corresponding values of f(x) using Eq. (1.140), the third column gives the values of M_x which are determined by applying statics, and the last column of the table gives

Table 1.3. Values of f(x), M_x and M_e for the indicated values of x (1 in. = 0.0254 m, 1 kip in. = 113 N m)

(1) x (in)	(2) f(x)	(3) M_x (kip-in)	(4) $M_e = \frac{M_x}{f(x)}$ (kip in.)
0	1	0	0
100	1.1576	−101.67	−87.83
200	1.3310	−213.33	−160.28
300	1.5209	−345.00	−226.84
400	1.7280	−506.67	−293.21
500	1.9531	−708.33	−362.67
600	2.1970	−960.00	−436.96
700	2.4604	−1,271.67	−516.85
800	2.7440	−1,653.33	−602.44
900	3.0486	−2,115.00	−693.76
1,000	3.3750	−2,666.67	−790.12

the values of M_e at the selected values of x, which are obtained by using Eq. (1.136).

The M_e diagram is plotted as shown by the solid line in Fig. 1.12b. A very reasonable approximation of this diagram would be the straight dahed line BD drawn as shown in the same figure. By using the straight line approximation and applying statics, we obtain the simplified nonlinear equivalent system of uniform stiffness EI_B shown in Fig. 1.12c. This system is a great deal simpler when we compare it to the original system in Fig. 1.12a. To solve the simplified nonlinear system we can use the pseudolinear analysis discussed earlier in this section Example 1.4. We can also do it by direct integration of the Euler–Bernoulli equation given by Eq. (1.135), or by using tables which are prepared using elliptic integral solution and are available in the literature [85].

In [85, p. 516], a table is provided by the author that can be used to determine the horizontal, vertical, and rotational displacements at the free end of a uniform flexible cantilever beam, loaded by a vertical concentrated load P at the free end. The solution is based on elliptic integrals, and provides reliable answers for practical applications. By using this table we find that at the free end of the simplified nonlinear equivalent system shown in Fig. 1.12c, we have, $\theta_B = 65.46^\circ$, vertical displacement $\delta_B = 679.78$ in. (17.27 m), and horizontal displacement $\Delta_B = 342.11$ in. (8.69 m).

In order to compare results, the original system in Fig. 1.12a was solved by the author and his students by deriving its Euler–Bernoulli nonlinear differential equation based on $x_o = x + \Delta$, where Δ is the horizontal displacement of its free end B. Then this equation was integrated twice to determine the vertical deflection at the free end B. The two constants of integration were determined by using the boundary conditions of zero rotation and zero deflection at the fixed end, and the required integrations were carried out by using Simpson's rule. The result obtained was $\delta_B = 701.0$ in. (17.81 m), giving a difference of 3.02%. The same problem was also solved by using the Runge–Kutta method, yielding $\theta_B = 67.54°$, a difference of 3.08%, and $\delta_B = 708.00$ in. (17.99 m), a difference of 4.0%. We note that all these approaches provided reliable results, but the use of simplified nonlinear equivalent systems is by far the simplest method to use.

Other cases of such problems can be treated in a similar manner. We should always remember that even with very crude approximations of the moment diagram with straight line segments, we will always obtain a reasonable solution to the problem for practical applications. Some caution, however, should be taken to retain the general shape of the moment diagram during its approximation with straight line segments. The approximations shown in Figs. 1.9a and 1.12b are considered to be excellent, and good accuracy, for practical applications, could be obtained using even fewer straight line segments in Fig. 1.9a.

1.8.3 Linear Theory of the Equivalent Systems

When the deformations of a member in bending are small, the y' in Eq. (1.110) is small when it is compared to one and it can be neglected for practical applications. On this basis, Eq. (1.110) reduces to the following second order differential equation:

$$y'' = -\frac{1}{E_1 I_1}\frac{M_x}{f(x)g(x)} \tag{1.141}$$

Also, from Eq. (1.120), we find $z_e = 1.0$, and from Eq. (1.119) we find

$$M'_e = M_e = \frac{M_x}{f(x)g(x)} \tag{1.142}$$

Therefore, Eq. (1.141) can be written as

$$y'' = -\frac{M_e}{E_1 I_1} \tag{1.143}$$

where in this equation the moment M_e is given by Eq. (1.142).

Equation (1.142) provides the bending moment of a linear equivalent system of constant stiffness $E_1 I_1$ at any location x of the member. With known M_e, the equivalent shear force V_e and the equivalent loading w_e of the constant stiffness equivalent system can be obtained from Eq. (1.142) by differentiation. That is,

$$V_e = \frac{d}{dx}(M_e) = \frac{d}{dx}\left[\frac{M_x}{f(x)g(x)}\right] \tag{1.144}$$

$$w_e = -\frac{d}{dx}(V_e)\cos\theta = -\frac{d^2}{dx^2}\left[\frac{M_x}{f(x)g(x)}\right]\cos\theta \tag{1.145}$$

where $\cos\theta \approx 1$ when the rotations are small. For more information on this subject you may refer to the work of the author and his collaborators given in the references at the end of this book.

The following examples illustrate the application of the theory.

Example 1.6 A tapered cantilever beam of rectangular cross section is loaded by a vertical load P at the free end B, as shown in Fig. 1.13a. The depth h_x of the member, for convenience, is selected to be

$$h_x = \left(\frac{3L - 2x}{L}\right)^{1/3} h \tag{1.146}$$

If the deformations of the member are assumed to be small, determine an exact equivalent system of constant stiffness EI_B. Also determine an approximate

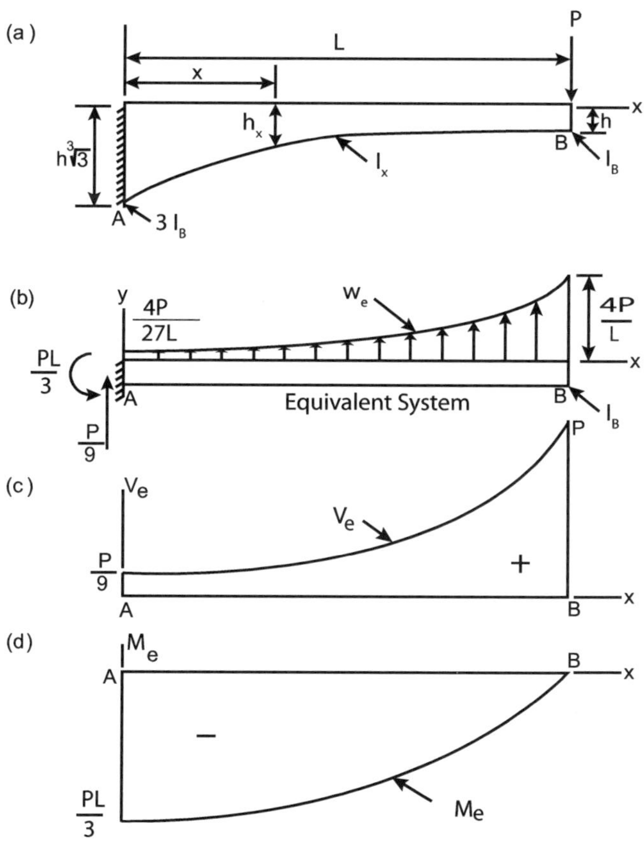

Fig. 1.13. (**a**) Tapered cantilever beam loaded as shown. (**b**) Exact equivalent system of constant stiffness EI_B. (**c**) Shear force diagram of the equivalent system. (**d**) Moment diagram M_e of the equivalent system

equivalent system loaded with vertical concentrated loads. The modulus of elasticity E and the width b of the member are constant.

Solution: For the assumed variation of h_x, the moment of inertia I_x of the member is

$$I_x = \frac{bh_x^3}{12} = \frac{3L - 2x}{L}\frac{bh^3}{12} = I_B f(x) \tag{1.147}$$

where

$$f(x) = \frac{3L - 2x}{L} \tag{1.148}$$

$$I_B = \frac{bh^3}{12} \tag{1.149}$$

At any x from the fixed end of the member, the bending moment M_x is

$$M_x = -P(L - x) \tag{1.150}$$

From Eq. (1.142), by substituting for f(x) which is given by Eq. (1.148) and noting that the function $g(x) = 1$, we find that the bending moment M_e of the equivalent system of constant stiffness EI_B is

$$M_e = \frac{M_x}{f(x)} = -\frac{PL(L - x)}{3L - 2x} \tag{1.151}$$

where M_x is given by Eq. (1.150).

By differentiating M_e once, we find the shear force V_e of the equivalent system, which is

$$V_e = \frac{dM_e}{dx} = \frac{PL^2}{(3L - 2x)^2} \tag{1.152}$$

Also, by differentiating V_e once, we obtain the loading w_e of the equivalent system of constant stiffness EI_B. That is

$$w_e = \frac{dV_e}{dx} = \frac{4PL^2}{(3L - 2x)^3} \tag{1.153}$$

Equations (1.151), (1.152), and (1.153) are plotted as shown in Figs. 1.13d, 1.13c, and 1.13b, respectively. Figure 1.13b illustrates the exact equivalent system of constant stiffness EI_B, loaded as shown, which produces a deflection curve that is identical to the deflection curve produced by the original system in Fig. 1.13a.

We proceed now with the second part of the problem which requires to determine an approximate equivalent system of constant stiffness EI_B which will be loaded with vertical concentrated loads.

An accurate approximation may be obtained by approximating the shape of M_e in Fig. 1.13d with a few straight-line segments, as it was done earlier for the nonlinear analysis. In order to get numerical results, we assume that

P = 10 kip (44,480 N) and L = 300 in. (7.62 m). On this basis, Eq. (1.151) yields

$$M_e = -\frac{1,500(300 - x)}{450 - x} \tag{1.154}$$

Equation (1.154), which represents the exact variation of the moment M_e of the equivalent system of constant stiffness EI_B, is plotted as shown by the solid line in Fig. 1.14a. In this figure, the shape of M_e is approximated with two straight-line segments as shown by the dashed line. By using statics we plot the shear force diagram as shown in Fig. 1.14b, and the simplified equivalent system of constant stiffness EI_B shown in Fig. 1.14c. If we solve the approximate equivalent system in Fig. 1.14c, we will find out that its deflections and rotations are closely identical to those of the original system in

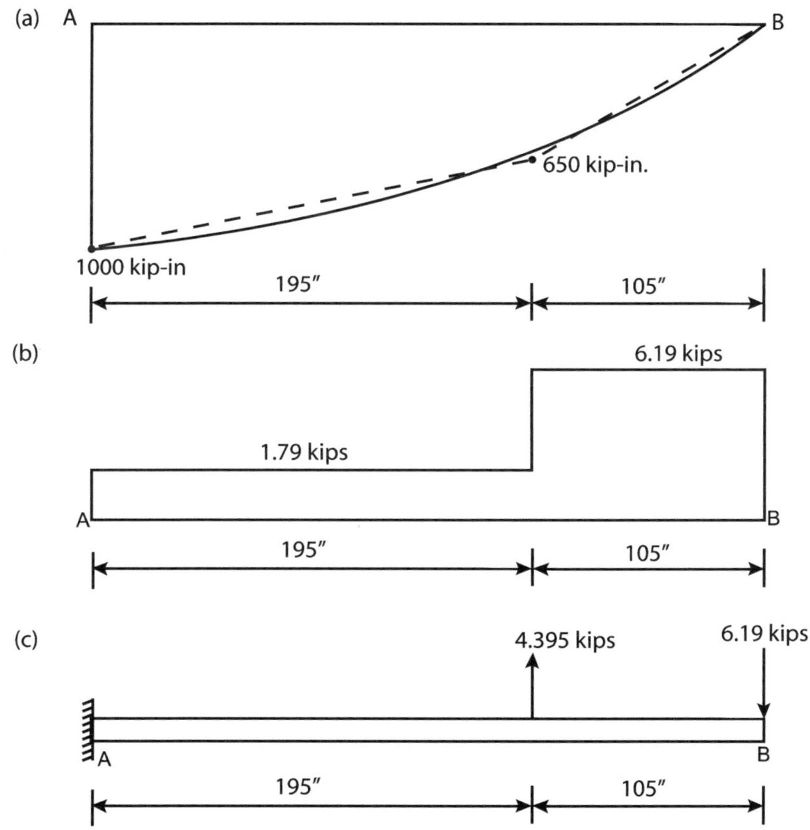

Fig. 1.14. (**a**) M_e diagram with its shape approximated with two straight-line segments. (**b**) Shear force diagram V_e. (**c**) Approximate equivalent system of constant stiffness EI_B loaded with two concentrated loads as shown (1 in. = 0.0254 m, 1 kip = 4,448 N, 1 kip in. = 112.9848 N m)

Fig. 1.13a and to those of the exact equivalent system in Fig. 1.13b. However, the solution of the approximate equivalent system in Fig. 1.14c is very simple when it is compared to the solution of the systems in Fig. 1.13a and 1.13b.

By using the approximate equivalent system in Fig. 1.14c and applying the moment–area method, we find that the deflection δ_B at the free end B is $105,022.44/EI_B$, and the rotation θ_B at the same end is $999.975/EI_B$ radians. The units of E and I_B are kip in.$^{-2}$ and in.^{4}, respectively. Compared to the exact solution, the error by using the two straight-line approximation for M_e, would be less than 1.5% for the deflection, and much less for the rotation. However, if required, we can improve the accuracy by using more straight-line segments for the approximation of M_e.

Example 1.7 A simply supported rectangular stepped beam of variable cross section, is loaded as shown in Fig. 1.15a. By approximating the shape of the M_e diagram with straight-line segments, determine an approximate equivalent system of uniform stiffness EI_I, where I_I, is the moment of inertia of the member at its end B. The width b of the member and its modulus E are constant.

Solution: At various locations x, where x is measured from the left support A, the values of M_x, f(x), and M_e are calculated and they are shown in Table 1.4. The moment diagram M_x of the original system in Fig. 1.15a, is plotted in Fig. 1.15c using the values of M_x in Table 1.4. The variation f(x) of the moment of inertia I_x in terms of I_1, where I_1 is the moment of inertia at the end B, is plotted as shown in Fig. 1.15b.

By using the values of $M_e = M_x/f(x)$ shown in the last column of Table 1.4, the moment diagram M_e of the equivalent system of constant stiffness EI_1 is shown plotted by the solid line in Fig. 1.16a. The shape of the M_e diagram is now approximated with four straight-line segments as shown by the dashed line in the same figure. By applying statics, using the approximated M_e diagram, we plot the equivalent shear force diagram shown in Fig. 1.16b. By

Table 1.4. Calculated values of M_x, f(x), and M_e at various locations x from the left support A (1 ft = 0.3048 m, 1 kip ft = 1,355.75 N m)

x (ft)	M_x (kip ft)	f(x)	$M_e = M_x/f(x)$ (kip ft)
0	0	17.560	0
5	37.5	13.820	2.71
10	50.0	10.650	4.69
10	50.0	2.745	18.22
12	48.0	2.300	20.85
14	42.0	1.908	22.00
15	37.5	1.728	21.70
18	18.0	1.260	14.27
20	0	1.000	0

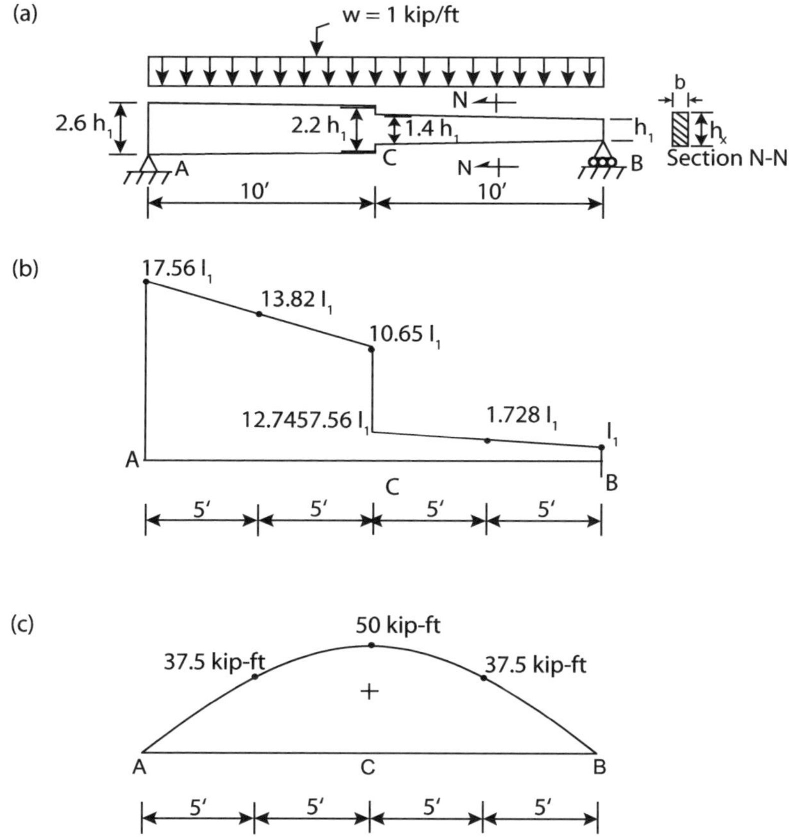

Fig. 1.15. (**a**) Simply supported stepped beam of variable stiffness. (**b**) Moment of inertia variation. (**c**) Moment diagram M_x of the initial system in Fig. 1.15a. (1 ft $= 0.3048$ m, 1 kip ft$^{-1} = 14,593$ N m^{-1}, 1 kip ft $= 1,355.75$ N m)

using this shear force diagram and applying static, we obtain the approximate equivalent system of constant stiffness EI_1 shown in Fig. 1.16c.

Note that in addition to the concentrated vertical loads, a concentrated moment equal to 12.72 kip ft $(18.22 - 5.5 = 12.72)$ (16,990.74 N m) must also act at point C of the equivalent system in Fig. 1.16c. This is justified because in this case there is an abrupt change of the bending moment M_e at point C. This means that a step in the original system in Fig. 1.15a requires a concentrated moment at the analogous point of the uniform stiffness equivalent system, in order to compensate for the elimination of the step.

By using the equivalent system in Fig. 1.16, deflections and rotations can be determined by using linear methods of analysis or handbook formulas.

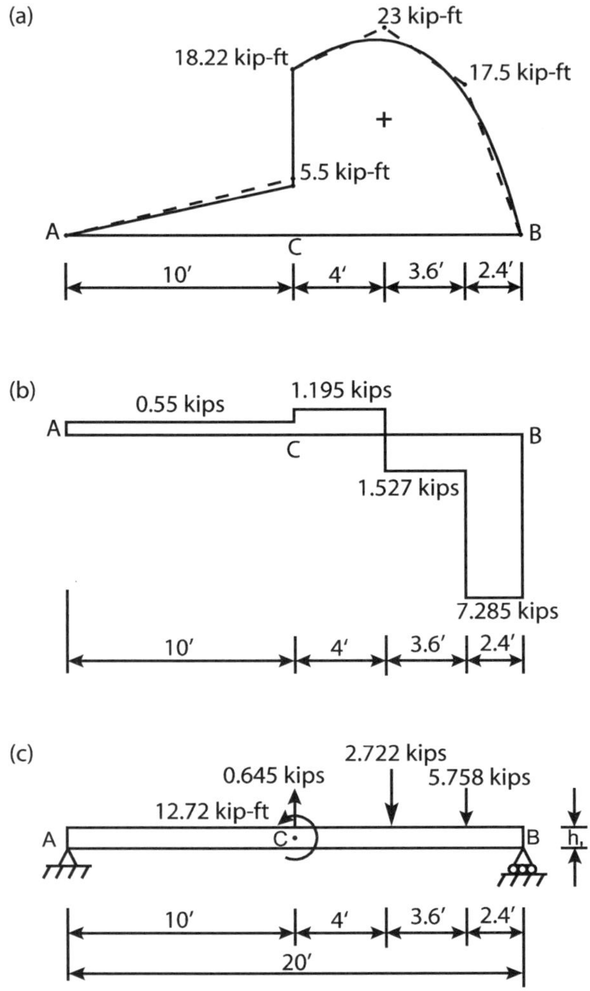

Fig. 1.16. (**a**) Moment diagram M_e with its shape approximated with four straight-line segments. (**b**) Equivalent shear force diagram. (**c**) Equivalent system of constant stiffness $EI_1(1 \text{ ft} = 0.3048\,\text{m}, 1 \text{ kip ft} = 1,335.75\,\text{N m})$

These rotations and deflections will be closely identical with the ones at corresponding points of the original system in Fig. 1.15a.

Example 1.8 The depth d_x and the moment of inertia variation I_x of the elastically supported beam in Fig. 1.17a, are

$$d_x = (\frac{L + 3x}{L})^{1/3} d_1 \tag{1.155}$$

$$I_x = \frac{L + 3x}{L} I_1 = f(x) I_1 \tag{1.156}$$

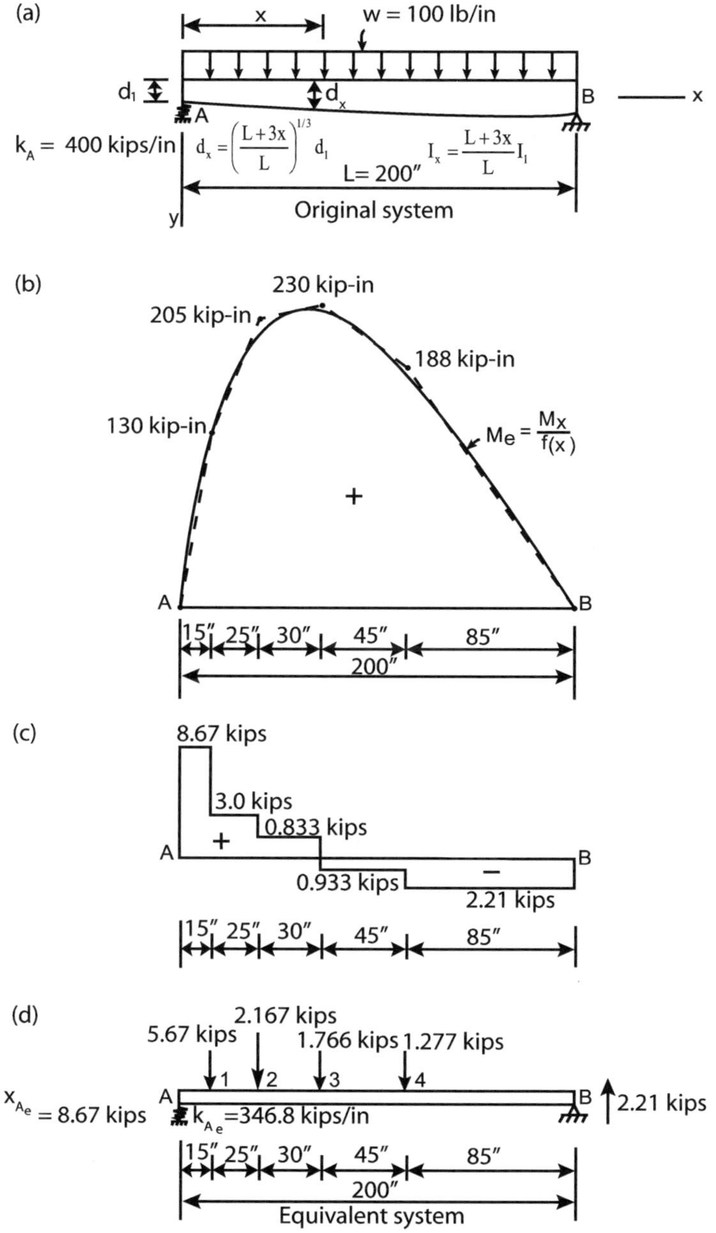

Fig. 1.17. (**a**) Original variable stiffness member. (**b**) M_e diagram with its shape approximated with five straight-line segments. (**c**) Equivalent shear force diagram. (**d**) Equivalent system of constant stiffness EI_1(1 ft $= 0.3048\,\mathrm{m}, 1$ kip $= 4{,}448.222\,\mathrm{N}, 1$ kip in. $= 113\,\mathrm{N\ m}, 1$ in. $= 0.0254\,\mathrm{m}, 1$ lb in^{-1} $= 175.118\,\mathrm{N\ m^{-1}}$)

where d_1 is the depth of the member at support A, $I_1 = bd_1^3/12$ (where b is the constant width of the member), is the moment of inertia at support A, and x is measured from support A. The modulus of elasticity E of the member is constant, the stiffnesss $EI_1 = 90 \times 10^6$ kip in.^{2}, and at support A the spring stiffness $k_A = 400$ kip in^{-1} (70,050.74 kN). Determine the equivalent system of constant stiffness EI_1. The length L = 200 in. (5.08 m).

Solution: If x_A is the elastic reaction from the spring at support A, then by considering the free body diagram of the beam and taking moments about the end point B, we obtain

$$\sum M_B = -200X_A + (100)(200)(100) = 0$$
$$X_A = 10 \text{ kip in. } 44,482.22$$

The values of f(x), M_x, and $M_e = M_x/f(x)$ are calculated at various positions x from support A, and they are as shown in Table 1.5.

The values of M_e given in the last column of Table 1.5 are used to plot the M_e diagram shown by the solid line in Fig. 1.17b. The shape of M_e is then approximated in the usual way with five straight-line segments as shown in the same figure. Note the values of the approximated M_e at the juncture points of the straight-line segments. Using these values of the approximated M_e and applying statics, we determine the equivalent shear force diagram shown in Fig. 1.17c. From the equivalent shear force diagram in Fig. 1.17c, by using statics, we determine the equivalent system of constant stiffness EI_1, shown in Fig. 1.17d.

There is however, an additional boundary condition at support A that must be satisfied. At this support, the vertical displacement δ_A of the spring of stiffness k_A in Fig. 1.17a, must be equal to the vertical displacement δ_{A_e} of the spring of stiffness k_{A_e} of the equivalent system in Fig. 1.17d. We know that

$$\delta_A = \frac{X_A}{k_A} \tag{1.157}$$

Table 1.5. Values of f(x), M_x, and M_e at various positions x from support A (1 in. = 0.0254 m, 1 kip in. = 113 N m)

x (in.)	f(x)	M_x (kip in.)	$M_e = M_x/f(x)$ (kip in.)
0	1.000	0	0
25	1.375	218.8	159.0
50	1.750	375.0	214.0
75	2.125	468.0	221.0
100	2.500	500.0	220.0
125	2.875	468.0	163.0
150	3.250	375.0	115.0
175	3.625	218.8	60.4
200	4.000	0	0

and

$$\delta_{A_e} = \frac{X_{A_e}}{k_{A_e}} \tag{1.158}$$

Thus, from Eqs. (1.157) and (1.158), for $\delta_A = \delta_{A_e}$, we must have

$$\frac{X_A}{k_A} = \frac{X_{A_e}}{k_{A_e}} \tag{1.159}$$

Thus, Eq. (1.159) yields

$$k_{A_e} - k_A \frac{X_{A_e}}{X_A} \tag{1.160}$$

By using again statics, the elastic reaction X_{A_e} of the equivalent spring at support A of the equivalent system in Fig. 1.17d, was found to be equal to 8.67 kip (38,566.085 N). On this basis, Eq. (1.160) yields

$$k_{A_e} = (400)\frac{8.67}{10} = 346.8 \text{ kip in.}^{-1}(60.731 \times 10^6 \text{ N m}^{-1})$$

Now the derivation of the equivalent system of constant stiffness EI_1 is completed and shown in Fig. 1.17d. The elastic line of this system will be practically identical to the one of the original system in Fig. 1.17a. The only approximation we have introduced is the approximation of the M_e diagram with five straight-line segments as shown in Fig. 1.17b. If you observe this diagram closely, you note that its general shape is well retained and, therefore, the error we introduce to the actual deflections and rotations is very small. Even a two- or a three-line approximation would give very reasonable results for practical applications.

By using the equivalent system in Fig. 1.17d and applying the conjugate beam method, we find that the vertical deflection at the distance $x = 70$ in. (0.778 m) from support A is 0.02542 in. (0.000646 m). The exact value is 0.02535 in. (0.000644 m), yielding an error of 0.28%. For practical purposes this error is considered to be negligible. In many cases, much higher errors are permissible for practical applications.

Example 1.9 A statically indeterminate tapered beam made out of steel is loaded as shown in Fig. 1.18a. The width b of the member is 6 in. (0.1524 m), and the depth $h = 10$ in. (0.254 m). By applying the method of the equivalent systems, determine the vertical reaction at support A. The constant modulus of elasticity $E = 30 \times 10^6$ psi (206.84×10^9 Pa).

Solution: The variation of the depth h_x at any distance x from support A is

$$h_x = h(1 + \frac{x}{L}) \tag{1.161}$$

and the expression for the variation of the moment of inertia I_x is

$$I_x = \frac{bh_x^3}{12} = I_A(1 + \frac{x}{L})^3 \tag{1.162}$$

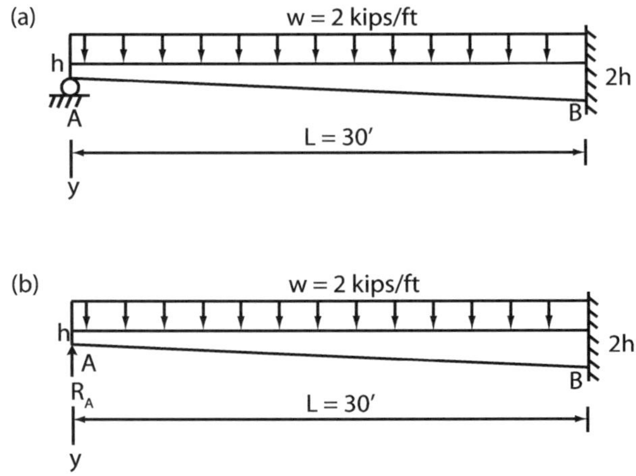

Fig. 1.18. (**a**) Tapered statically indeterminate steel beam. (**b**) Tapered cantilever beam loaded with the distributed loading and the redundant reaction R_A at the free end A ($1\,\text{ft} = 0.3048\,\text{m}, 1\text{ kip ft}^{-1} = 14,593.18\,\text{Nm}^{-1}$)

or

$$I_x = I_A f(x) \tag{1.163}$$

where

$$I_A = \frac{bh^3}{12}$$

is the moment of inertia at the left support A, and

$$f(x) = (1 + \frac{x}{L})^3 \tag{1.164}$$

represents the variation of the moment of inertia I_x.

Since the beam is statically indeterminate, the reaction R_A at the support A is taken as the redundant quantity. On this basis, we now have a cantilever beam loaded with the distributed load w and the reaction R_A as shown in Fig. 1.18b. The linear method of the equivalent systems may be applied here by using the cantilever beam in Fig. 1.18b and deriving two equivalent systems–one by using only R_A as the applied loading, and a second one by using only the distributed load w as the applied load.

The procedure is illustrated in Table 1.6. The first column of the table includes selected values of x, the second column gives the values of f(x) at these points, the third column gives the values of M_x produced only by the application of R_A, and the fourth column shows the values of the moment M_e of the equivalent system of constant stiffness EI_A produced by the application of R_A. The fifth column of the table gives the values of M_x produced only by the application of the distributed load w, and the last column of the table gives the values of the moment M_e of the equivalent system of constant stiffness EI_A produced by the application of the load w. The reader may verify the values in the table in order to become familiar with the methodology.

Table 1.6. Values of f(x), M_x, and M_e caused independently by the reaction R_A and loading w, for the cantilever beam in Fig. 1.18b (1 ft = 0.3048 m, 1 kip ft = 1,355.75 N m, 1 kip ft^{-1} = 14,593.18 N m^{-1})

(1) x (ft)	(2) f(x)	(3) M_x due to R_A (kip ft)	(4) M_e due to R_A (kip ft)	(5) M_x due to w (kip ft)	(6) M_e due to w (kip ft)
0	1.0000	0	0	0	0
5	1.5880	5 R_A	3.1486 R_A	−25	−15.7431
10	2.3704	10 R_A	4.2187 R_A	−100	−42.1870
15	3.3750	15 R_A	4.4444 R_A	−225	−66.6667
20	4.6296	20 R_A	4.3200 R_A	−400	−86.4006
25	6.1620	25 R_A	4.0571 R_A	−625	−101.4281
30	8.0000	30 R_A	3.7500 R_A	−900	−112.5000

By using the values of M_e in the fourth column of Table 1.6, the equivalent moment diagram M_e for the reaction R_A is plotted as shown in Fig. 1.19a. The approximation of its shape with four straight-line segments leads to the equivalent system for R_A shown in Fig. 1.19c. In a similar manner, by using the values of M_e in the last column of the table, the equivalent moment diagram M_e for the distributed load w is shown plotted in Fig. 1.20a. The approximation of its shape with three straight-line segments leads to the equivalent system for w shown in Fig. 1.20c.

If we use the equivalent systems in Figs. 1.19c and 1.20c and determine each time the vertical displacements $\delta_A{}'$ and $\delta_A{}''$, respectively, the vertical reaction R_A at the support A of the original system in Fig. 1.18a may be determined by satisfying the boundary condition

$$\delta_A = \delta_A{}' + \delta_A{}'' = 0 \tag{1.165}$$

where δ_A is the deflection at A of the original system, which is zero.

The deflection $\delta_A{}'$ may be determined by applying the moment–area method. It can be accomplished by using the approximated M_e in Fig. 1.19a, dividing it by EI_A, and taking the first moment of the M_e/EI_A area between points A and B, about point A. On this basis, we find

$$\begin{aligned}
\delta_A{}' = {} & \left[\frac{1}{2}(3R_A)(4)\left(\frac{2}{3}\right)(4) + (3R_A)(6)(7)\right. \\
& + \frac{1}{2}(1.4R_A)(6)(8) + (4.4R_A)(10)(15) \\
& \left. + \frac{1}{2}(0.65R_A)(10)(23.3333) + (3.75R_A)(10)(25)\right](25)^3 \\
= {} & \frac{1}{EI_A}\left[16R_A + 126R_A + 33.6R_A + 600R_A + 75.8332R_A + 937.5R_A\right](12)^3 \\
= {} & \frac{1,848.93R_A}{EI_A}(12)^3
\end{aligned}$$

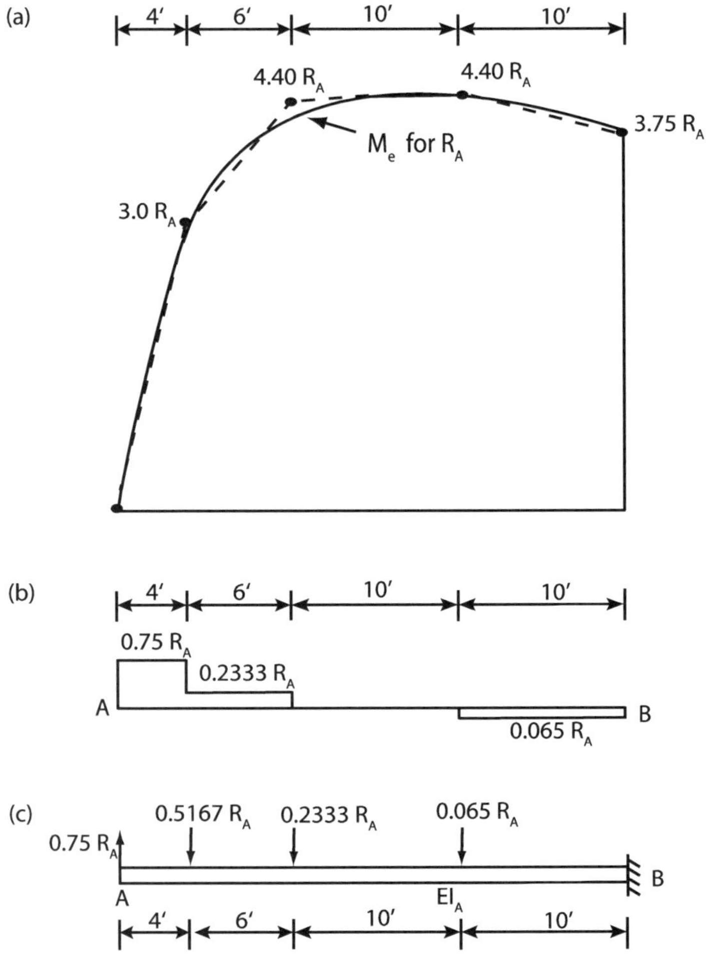

Fig. 1.19. Equivalent system for R_A. (**a**) M_e diagram with its shape approximated with four straight-line segments. (**b**) Equivalent shear-force diagram. (**c**) Equivalent system of constant stiffness EI_A (1 ft = 0.3048 m)

In a similar manner, by using the approximated M_e diagram in Fig. 1.120a, dividing it by EI_A, and taking its first moment about A, we find

$$\delta_A'' = -\frac{1}{EI_A}\left[\frac{1}{2}(25)(7)\left(\frac{2}{3}\right)(7) + (25)(12)(13) + \frac{1}{2}(60)(12)(15) + (85)(11)(24.5) + \frac{1}{2}(27.5)(11)(26.3333)\right](12)^3$$

$$= -\frac{35,271.11}{EI_A}(12)^3$$

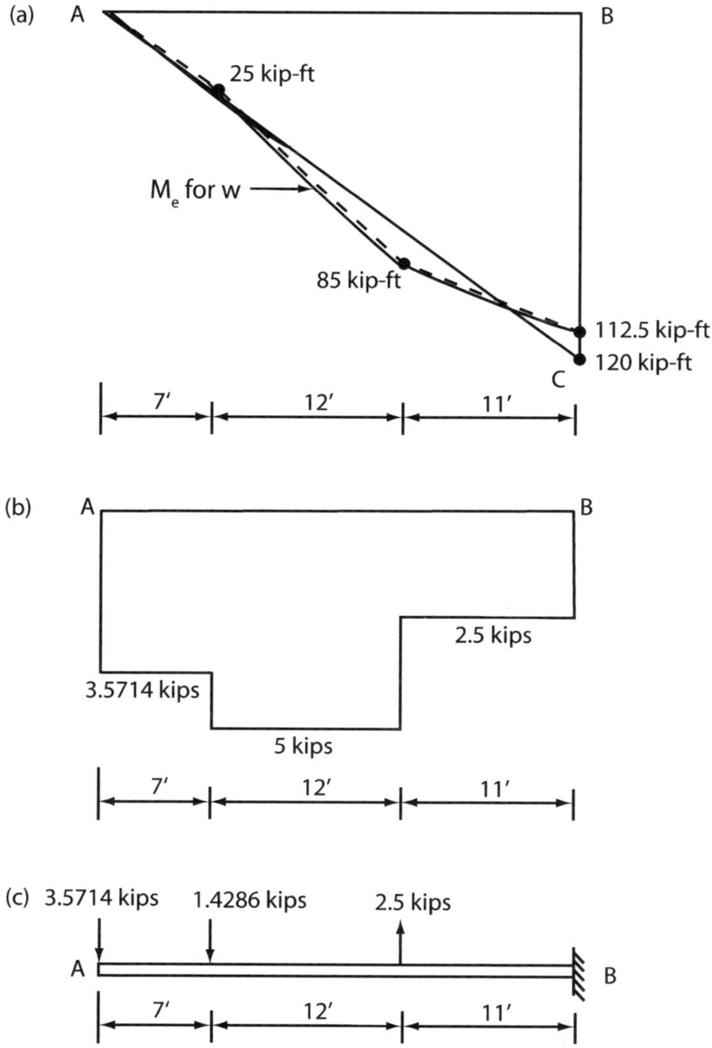

Fig. 1.20. Equivalent system for distributed load w. (**a**) M_e diagram with its shape approximated with three straight-line segments. (**b**) Equivalent shear-force diagram. (**c**) Equivalent system of constant stiffness EI_A (1 ft = 0.3048 m, 1 kip ft = 1,355.75 N m, 1 kip = 4,448 N)

By substituting the values of $\delta_A{}'$ and $\delta_A{}''$ into Eq. (1.165) and solving for R_A, we find

$$\frac{1,848.93R_A}{EI_A}(12)^3 - \frac{35.271.11}{EI_A} = 0$$

or

$$R_A = 19.08 \text{ kip } (84.85 \times 10^3 \text{ N})$$

The value of R_A obtained here by using the indicated approximations for M_e should be very accurate–well within one percent. Better accuracy may be obtained by using more straight lines to approximate M_e.

Very reasonable results may be obtained even with crude approximations of the shape of M_e. For example, the approximation of the M_e diagram in Fig. 1.20a with one straight line AC, yields

$$\begin{aligned}\delta''_A &= \frac{1}{EI_A}\left[\frac{1}{2}(120)(30)(20)\right](12)^3 \\ &= -\frac{36{,}000}{EI_A}(12)^3\end{aligned}$$

If it is compared to the value obtained earlier by using three straight-line segments, we find that the difference in δ_A'' is only 2.07%, and the difference in R_A would be 2.04%. Similar accuracy may be obtained if we approximate the M_e in Fig. 1.19a with two or three straight-line segments instead of four.

Other problems of this nature may be solved in a similar manner. See also [2, 3, 5, 6, 84] at the end of this text.

Problems

1.1 By using Simpson's One-Third rule, evaluate the integrals

$$\delta = \int_0^{20} (x+1)^2 dx, \qquad \delta = \int_0^{20} (x-5)^2 dx,$$

$$\delta = \int_0^{100} \frac{\sqrt{x}}{2} dx, \qquad \delta = \int_0^{100} \left(\frac{\sqrt{x}}{2}\right)^3 dx$$

1.2 Repeat Example 1.1 by assuming that $P = 0.6$ kip (2.67 kN) and compare the results.
Answer: $\Delta_B = 281.29$ in. (7.145 m), $\theta_B = 59.42°$.

1.3 Repeat Example 1.1 by assuming that $P = 1.4$ kip (6.23 kN) and compare the results.
Answer: $\Delta_B = 557.22$ in. (14.66 m), $\theta_B = 83.23°$.

1.4 Repeat Example 1.2 by assuming that the beam in Fig. 1.6 has a uniform cross section throughout its length and compare the results.

1.5 Repeat Example (1.3) by assuming that $\Delta(x)$ is given, (a) by Eq. (1.83), and (b) by Eq. (1.84). Compare the results.

1.6 For the uniform flexible beam shown in Fig. 1.1a, determine a pseudolinear system of constant stiffness EI. Assume that $P = 0.6$ kip (2.67 kN), $L = 1{,}000$ in. (25.4 m), and $EI = 180 \times 10^3$ kip $n^2 (516.54 \times 10^3\ N\ m^2)$. By using the pseudolinear system, determine the rotation and vertical deflection at the free end B. Also determine the horizontal displacement of the free end B.
Answer: $\delta_B = 629.0$ in. (15.98 m), $\Delta_B = 281.29$ in. (7.145 m), and $\theta_B = 59.42°$.

1.7 By using the pseudolinear system derived in Problem 1.6, determine the rotations and vertical displacements at x = 100 in. (2.54 m) and x = 300 in. (7.62 m). Also determine what values of x_o, in the original system, correspond to the indicated values of x.

1.8 Repeat Problem 1.6 by assuming that P = 1 kip (4.448 kN) and compare the results.

1.9 The tapered cantilever beam in Fig. 1.6 is loaded with a concentrated load P = 2.5 kip (11.12 kN) at the free end. By using a pseudolinear equivalent system of constant stiffness, determine the vertical and horizontal displacements δ_B and Δ_B, respectively, at the free end B of the beam, as well as the rotation θ_B at the same end. The length L = 1,000 in. (25.4 m), EI_B = 180,000 kip in^2(516.54 × 10^3 N m^2), and taper n = 1.5.
Answer: δ_B = 720.82 in. (18.31 m), Δ_B = 436.60 in. (11.10 m), and θ_B = 78.44°.

1.10 Repeat Problem 1.9 by assuming that P = 1.5 kip (6.67 kN), and compare the results.
Answer: δ_B = 622.75 in. (15.82 m), Δ_B = 298.0 in. (7.145 m), and θ_B = 65.86°.

1.11 Repeat Problem 1.9 by assuming that taper n = 2 and P = 1 kip (4.448 kN). Compare the results.
Answer: δ_B = 334.47 in. (8.50 m), Δ_B = 78.42 in. (1.99 m), and θ_B = 36.14°.

1.12 Solve Problem 1.9 by using a simplified nonlinear equivalent system of constant stiffness EI_B that is loaded with one equivalent concentrated load P_e at the free end B of the beam. Apply pseudolinear analysis to solve the simplified nonlinear equivalent system. Compare the results.

1.13 Solve Problem 1.9 with P = 2.0 kip (8.889 kN).

1.14 Solve Problem 1.9 with P = 1 kip (4.448 kN), and n = 1.80.
Answer: δ_B = 401.23 in. (10.19 m), Δ_B = 113.26 in. (2.88 m), and θ_B = 42.39°.

1.15 Solve the problem in Example 1.5 by assuming that P = 1.5 kip (6.672 kN) and w_o = 0.02 kips in.$^{-1}$(3, 502.54 N m^{-1}).

1.16 The uniform flexible cantilever beam in Fig. P1.16 is loaded by two concentrated loads located as shown in the figure. Determine a simplified nonlinear equivalent system of constant stiffness that is loaded with only

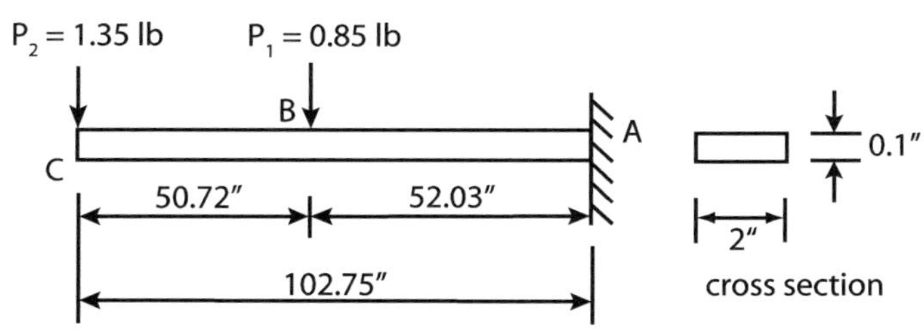

Fig. P1.16.

one concentrated equivalent vertical load at its free end C. The modulus of elasticity $E = 30 \times 10^6$ psi (206.84×10^6 kPa).

1.17 By using the simplified nonlinear equivalent system obtained in Problem 1.16 and applying pseudolinear analysis, determine the vertical deflection δ_C, the horizontal displacement Δ_C, and rotation θ_C, at its free end C. *Answer*: $\delta_C = 68.84$ in. (1.7485 m), $\Delta_C = 29.51$ in. (0.7496 m), and $\theta_C = 60°$ (1.0462 rad).

1.18 Repeat the problem in Example 1.6 by assuming that the length $L = 600$ in. (15.24 m) and compare the results.

1.19 Repeat the problem in Example 1.7 by assuming that $w = 2$ kips ft^{-1} (29, 186.36 N m^{-1}), and compare the results.

1.20 By using the constant stiffness equivalent system obtained in Example 1.7, determine the rotation θ_C and vertical displacement δ_C at midspan C of the member.

1.21 By using the constant stiffness equivalent system obtained in Example 1.8, determine its deflection and rotation at midspan.

1.22 Repeat the problem in Example 1.8 by assuming that $w = 200$ lb in.$^{-1}$ (35, 026 N m^{-1}).

1.23 Repeat the problem in Example 1.9 by assuming that $w = 4$ kips ft^{-1} (58, 372.72 N m^{-1}).

1.24 Repeat the problem in Example 1.8 by approximating the M_e diagram (a) with three straight-line segments, and (b) with only two straight-line segments appropriately selected. In each case, solve for the deflection and rotation at midspan and compare the results.

1.25 The variable stiffness steel beams in Fig. P1.25 are loaded as shown. The cross section of each member is rectangular with width $b = 8$ in.

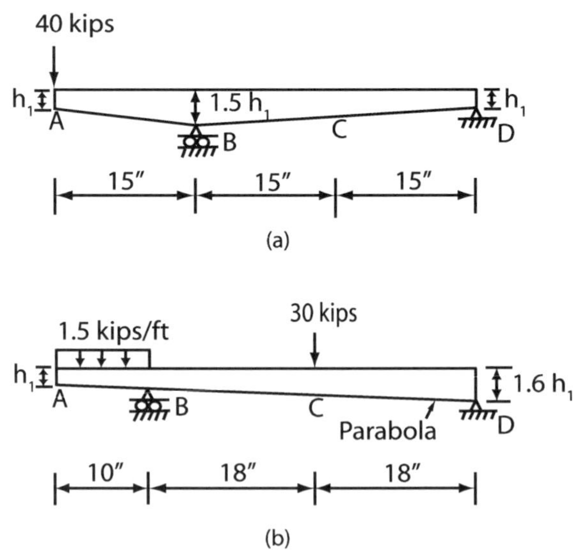

Fig. P1.25.

(0.2032 m) and depth $h_1 = 16$ in. (0.4064 m). By applying the linear theory of equivalent systems, determine in each case an equivalent system of uniform stiffness EI_A, where $E = 30 \times 10^6$ psi $(206.84 \times 10^9$ Pa) is the constant modulus of elasticity, and I_A is the moment of inertia at the free end A. By using in each case the equivalent system, determine the vertical deflections at points A and C.

2

Solution Methodologies for Uniform Flexible Beams

2.1 Introduction

In the first chapter, the basic aspects of nonlinear deformations of structural components such as beam, were extensively discussed, and various methods of analysis, together with their associated difficulties, or simplifications, were explored. This preliminary analysis was concentrated on flexible members of uniform and variable stiffness EI, with emphasis in the derivation of equivalent pseudolinear systems and equivalent simplified nonlinear systems in order to provide a convenient and relatively easy solution to such extremely complicated nonlinear problems. The purpose of the pseudolinear analysis is to linearize mathematically the given complex nonlinear system, and then apply known methods of linear analysis to solve the derived pseudolinear system. Very complicated problems of nonlinear analysis can be solved conveniently by following this methodology. In many cases, before we apply the pseudolinear analysis, we reduce mathematically the initially complex nonlinear system into a simpler nonlinear equivalent system and then apply to it pseudolinear analysis. This procedure is particularly useful to practicing engineers, because they can obtain easy and accurate solutions for their important practical problem that requires reliable nonlinear analysis.

In this chapter, various cases of statically determinate and statically indeterminate flexible members with various loading conditions are examined in great detail, and convenient solution methodologies based on equivalent systems are obtained. In all cases in this chapter, the modulus of elasticity E and the moment of inertia I are assumed to be constant along the length of the member. However, in Chaps. 3 and 4, both E and I are permitted to vary in any arbitrary manner along the length of the member, and in Chap. 5, vibration theories with applications are derived for such nonlinear problems. It should be emphasized here that these methods are mathematically derived, and they are designed to reduce the enormous complexity of the initial complex nonlinear problem, and they are general.

2.2 Pseudolinear Analysis for Uniform Flexible Cantilever Beams Loaded with Uniformly Distributed Loading Throughout their Length

Consider the uniform flexible cantilever beam in Fig. 2.1, which is loaded with a uniformly distributed vertical load w_o as shown in the figure. During the large deformation configuration of the member, the distributed load retains its vertical configuration. Such large deformation position is shown in Fig. 2.1.

At any distance $0 \leq x \leq L_o$ from the end B, where $L_o = (L - \Delta)$, the bending moment M_x is

$$M_x = -w_0 x_0 \frac{x}{2} \tag{2.1}$$

From Chap. 1, Eq. (1.75), we know that

$$x_0(x) = \int_0^x \sqrt{1 + [y'(x)]^2} dx \tag{2.2}$$

By substituting Eq. (2.2) into Eq. (2.1), we find

$$M_x = -\frac{w_0 x}{2} \int_0^x \sqrt{1 + [y'(x)]^2} dx \tag{2.3}$$

We assume in this case that the stiffness EI of the member, where E is its modulus of elasticity and I is its moment of inertia, is constant throughout the length of the member. The cases where both E and I are variable are examined in detail in later parts of this book and in Chap. 1.

By substituting Eq. (2.3) into the Euler–Bernoulli equation, which is given by Eq. (1.7), or Eq. (1.86) with f(x) = g(x) = 1, we find

$$\frac{y''}{[1 + (y')^2]^{3/2}} = \frac{w_0 x}{2EI} \int_0^x \left\{1 + [y'(x)]^2\right\}^{1/2} dx \tag{2.4}$$

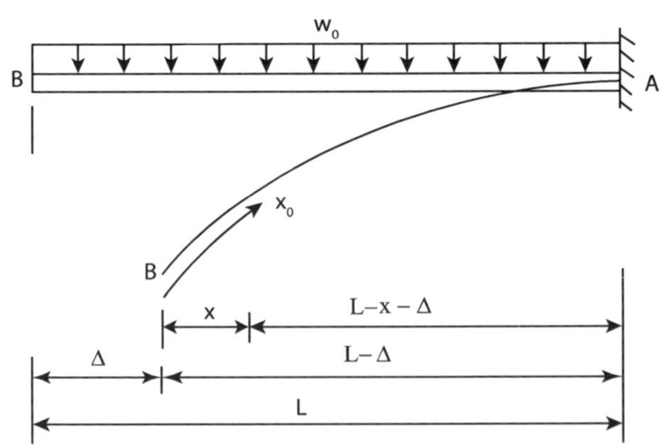

Fig. 2.1. Uniform flexible cantilever beam subjected to a uniformly distributed loading w_o as shown

Equation (2.4) is the exact nonlinear integral differential equation that is very difficult to solve. An accurate and simpler solution of this equation, however, may be obtained by using the expression for x_o given by Eq. (1.74), and assuming that the horizontal displacement $\Delta(x)$ can be expressed as shown by Eqs. (1.81) – (1.84). All these four expression for $\Delta(x)$ eliminate the integral function in Eq. (2.4), and at the same time they provide an accurate and much easier solution to the problem. For example, if we use Eq. (1.81) and assume that $\Delta(x) = \text{constant} = \Delta$, where Δ is the horizontal displacement at the free end of the member (see Fig. 2.1), then Eq. (2.2) yields

$$x_0\,(x) = x + \Delta \tag{2.5}$$

By substituting Eq. (2.5) into Eq. (2.1), we obtain

$$M_x = -\frac{w_0 x}{2}(x + \Delta) \tag{2.6}$$

On this basis, the Euler–Bernoulli nonlinear differential equation yields

$$\frac{y''}{[1 + (y')^2]^{3/2}} = \frac{w_0 x}{2EI}(x + \Delta) \tag{2.7}$$

which is much easier to solve compared to the one given by Eq. (2.4). Note that the integral in Eq. (2.4) is replaced by the expression given by Eq. (2.5) to yield Eq. (2.7).

By following the procedure discussed in Sect. 1.4, we note that

$$\lambda(x) = -\frac{M_x}{EI} = \frac{w_0 x}{2EI}(x + \Delta) \tag{2.8}$$

and

$$\phi(x) = \int \lambda(x)dx = \frac{w_0}{2EI}\left[\frac{x^3}{3} + \frac{x^2\Delta}{2}\right] \tag{2.9}$$

Thus, by using Eq. (1.32), we find

$$\frac{y''}{\left[1 + (y')^2\right]^{1/2}} = \phi\,(x) + C \tag{2.10}$$

where C is the constant of integration.

The constant of integration C can be determined by applying the boundary condition of zero rotation ($y' = 0$) at $x = L_o = (L - \Delta)$. By applying this boundary condition, we obtain

$$0 = \frac{w_0}{2EI}\left[\frac{(L-\Delta)^3}{3} + \frac{(L-\Delta)^2\Delta}{2}\right] + C \tag{2.11}$$

or

$$C = -\frac{w_0}{2EI}\left[\frac{(L-\Delta)^3}{3} + \frac{(L-\Delta)^2\Delta}{2}\right] \tag{2.12}$$

Thus,

$$G(x) = \phi(x) + C = \frac{w_0}{12EI}\left[2x^3 + 3\Delta x^2 - 2(L-\Delta)^3 - 3\Delta(L-\Delta)^2\right] \quad (2.13)$$

By using Eqs. (1.33) and (2.13), we find

$$y'(x) = \frac{G(x)}{\left\{1 - [G(x)]^2\right\}^{1/2}} \quad (2.14)$$

where G(x) is as shown by Eq. (2.13). Since the horizontal displacement Δ in Eq. (2.13) is not known, we can determine it by using Eq. (1.59) and applying a trial-and-error procedure, as discussed in Sect. 1.6 in the first chapter of this book. When Δ is evaluated, then $y'(x)$ at any $0 \leq x \leq L_o$ can be obtained by using Eq. (2.14). With known $y'(x)$, the pseudolinear analysis discussed in Sect. 1.8.1 may be applied to this problem to determine the large deflections and rotations of the member.

The following numerical example illustrates the application of the above methodology.

Example 2.1 Assume that the uniformly distributed loading w_o in Fig. 2.1 is 1.5 lb in.$^{-1}$ (262.69 N m^{-1}), the uniform stiffness EI is 180 $\times$ 10^3 kip in.^{2} (516, 541 N m^2), and the length L is 1,000 in (25.4 m). By using pseudolinear analysis as discussed in Sect. 1.8.1 in the first chapter, determine the rotation θ_B, the vertical displacement δ_B, and the horizontal displacement Δ of the free end B of the flexible member. Assume that $\Delta(x) = \Delta x/L_o$.

Solution: For the assumed function for $\Delta(x)$, we have

$$x_0 = x + \Delta(x) = x + \frac{\Delta x}{L_0} = x + \frac{\Delta x}{L-\Delta} \quad (2.15)$$

Therefore,

$$M_x = -\frac{w_0 x}{2}x_0 = -\frac{w_0 x}{2}\left(x + \frac{\Delta x}{L-\Delta}\right) \quad (2.16)$$

and

$$\frac{y''}{[1+(y')^2]^{3/2}} = \frac{w_0 x}{2}\left(x + \frac{\Delta x}{L-\Delta}\right) \quad (2.17)$$

On this basis, we find that

$$\lambda(x) = -\frac{M_x}{EI} = \frac{w_0 x}{2EI}\left(x + \frac{\Delta x}{L-\Delta}\right) \quad (2.18)$$

$$\phi(x) = \int \lambda(x)dx = \frac{w_0}{2EI}\left[\frac{x^3}{3} + \frac{\Delta x^3}{3(L-\Delta)}\right] \quad (2.19)$$

Thus, by using Eq. (1.32), we find

$$\frac{y'}{[1+(y')^2]^{1/2}} = \frac{w_0}{2EI}\left[\frac{x^3}{3} + \frac{\Delta x^3}{3(L-\Delta)}\right] + C \tag{2.20}$$

where C is the constant of integration. We can determine C by applying the boundary condition $y' = 0$ at $x = (L - \Delta)$. Using Eq. (2.20), this boundary condition for $x = (L - \Delta)$, yields

$$\frac{w_0}{2EI}\left[\frac{(L-\Delta)^3}{3} + \frac{\Delta(L-\Delta)^3}{3(L-\Delta)}\right] + C = 0$$

or,

$$C = \frac{w_0}{6EI}\left[(L-\Delta)^3 + \Delta(L-\Delta)^2\right] \tag{2.21}$$

We also have,

$$G(x) = \phi(x) + C = \frac{w_0}{6EI}\left[x^3 + \frac{\Delta x^3}{(L-\Delta)} - (L-\Delta)^3 + \Delta(L-\Delta)^2\right] \tag{2.22}$$

and by referring to Eq. (2.14), we can write

$$y'(x) = \frac{G(x)}{\{1-[G(x)]^2\}^{1/2}} \tag{2.23}$$

where G(x) in Eq. (2.23) is given by Eq. (2.22).

In order to apply Eq. (2.23) to determine $y'(x)$, we need to know first the value of the horizontal displacement Δ of the free end B of the flexible member. This can be determined as explained in Example 1.1, by using the equation

$$L = \int_0^{L_0} \left[1+(y')^2\right]^{1/2} dx \tag{2.24}$$

and applying a trial-and-error procedure. That is, we assume a value of Δ in Eq. (2.23) and then carry out the integration in Eq. (2.24) to determine the length L of the member. The Simpson's One-Third rule given by Eq. (1.36) in Sect. 1.5 can be used to carry out the integrations in Eq. (2.24). The procedure may be repeated until the correct length L is obtained. See Examples 1.1 and 1.4. The value of Δ that satisfies Eq. (2.24) is $\Delta = 251.16$ in. (6.38 m).

By substituting the values of w_o, L, EI, and Δ into Eq. (2.22), we find

$$\begin{aligned} G(x) &= \frac{1.5}{(6)(180)(10)^3(10)^3}\left[x^3 + \frac{251.16}{1{,}000-251.16}x^3 - (1{,}000-251.16)^3\right. \\ &\qquad \left. -251.16(1{,}000-251.16)^2\right] \\ &= 1.388889(10)^{-9}\left(1.335399x^3 - 560{,}761{,}345.60\right) \end{aligned} \tag{2.25}$$

For $x = 0$, Eq. (2.25) yields

$$G(0) = -0.778835$$

and from Eq. (2.23), we find

$$y'(0) = \frac{-0.778835}{\sqrt{1 - (-0.778835)^2}} = -1.241708$$

Thus, the rotation of the free end B, is

$$\begin{aligned} \theta_B = \tan^{-1}[y'(0)] &= 0.892806\,\text{rad} \\ &= 51.15^\circ \end{aligned}$$

Table 2.1 gives the values of $M_x, M'_e, y'(x)$, and z_e for the indicated values of $0 \leq x \leq L_0$. Equation (2.16) is used to find M_x, and Eq. (2.23) for $y'(x)$. The values of M'_e and z_e of the pseudolinear system are determined by using the expression

$$M'_e = z_e M_x \tag{2.26}$$

$$z_e = \left\{1 + [y'(x)]^2\right\}^{3/2} \tag{2.27}$$

By using the values of M'_e shown in the last column of Table 2.1, the moment diagram M'_e of the equivalent pseudolinear system is plotted as shown by the solid line in Fig. 2.2a. In the same figure, the approximation of its shape with four straight-line segments is also shown. By using the approximated M'_e diagram and applying statics, we obtain the pseudolinear equivalent system shown in Fig. 2.2b.

The moment–area method (or any other known method of linear mechanics), may be applied here, using the equivalent pseudolinear system in

Table 2.1. Values of $M_x, y', (x), z_e$, and M'_e, corresponding to the assumed vales of x(1 in. = 0.0254 m, 1 kip in. = 113.0 N m)

(1) x (in.)	(2) M_x (kip in.)	(3) $y'(x)$	(4) z_e	(5) M'_e (kip in.)
0	0	−1.241708	4.052490	0
100	−10.0155	−1.234236	4.008326	−40.1454
200	−40.0620	−1.184089	3.722870	−149.1456
300	−90.1394	−1.064236	3.114323	−280.7232
400	−160.2478	−0.878832	2.359515	−378.1071
500	−250.3873	−0.653412	1.704559	−426.8000
600	−360.5576	−0.408565	1.260563	−454.5056
700	−490.7590	−0.144140	1.031326	−506.1325
748.84	−561.6300	0	1.000000	−561.6300

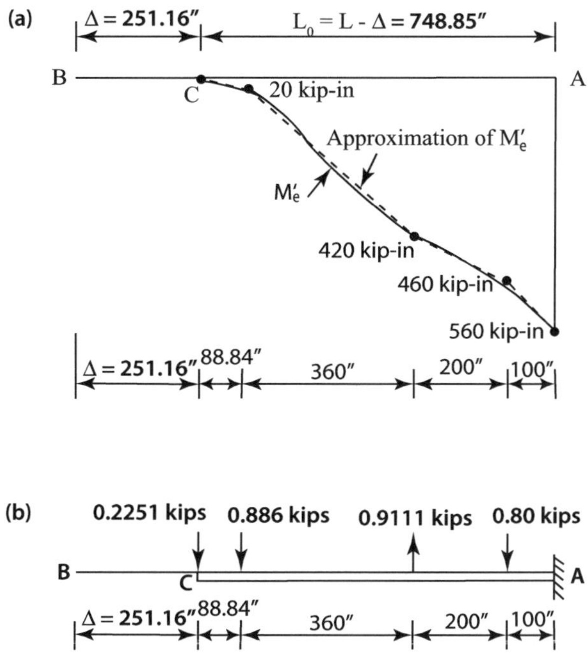

Fig. 2.2. (**a**) Moment diagram M'_e of the pseudolinear system approximated with four straight-line segments. (**b**) Pseudolinear equivalent system (1 in. = 0.0254 m, 1 kip in. = 113.0 N m, 1 kip = 4,448 N)

Fig. 2.2b, to determine the vertical deflection δ_C at the free end C of the pseudolinear system. This value should be closely identical to the deflection δ_B at the free end B of the original system in Fig. 2.1. The only approximation we have introduced here is the approximation with four straight-line segments of the exact M'_e diagram shown by the solid line in Fig. 2.2a. The accuracy can be improved by using more straight-line segments, but very accurate values for practical purposes can be obtained by using only a few straight-line segments.

We determine here the value of δ_C (which is equal to δ_B) by using the moment–area method. That is, we divide the approximated M'_e in Fig. 2.2a by EI, and we take the first moment of its area about point C. This yields

$$\begin{aligned}\delta_C = \delta_B &= \frac{1}{EI}\left[\frac{1}{2}(20)(88.84)\left(\frac{2}{3}\right)(88.84) + (20)(360)(268.84)\right.\\ &\quad + \frac{1}{2}(400)(360)(328.84) + (420)(200)(548.84) + \frac{1}{2}(40)(200)(582.17)\\ &\quad \left. + (460)(100)(698.84) + \frac{1}{2}(100)(100)(715.51)\right]\\ &= \frac{109,820,175}{EI}\end{aligned}$$

By substituting for EI, we obtain

$$\delta_C = \delta_B = \frac{109,820,175}{(180)(10)^3} = 610.11\,\text{in.}\ (15.50\ \text{m})$$

For comparison purposes, the same problem was solved by assuming in each case that $\Delta(x) = \Delta, \Delta(x) = \Delta x/L_o, \Delta(x) = \Delta(x/L_o)^{1/2}, \Delta(x) = \Delta\sin(\pi x/2L_o)$, and $x_o = x$. The results are shown in Table 2.2. The results are also compared with the ones obtained by using the fourth order Runge–Kutta method, as shown in the last column of the table. The large error is obtained when it was assumed that $x_o = x$, which means that the horizontal displacement $\Delta(x)$ is zero. All other cases examined provided very reasonable results for practical applications.

Another comparison of results was made here by solving the same problem for various values of w_o and applying the method of the equivalent pseudolinear systems with $x_o = x + \Delta$, the Power Series method, and the Fourth Order Runge–Kutta method. The results are shown in Table 2.3 and they are in close agreement for practical applications.

Table 2.2. Values of Δ_B, θ_B, and δ_B for various assumed cases of $\Delta(x)$ and with $w_o = 1.5$ lb in.$^{-1}$(1 in. = 0.0254 m)

displacement	assumed cases of $\Delta(x)$					Runge–Kutta method
	$\Delta(x) = \Delta$	$\Delta(x) = \Delta x/L_o$	$\Delta(x) = \Delta(x/L_o)^{1/2}$	$\Delta(x) = \Delta \sin \pi x/2L_o$	$x_o = x$	
Δ_B (in.)	277.25	251.16	261.64	263.65	199.24	287.74
θ_B (deg.)	55.70	51.15	52.83	52.74	45.49	55.07
δ_B (in.)	637.96	610.11	626.27	628.64	558.46	652.06

Table 2.3. Variation of θ_B and δ_B for various values of the distributed load w_o (1 in. = 0.0254 m, 1 lb in.$^{-1}$ = 175.13 N m^{-1})

	pseudolinear equivalent systems method		power series method		fourth order Runge–Kutta method		% difference base on δ_B
w_o (lb in.$^{-1}$)	θ_B (deg.)	δ_B (in.)	θ_B (deg.)	δ_B (in.)	θ_B (deg.)	δ_B (in.)	
1	43.41	522.42	43.14	531.48	43.01	529.53	−1.34
1.5	55.69	638.00	55.37	656.38	55.07	652.07	−2.15
2.0	66.30	705.72	63.62	731.04	63.23	725.34	−2.70
2.5	70.88	748.75	69.36	778.08	68.95	771.99	−3.01
3.0	73.88	777.86	73.49	809.39	73.12	803.57	−3.20

2.3 Pseudolinear Analysis for Uniform Simply Supported Beams Loaded with a Uniformly Distributed Loading Throughout their Length

We consider the uniform simply supported beam in Fig. 2.3 which is loaded by a uniformly distributed load w as shown. The large deflection configuration of the member is shown in the same figure. At any distance $0 \leq x \leq L_o$ from support B, the bending moment M_x is

$$\begin{aligned} M_x &= \frac{wLx}{2} - \frac{wxx_0}{2} \\ &= \frac{wLx}{2} - \frac{wx}{2}\int_0^x \left\{1 + [y'(x)]^2\right\}^{1/2} dx \end{aligned} \tag{2.28}$$

By using the exact expression for x_o and substituting Eq. (2.28) into the Euler–Bernoulli equation, we find

$$\frac{y''}{[1+(y')^2]^{3/2}} = \frac{wx}{2EI}\left\{-L + \int_0^x \left\{1 + [y'(x)]^2\right\}^{1/2} dx\right\} \tag{2.29}$$

Again, Eq. (2.29) is the exact complex nonlinear integral differential equation for the uniform simply supported beam that is very difficult. However, we can simplify a great deal its solution by assuming that $x_o = x + \Delta$, where Δ is the horizontal displacement of the support A of the member, as shown in Fig. 2.3. $\Delta(x)$, and consequently x_o, could involve any of the expressions given by Eqs. (1.81)–(1.84). With this in mind, the bending moment M_x at any $0 \leq x \leq L_o$ is given by the expression

$$M_x = \frac{wLx}{2} - \frac{wx}{2}x_0 = \frac{w(L-\Delta)x}{2} - \frac{wx^2}{2} \tag{2.30}$$

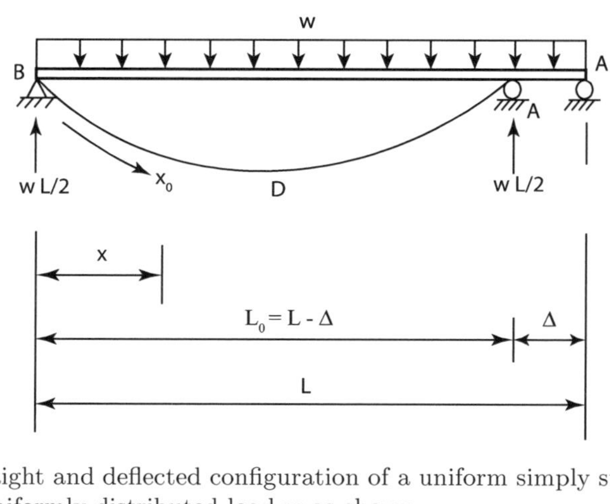

Fig. 2.3. Straight and deflected configuration of a uniform simply supported beam loaded by a uniformly distributed load w as shown

By substituting Eq. (2.30) into the Euler–Bernoulli equation, we find

$$\frac{y''}{[1+(y')^2]^{3/2}} = \frac{w}{2EI}\left[-(L-\Delta)x + x^2\right] \tag{2.31}$$

which is a much easier nonlinear differential equation to solve compared to the one given by Eq. (2.29).

Proceeding with the integration of Eq. (2.31), we note that in this case we have

$$\lambda(x) = \frac{w}{2EI}[x^2 - (L-\Delta)x] \tag{2.32}$$

$$\phi(x) = \int \lambda(x)dx = \frac{w}{12EI}[2x^3 - 3(L-\Delta)x^2] \tag{2.33}$$

Thus, by using Eq. (1.32), we find

$$\frac{y'}{[1+(y')^2]^{1/2}} = \phi(x) + C \tag{2.34}$$

where C is the constant of integration and $\Phi(\chi)$ is as shown by Eq. (2.33).

We can determine C by using Eq. (2.34) and applying the boundary condition of zero y' at $x = L_o/2$, where $L_o = L - \Delta$. This yields,

$$\frac{w}{12EI}\left[2\left(\frac{L-\Delta}{2}\right)^3 - 3(L-\Delta)\left(\frac{L-\Delta}{2}\right)^2\right] + C = 0$$

$$C = \frac{w}{12EI}\left[\frac{(L-\Delta)^3}{2}\right] \tag{2.35}$$

On this basis, we find that

$$G(x) = \phi(x) + C = \frac{w}{24EI}\left[4x^3 - 6(L-\Delta)x^2 + (L-\Delta)^3\right] \tag{2.36}$$

Thus, by using Eq. (1.33), we write

$$y'(x) = \frac{G(x)}{\left\{1 - [G(x)]^2\right\}^{1/2}} \tag{2.37}$$

where G(x) is given by Eq. (2.36).

The unknown horizontal displacement Δ of the end support A of the beam, may be determined as in previous cases by using the expression

$$L = \int_0^{L_0} \left\{1 + [y']^2\right\}^{1/2} dx \tag{2.38}$$

and applying a trial-and-error procedure. With known Δ, y' can be determined by using Eq. (2.37), and the pseudolinear analysis can be carried out as discussed in Sect. 1.8.1 and in the examples of preceding sections.

The following numerical example illustrates the procedure:

Example 2.2 The flexible simply supported beam in Fig. 2.3 has a uniform cross section throughout its length and it is loaded as shown. By applying the methodology discussed above, determine the horizontal displacement Δ at the support A of the member, the rotation θ_B of the end B, and the vertical displacement δ_D at $x = L_o/2$. The length L of the beam is 1,000 in. (25.4 m), $EI = 75 \times 10^3$ kip in.^{2}(215, 224 N m^2), and $w = 10$ lb in.$^{-1}$(1, 751.27 N m^{-1}).

Solution: By substituting into Eq. (2.36) the assigned values for w, EI and L, we find

$$\begin{aligned} G(x) &= \frac{10}{24(75)(10)^3(10)^3}\left[4x^3 - 6(L-\Delta)x^2 + (L-\Delta)^3\right] \\ &= 5.555556(10)^{-9}\left[4x^3 - 6(L-\Delta)x^2 + (L-\Delta)^3\right] \end{aligned} \tag{2.39}$$

The horizontal displacement Δ of support A is an important part for G(x) in Eq. (2.39) and needs to be determined. This is accomplished, as in preceding examples, by using Eq. (2.38) and applying a trial-and-error procedure to find the value of Δ that satisfies this equation. By applying Simpson's One-Third rule to carry out the required integrations, we found that the value of Δ that satisfies Eq. (2.38) is 440.53 in. (11.19 m). Using a digital computer will facilitate the procedure.

By substituting into Eq. (2.30), we find that

$$\begin{aligned} M_x &= \frac{(10)(1{,}000 - 440.53)x}{2} - 5x^2 \\ &= 2,797.35x - 5x^2 \end{aligned} \tag{2.40}$$

and from Eq. (2.39), by substitution, we find

$$\begin{aligned} G(x) &= 5.555556(10)^{-9}\left[4x^3 - 6(1{,}000 - 440.53)x^2 + (1{,}000 - 440.53)^3\right] \\ &= 5.555556(10)^{-9}\left[4x^3 - 3{,}356.82x^2 + 175,117.76\right] \end{aligned} \tag{2.41}$$

At $x = 0$, we have $G(0) = 0.972877$. Thus, Eq. (2.37) yields

$$y'(0) = \frac{0.972877}{\sqrt{1-(0.972877)^2}} = 4.205726$$

Therefore, the rotation θ_B of the end B of the flexible member, is

$$\begin{aligned} \theta_B = \tan^{-1}[y'(0)] &= 3.141593\,\text{rad} \\ &= 76.625^\circ \end{aligned}$$

At the selected values of $0 \le x \le L_o$, where $L_o = L - \Delta = 1{,}000 - 440.53 = 559.47$ in. (14.21 m), the values of $M_x, y'(x), z_e$, and M'_e were computed and

Table 2.4. values of $M_x, y'(x), z_e$, and M'_e, corresponding to the assumed values of x (1 in. = 0.0254 m, 1 kip in. = 113.0 N)

(1) x (in.)	(2) M_x (kip in.) Eq. (2.40)	(3) $y'(x)$ Eqs. (2.37) and (2.41)	(4) z_e Eq. (2.27)	(5) M'_e (kip in.) Eq. (2.26)
0	0	4.205726	80.78836	0
50	127.368	2.510904	19.74232	2,514.54
100	229.735	1.374377	4.910224	1,128.05
200	359.470	0.442555	1.307728	470.09
300	389.205	−0.106126	1.016941	395.80
400	318.940	−0.728350	1.893421	603.89
450	246.308	−1.240537	4.045540	996.45
500	148.675	−2.217513	14.394354	2,140.08
550	26.043	−4.077856	74.018149	1,927.65
559.47	0	−4.205726	80.788364	0

Table 2.4 was prepared. The values of M_x were calculated using Eq. (2.40), $y'(x)$ by using Eq. (2.37), and M'_e and z_e by using Eqs. (2.26) and (2.27), respectively, or Eqs. (1.119) and (1.120), respectively.

The moment diagram M'_e of the pseudolinear equivalent system is plotted by using the values in the last column of Table 2.4, and it is shown by the solid line in Fig. 2.4a. The approximation of its shape with eight straight-line segments is shown by the dashed line in the same figure. By using the approximated M'_e diagram and applying statics, the equivalent pseudolinear system in Fig. 2.4b is obtained.

At midpoint D, the vertical deflection δ_D can be determined by using the moment–area method, or any other method or hand book formulas of linear analysis, because the system is linearized. If we use the moment–area method, we note that the vertical deflection δ_D at D is equal to the first moment about point B of the M'_e/EI area between points B and D. Note that the rotation θ_D at midpoint D is zero. By using the approximated M'_e in Fig. 2.4a and dividing it by the stiffness EI, the indicated first moment about point B, and consequently δ_D, is

$$\delta_D = \frac{1}{EI}\left[\frac{1}{2}(2{,}600)(40)(26.667) + (750)(80)(80) + \frac{1}{2}(1850)(80)(66.667)\right.$$
$$\left. + (400)(120)(180) + \frac{1}{2}(350)(120)(160) + (400)(39.785)(259.8925)\right]$$
$$= \frac{27,255,971}{EI}$$

or, by substituting for EI, we find

$$\delta_D = \frac{27,255,971}{(75)(10)^3} = 363.41 \text{ in.} (9.23 \text{ m})$$

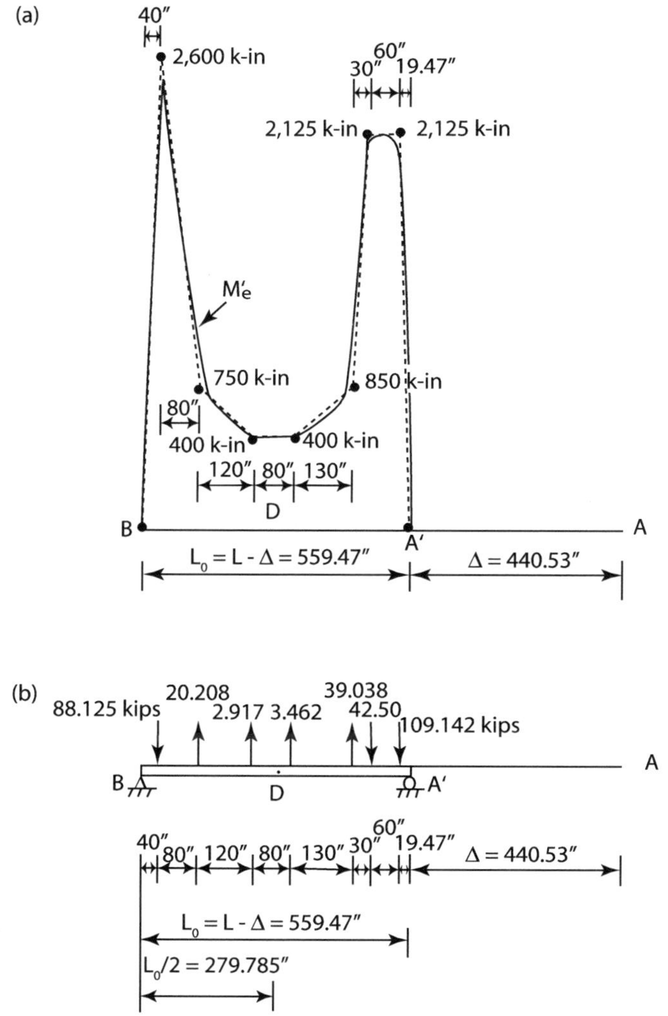

Fig. 2.4. (**a**) Moment diagram M'_e of the pseudolinear system approximated with eight straight-line segments. (**b**) Pseudolinear equivalent system loaded with seven concentrated loads (1 in. = 0.0254 m, 1 kip in. = 113.0 N m, 1 kip = 4, 448 N)

In order to compare results, Table 2.5 was prepared. In this table, the value of the horizontal displacement Δ_A at support A, the rotation θ_B at the end B, and the vertical displacement δ_D at midpoint D, were calculated for various values of the uniformly distributed load w, by using pseudolinear equivalent systems and the Fourth Order Runge-Kutta method. In terms of δ_D, the percentage difference between these two methods is shown in the last column of the table. With respect to Δ_A and θ_B, the percentage differences are larger. However, the values that are obtained by using the pseudolinear equivalent system should be more accurate, because the only approximations we have

Table 2.5. Variation of the horizontal displacement Δ_A, rotation θ_B, and vertical displacement δ_D for various values of the distributed load $w(1\,\text{in.} = 0.0254\,\text{m}, 1\,\text{lb in.}^{-1} = 175.118\,\text{N m}^{-1})$

w	pseudolinear equivalent systems			fourth order Range–Kutta method			% difference
(lb in.$^{-1}$)	Δ_A (in.)	θ_B (deg.)	δ_D (in.)	Δ_A (in.)	θ_B (deg.)	δ_D (in.)	in δ_D
2	143.60	44.26	230.07	161.56	47.21	242.19	−5.00
5	311.43	65.08	321.25	358.65	70.59	337.72	−4.87
10	440.53	76.63	367.65	496.17	81.93	379.62	−3.15
15	508.55	81.54	389.87	562.53	85.81	396.42	−1.68
20	552.62	84.22	404.08	603.66	87.57	406.32	−0.55

introduced to this problem is the approximation with straight-line segments of the M'_e diagram in Fig. (2.4a), and the function we have used for $\Delta(x)$.

These two approximations are very reasonable, and if we require better accuracy, we can approximate the M'_e with more straight-line segments, and we can also use a better function to approximate $\Delta(x)$, instead of assuming it to remain constant and equal to the horizontal displacement Δ of support A. For many practical applications the results obtained using these two approximations are very reasonable and further refinement is not required. We should not forget that the derivation of the pseudolinear equivalent system is mathematically exact, and any simplifications we introduce to any of its applications we are in a very excellent position to know their effect on accuracy. This problem was examined in a great detail by the author and his collaborators, and a great deal of these results are shown in the references at the end of this text. Refer also to Sect. 1.7 of the first chapter of this text.

2.4 Flexible Uniform Simply Supported Beam Loaded with a Vertical Concentrated Load

The uniform flexible simply supported beam in Fig. 2.5, is assumed to be loaded with a concentrated vertical load P, located at any point along the length of the member. Its large deformation configuration is also shown in the same figure. The modulus of elasticity E and moment of inertia I are assumed to be constant throughout the length of the member. In following chapters of this text, both E and I are permitted to vary in any arbitrary manner, and detailed solutions are obtained.

For the portions a_1 and a_2 in Fig. 2.5, the bending moment expressions $M_1(x)$ and $M_2(x)$, respectively, are obtained by using simple statics, and they are as follows:

$$M_1(x) = \frac{Pa_2x}{L_0}, \qquad 0 \leq x \leq a_1 \tag{2.42}$$

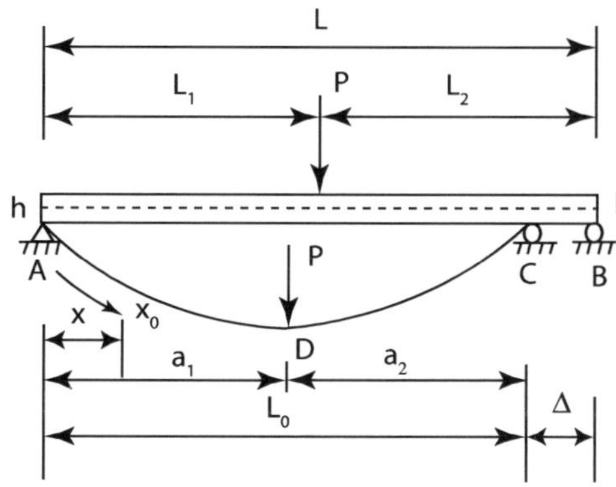

Fig. 2.5. Uniform simply supported beam loaded with a concentrated load P at any point along its length

$$M_2(x) = \frac{Pa_1}{L_0}(L_0 - x), \qquad a_1 \le x \le (a_1 + a_2) \tag{2.43}$$

Note that $L_o = a_1 + a_2$, and $L = L_1 + L_2$, as shown in Fig. 2.5.

In this case, the Euler–Bernoulli nonlinear differential equation must be applied two times, once for portion a_1, and secondly for portion a_2. Thus, by using Eqs. (2.42) and (2.43) and substituting into the Euler–Bernoulli equation, we obtain the following two nonlinear differential equations for portions a_1 and a_2:

$$\frac{y_1''}{\left[1 + (y_1')^2\right]^{3/2}} = -\frac{Pa_2 x}{EIL_0}, \qquad 0 \le x \le a_1 \tag{2.44}$$

$$\frac{y_1''}{\left[1 + (y_1')^2\right]^{3/2}} = -\frac{Pa_1}{EIL_0}(L_0 - x), \qquad a_1 \le x \le (a_1 + a_2) \tag{2.45}$$

By integrating each of the Eqs. (2.44) and (2.45) once, as it was done in the preceding sections of this text, we find

$$\frac{y_1''}{\left[1 + (y_1')^2\right]^{3/2}} = -\frac{Pa_2 x^2}{2EIL_0} + C_1, \qquad 0 \le x \le a_1 \tag{2.46}$$

$$\frac{y_1''}{\left[1 + (y_1')^2\right]^{3/2}} = -\frac{Pa_1}{EIL_0}\left(-\frac{x^2}{2} + L_0 x\right) + C_2, \qquad a_1 \le x \le (a_1 + a_2) \tag{2.47}$$

where C_1 and C_2 are the constants of integration.

We determine C_1 and C_2 by applying the boundary conditions $y_1'(0) = y_A'$, and $y_1'(a_1) = y_2'(a_1)$, where y_A' is the value of $y'(x)$ at the left support of the member where $x = 0$. Thus, by using Eqs. (2.46) and (2.47) and applying the indicated boundary conditions, we obtain

$$C_1 = \frac{y'_A}{\left[1 + (y'_A)^2\right]^{1/2}} \tag{2.48}$$

$$C_2 = \frac{Pa_1^2}{2EI} + \frac{y'_A}{\left[1 + (y'_A)^2\right]^{1/2}} \tag{2.49}$$

We note here that Eqs. (2.46) and (2.47) contain the horizontal displacement Δ of the right support B of the member, because $L_o = L - \Delta$, and that the two unknown quantities in these two equations are Δ and y'_A. Also unknown are the two horizontal displacements Δ_1 and Δ_2 of the lengths L_1 and L_2, respectively, where $\Delta_1 + \Delta_2 = \Delta$.

The quantities Δ_1, Δ_2, and Δ may be determined by trial-and-error, as in preceding sections, using the following equations:

$$L_1 = \int_0^{a_1} \left[1 + (y'_1)^2\right]^{1/2} dx \tag{2.50}$$

$$L_2 = \int_{a_1}^{(a_1+a_2)} \left[1 + (y'_2)^2\right]^{1/2} dx \tag{2.51}$$

$$L = \int_0^{a_1} \left[1 + (y'_1)^2\right]^{1/2} dx + \int_{a_1}^{(a_1+a_2)} \left[1 + (y'_2)^2\right]^{1/2} dx \tag{2.52}$$

By considering Eqs. (2.46)–(2.49), and consulting Eqs. (1.32) and (1.33), we conclude that

$$y'_1 = \frac{G_1(x)}{\sqrt{1 - [G_1(x)]^2}}, \quad 0 \le x \le a_1 \tag{2.53}$$

$$y'_2 = \frac{G_2(x)}{\sqrt{1 - [G_2(x)]^2}}, \quad a_1 \le x \le (a_1 + a_2) \tag{2.54}$$

where

$$G_1(x) = \phi_1(x) + C_1 = -\frac{Pa_2x^2}{2EIL_0} + \frac{y'_A}{[1 + (y'_A)^2]^{1/2}} \tag{2.55}$$

$$G_2(x) = \phi_2(x) + C_2 = -\frac{Pa_1}{2EIL_0}\left[\frac{x^2}{2} - L_0x\right] + \frac{Pa_1^2}{2EI} + \frac{y'_A}{\left[1 + (y'_A)^2\right]^{1/2}} \tag{2.56}$$

In the above equations the rotation y'_A at the left end of the beam is not also known and it has to be determined. This can be accomplished by using a trial-and-error procedure with assigned values for the slope y'_A. For example, we assume first a value for y'_A, and then we move to Eqs. (2.50) and (2.51) and assign values for a_1 and a_2 until these two equations are satisfied. When this procedure is accomplished, we check if the condition of zero displacement at $x = L_o$ or $x = 0$ is satisfied. If this is not satisfied, we repeat the procedure with new assigned values of y'_A until the displacement condition is satisfied. When we will complete this trial-and-error procedure, then the values of θ_A, Δ_1, Δ_2, Δ, a_1, a_2, and $L_o = (a_1 + a_2)$ will all be known, and the values of y' at any $0 \leq x \leq L_o$ would be evaluated by using Eqs. (2.53) and (2.54).

With known y', the pseudolinear analysis can be carried out as it was done in preceding sections of this text. The following numerical example illustrates the application of the above methodology.

Example 2.3 Consider the flexible simply supported beam in Fig. 2.5 that has a uniform stiffness $EI = 312.50$ lb in.^{2} ($0.8968\,\mathrm{N\,m^2}$) throughout its length, and loaded with a vertical load $P = 0.50$ lb (2.4482 N) located as shown. Its length $L = L_1 + L_2 = 104.60$ in. (2.6568 m), where $L_1 = 40.0$ in. (1.016 m) and $L_2 = 64.60$ in. (1.6408 m). The cross section of the member is rectangular, with depth $h = 0.05$ in. (0.00127 m), width $b = 1.0$ in. (0.0254 m), and modulus of elasticity $E = 30 \times 10^6$ psi ($206,844 \times 10^6$ Pa). Determine the rotation θ_A at the end A of the member, the horizontal displacement Δ_B at the end B, and the vertical displacement under the concentrated load P.

Solution: We follow the procedure discussed above, and for a start we assume that $y'_A = 1.1505$ rad, which corresponds to a rotation $\theta_A = 49^o$ of the left end of the member. On this basis, Eq. (2.55) yields

$$G_1(x) = -\frac{0.8(10)^{-3}}{L_0} a_2 x^2 + 0.754747 \tag{2.57}$$

As a next step, we assume values for a_1 and a_2. We note, however, that the ratio a_1/a_2 should be close to the ratio L_1/L_2, which is known. Therefore, we assume $a_1 = 33.58$ in. (0.853 m) and $a_2 = 54.22$ in. (1.377 m), yielding a ratio a_1/a_2 approximately equal to the ratio L_1/L_2. On this basis, $L_o = a_1 + a_2 =$ 87.80 in. (2.23 m), and $\Delta = L - L_o = 16.80$ in. (0.427 m).

By substituting into Eq. (2.57), we find

$$G_1(x) = -0.494032\,(10)^{-3}\,x^2 + 0.754747 \tag{2.58}$$

With the assumed values for a_1 and a_2, we check to find out if Eq. (2.50) is satisfied. The integrations associated with this equation can be carried out by using Simpson's One-Third rule, as discussed in preceding examples and in Sect. 1.5 of the first chapter. Computer programs are usually available for this purpose. For illustration purpose, we perform this integration manually by using Simpson's rule.

By following the procedure in Sect. 1.5, we assume that $n = 5$, yielding

$$\lambda = \frac{33.58 - 0}{5} = 6.716 \tag{2.59}$$

On this basis, the application of Simpson's rule given by Eq. (1.36), yields

$$\begin{aligned} L_1 &= \int_0^{33.58} \left[1 + (y_1')^2\right]^{1/2} dx = \frac{6.716}{3}\Big[1.524353 + (4)(1.468849) \\ &\quad +(2)(1.339951) + (4)(1.201368) + (2)(1.090166) + 1.020128\Big] \\ &= 40.4876 \text{ in. } (1.028 \text{ m}) \end{aligned} \tag{2.60}$$

which is approximately equal to the actual value $L_1 = 40$ in. (1.016 m).

The required ordinates $y_o, y_1, y_2, \ldots, y_5$ for the application of Simpson's rule were determined as discussed in Sect. 1.5 and in Example 1.1 of the first chapter of this text. For example, at $x = 0$, Eq. (2.58) yields $G_1(0) = 0.754747$. Therefore, by applying Eq. (2.53) we obtain

$$y_1'(0)\frac{0.754747}{\sqrt{1 - (0.754747)^2}} = 1.150501$$

The function f(x) in Eq. (2.50), or Eq. (2.60), is

$$f(x) = \left[1 + (y_1')^2\right]^{1/2} \tag{2.61}$$

Therefore, the ordinate y_o is

$$y_0 = \left\{1 + [y_1'(0)]^2\right\}^{1/2} = \left\{1 + (1.150501)^2\right\}^{1/2} = 1.524353$$

In a similar manner, we determine the values of $y_1, y_2, \ldots, y_5$, ordinates corresponding to $x_1 = 6.716, 13.432, 20.148, 26.864$, and 33.58, respectively.

The final step is to find out if the zero displacement condition at $x = L_o$, which is the right support of the member, is satisfied. This can be accomplished by using Eq. (1.34) and applying Simpson's rule to carry out the required integrations. This task was carried out by using a digital computer, and it was found that this displacement condition is adequately satisfied. Therefore, we accept the assumed value of $y_A' = 1.1505$ ($\theta_A = 49°$), and the procedure will not be repeated.

Knowing y_A', a_1, and a_2, the rotation at any $0 \leq x \leq L_o$ can be determined by using Eqs. (2.53) and (2.54), and the pseudolinear analysis can be carried out as discussed in Sect. 1.8.1 of the first chapter and in other preceding parts of the text. For example, by using Eqs. (2.42) and (2.43) and making the appropriate substitutions for a_1, a_2, L_o, and P, we obtain

$$M_1(x) = \frac{(0.5)(54.22)}{87.80}x = 0.30877\ x, \quad 0 \leq x \leq a_1 \tag{2.62}$$

$$\begin{aligned} M_2(x) &= \frac{(0.5)(33.58)}{87.80}(87.80 - x) = 0.19123(87.80 - x), \\ &\quad a_1 \leq x \leq (a_1 + a_2) \end{aligned} \tag{2.63}$$

The moment diagram M'_e of the pseudolinear equivalent system can be determined from the equation

$$M'_e = z_e M_x \tag{2.64}$$

where

$$z_e = \left[1 + (y')^2\right]^{3/2} \tag{2.65}$$

The required values of M_x in Eq. (2.64) can be determined by using Eqs. (2.62) and (2.63) and y' in Eq. (2.65) can be determined by using Eqs. (2.53) and (2.54).

The calculated values of $M_x, y'(x), z_e$, and M'_e, for the assumed values of x, are given in Table 2.6. By using the values of M'_e in the last column of the table, we plot the moment diagram M'_e of the pseudolinear system shown by the solid line in Fig. 2.6a. The approximation of its shape with four straight-line segments is shown by the dashed line in the same figure. The approximated M'_e leads to the pseudolinear equivalent system shown in Fig. 2.6b.

By looking at the deformed configuration of the member, which is shown in Fig. 2.5, we see that the concentrated load P is located at a distance $a_1 =$ 33.58 in. (0.853 m) from support A. This position is marked in Fig. 2.6b by the capital letter D. The vertical deflection δ_D at point D is determined by using the pseudolinear system in Fig. 2.6b and applying handbook formulas [84]. Thus,

$$\begin{aligned}
\delta_D &= \frac{(0.9643)(6)(54.22)}{(6)(312.50)(87.80)}\left[(87.8)^2 - (6)^2 - (54.22)^2\right] \\
&\quad + \frac{(0.382)(20)(54.22)}{(6)(312.50)(87.80)}\left[(87.8)^2 - (20)^2 - (54.22)^2\right] \\
&\quad + \frac{(0.3602)(13.8)(33.58)}{(6)(312.50)(87.80)}\left[(87.8)^2 - (13.8)^2 - (33.58)^2\right] \\
&= 26.306 \text{ in. } (0.668 \text{ m})
\end{aligned}$$

Table 2.6. Values of $M_x, y'(x), z_e$, and M'_e, corresponding to the assumed values of x(1 in. = 0.0254 m, 1 lb in. = 0.113 N m)

x (in.)	M_x (lb in.) Eqs. (2.62) and (2.63)	$y'(x)$ Eqs. (2.53) and (2.54)	z_e Eq. (2.65)	M_e (lb in.) Eq. (2.64)
0	0	1.150500	3.542062	0
10	3.088	0.995033	2.807432	8.670
20	6.175	0.670904	1.746234	10.783
30	9.263	0.326200	1.163783	10.780
40	9.141	−0.002904	1.000004	9.141
50	7.228	−0.274647	1.115254	8.061
60	5.316	−0.526070	1.442642	7.669
70	3.404	−0.760057	1.981670	6.746
80	1.492	−0.936231	2.570591	3.835
87.8	0	−0.985890	2.769194	0

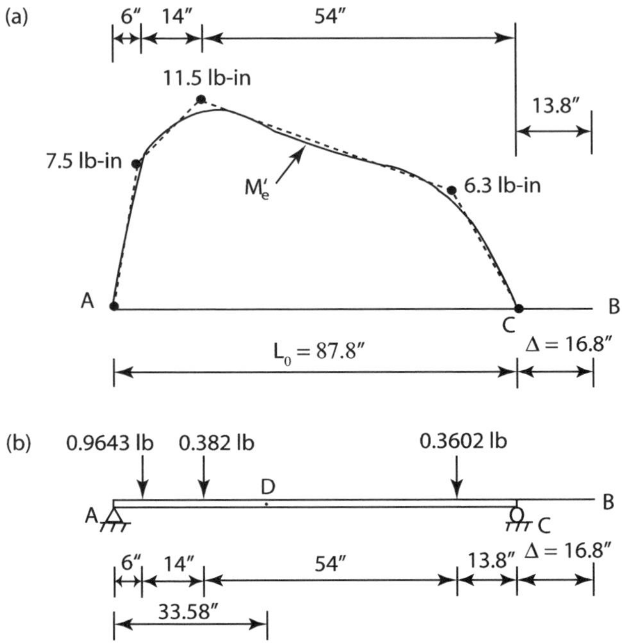

Fig. 2.6. (**a**) Moment diagram M'_e of the pseudolinear system with its shape approximated with four straight-line segments. (**b**) Pseudolinear equivalent system (1 in. = 0.0254 m, 1 lb in. = 0.113 N m, 1 lb = 4.448 N)

The same problem was solved by Frisch-Fay [63], by using the very complex method of elastic similarities. His results are $\theta_A = 50^\circ$, $\Delta = 16.10$ in. (0.409 m) and $\delta_D = 23.80$ in. (0.6045 m). Frisch-Fay used an iterative procedure to evaluate the elliptic integrals involved in his analysis, and a good part of the difference in the results may be attributed to the utilization of this procedure.

The computerized results of the pseudolinear analysis, which should be very accurate, yielded $\delta_D = 25.48$ in. (0.409m).

2.5 Uniform Statically indeterminate Single Span Flexible Beam Loaded with a Uniformly Distributed Load w_o on its Entire Span

The single span uniform beam in Fig. 2.7 is fixed at the end A, it is supported by a roller at the end B, and it is loaded with a uniformly distributed load w_o over its entire span length. Its modulus of elasticity E and its moment of inertia I are also constant over the whole length of the member. Since the end A is fixed and the member is statically indeterminate, the reaction R_B at the end B is considered to be the redundant. However, if desired, the moment M_A at

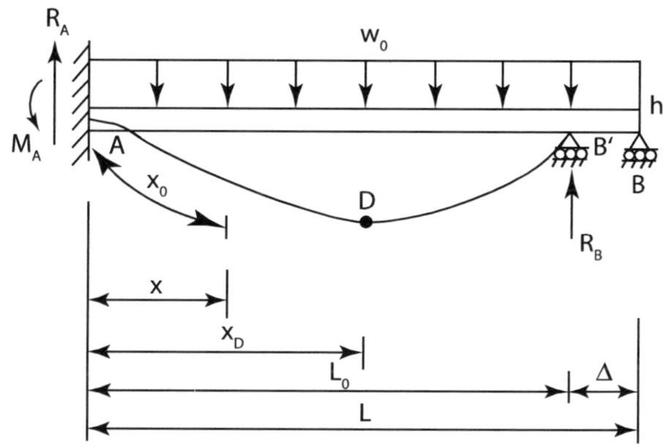

Fig. 2.7. Uniform statically indeterminate flexible beam subjected to a uniformly distributed load w_o

the end A could be taken as the redundant instead of R_B. By satisfying static equilibrium in the vertical direction, we can write the following equation:

$$R_A + R_B = Lw_0 \tag{2.66}$$

The bending moment at any $0 \le x \le L_o$ is obtained by using simple statics, and it is given by the expression

$$M_x = \frac{w_0}{2}\left[2Lx - LL_0 - xx_0\right] + R_B\left[L_0 - x\right] \tag{2.67}$$

By using Eq. (2.67), the Euler–Bernoulli equation is written as

$$\frac{y''}{\left[1+(y')^2\right]^{3/2}} = \frac{w_0}{2EI}\left[LL_0 - 2Lx + xx_0\right] - \frac{R_B}{EI}\left[L_0 - x\right] \tag{2.68}$$

where

$$x_0(x) = \int_0^x \left\{1 + \left[y'(x)\right]^2\right\}^{1/2} dx \tag{2.69}$$

By substituting Eq. (2.69) into Eq. (2.68), we obtain the following nonlinear integral differential equation:

$$\frac{y''}{\left[1+(y')^2\right]^{3/2}} = \frac{w_0}{2EI}\left\{LL_0 - 2Lx + x\left[\int_0^x \left\{1 + \left[y'(x)\right]^2\right\}^{1/2} dx\right]\right\} - \frac{R_B}{EI}\left[L_0 - x\right] \tag{2.70}$$

which is very difficult to solve. Its solution, however, may be simplified in the same way as it was done in preceding sections. That is, we replace the integral that represents $x_o(x)$ by a function that contains the horizontal displacement

$\Delta(x)$. On this basis we may write

$$x_0(x) = x + \Delta(x) \tag{2.71}$$

which is the same as Eq. (1.74) in Sect. 1.7. Various expressions for $\Delta(x)$ are suggested in the same section. See, for example, Eqs. (1.81)–(1.84).

A popular expression for $x_o(x)$ which was used extensively in preceding sections and provided accurate results, is to use $\Delta(x) = \Delta$ in Eq. (2.71), where Δ is the horizontal displacement of the end support B in Fig. 2.7. This yields

$$x_0(x) = x + \Delta \tag{2.72}$$

Thus, by using Eq. (2.72) instead of Eq. (2.69), the Euler–Bernoulli differential equation, and consequently Eq. (2.70), reduces to the following equation:

$$\frac{y''}{\left[1 + (y')^2\right]^{3/2}} = \frac{w_0}{2EI}\left[LL_0 - 2Lx + x\,(x + \Delta)\right] - \frac{R_B}{EI}\left[L_0 - x\right] \tag{2.73}$$

which is a much easier differential equation to solve.

By integrating Eq. (2.73) once, and determining the constant of integration by satisfying the boundary condition of zero rotation at $x = 0$, which is the fixed end of the member, we obtain

$$y'(x) = \frac{G(x)}{\left\{1 - [G(x)]^2\right\}^{1/2}} \tag{2.74}$$

where

$$G(x) = \frac{w_0}{12EI}\left[6LL_0x - 6Lx^2 + 3\Delta x^2 + 2x^3\right] - \frac{R_B}{2EI}\left[2L_0x - x^2\right] \tag{2.75}$$

By substituting the expression for $y'(x)$, given by Eq. (2.74), into the expression

$$L = \int_0^{L_0 = L - \Delta} \left[1 + (y')^2\right]^{1/2} dx \tag{2.76}$$

and performing some mathematical manipulations, we obtain

$$L = \int_0^{(L-\Delta)} \frac{1}{\left\{1 - [G(x)]^2\right\}^{1/2}} dx \tag{2.77}$$

The vertical displacement $y(x)$ of the member at any $0 \le x \le L_o$ may be obtained from equation

$$y(x) = \int_0^x y'(x)dx = \int_0^x \frac{G(x)}{\left\{1 - [G(x)]^2\right\}^{1/2}} dx \tag{2.78}$$

or by using pseudolinear analysis as discussed in preceding sections of this text and in Sect. 1.8.1, since $y'(x)$ can be determined using Eq. (2.74).

It should be noted, however, that Eqs. (2.74), (2.77), and (2.78) contain the unknown horizontal displacement Δ of the end B and the unknown redundant reaction R_B at the same end B of the member. These two quantities can be determined by using Eqs. (2.77) and (2.78) and applying a trial-and-error procedure in a way similar to the one used in preceding parts of this text. That is, we can assume a value for the redundant reaction R_B, and then apply a trial-and-error procedure to find the value of Δ that satisfies Eq. (2.77). Once this is done, we check if the boundary conditions at the ends of the member are also satisfied. If not, then we repeat the procedure by assigning new values of R_B until the boundary conditions are satisfied.

The example that follows illustrates the application of the methodology.

Example 2.4 The statically indeterminate flexible member shown in Fig. 2.7, has a uniform stiffness $EI = 75 \times 10^3$ kip in.^{2} (215, 224 N m^2) and a length $L = 1{,}000$ in. (25.4 m). Determine the horizontal displacement Δ of support B, the reaction R_B of support B, the rotation θ_B at the same support, the maximum displacement δ_D of the member, and the location x_D of the maximum displacement. Assume various values of w_o and compare the results by assuming $\Delta(x) = \Delta, \Delta(x) = \Delta x/L_o, \Delta(x) = \Delta[x/L_o]^{1/2}$, and $\Delta(x) = \Delta \sin(\pi x/2L_o)$.

Solution: We initiate the solution by considering first the case where $\Delta(x) = \Delta$, and assuming $w_o = 5$ lb in.$^{-1}$ (875.65 N m^{-1}). The trial-and-error procedure may be initiated by assuming a value for the redundant reaction R_B, and then look for the value of Δ that satisfies Eq. (2.77). We start with $R_B = 1.96$ kips (8,718 N) and assuming $\Delta = 150.6$ in. (3.83 m). Thus, $L_o = L - \Delta = 849.4$ in. (21.57 m).

By substituting for L_o and Δ into Eq. (2.75), we obtain

$$G(x) = 0.005556\,(10)^{-6}\left[5,096,400x - 5,548.2x^2 + 2x^3\right] \\ -0.013067\,(10)^{-3}\left[1,698.8x - x^2\right] \tag{2.79}$$

Note that at $x = 0$ we have $G(0) = 0$, indicating that the rotation is zero at the fixed end, as expected. We also observe that the assumed value of Δ satisfies Eq. (2.77). The integration of this equation was carried out by using Simpson's rule. By using Eq. (2.78) and applying Simpson's rule to carry out the integration, we find that the boundary condition of zero displacement at support A is satisfied. This means that the assumed value of R_B is correct and, consequently, we do not need to repeat the procedure.

The rotation $\theta_B = \tan^{-1}(y_B')$ may be determined by using Eqs. (2.74) and (2.79). Thus, at $x = L_o = L - \Delta = 849.4$ in. (21.57 m), Eq. (2.79) yields $G(L_o) = -0.806807$, and from Eq. (2.74) we obtain

$$y'(L_0) = \frac{-0.806807}{\sqrt{1-(-0.806807)^2}} = -1.365583$$

Thus,

$$\theta_B = \tan^{-1} y'(L_0) = 0.938728 \text{ rad}$$
$$= 53.79^\circ$$

The maximum vertical displacement δ_D occurs at point D in Fig. 2.7, which is the point where $y'(x) = 0$. From Eq. (2.74) we note that $y'(x)$ is zero when $G(x)$ is zero. Thus, for $G(x) = 0$, Eq. (2.79) yields $x = x_D = 502.6$ in. (12.766 m).

Since Δ and R_B are now known quantities, Eqs. (2.74) and (2.75) can be written as functions of $0 \leq x \leq L_o$ only, see for example Eq. (2.79). Thus, we can determine $y'(x)$ at any $0 \leq x \leq L_o$ by using Eq. (2.74), where in this equation $G(x)$ is given by Eq. (2.79). With known $y'(x)$, the pseudolinear analysis discussed in Sect. 1.8.1 and in other parts of this text, can be used to determine the vertical displacements of the member at any $0 \leq x \leq L_o$. The moment diagram M_e' of the pseudolinear system can be determined by using Eqs. (1.119) and (1.120), where M_x in Eq. (1.119) is given by Eq. (2.67) and $f(x) = g(x) = 1$.

With known values of M_e', we can plot the moment diagram M_e' of the pseudolinear equivalent system in the usual way. The approximation of its shape with three or four straight-line segments, as it was done in preceding examples of this book, and applying statics, we determine the equivalent pseudolinear system. As an exercise, the reader could perform this detail derivation of the pseudolinear system for better understanding of the problem.

By using the pseudolinear system and applying linear analysis or handbook formulas, we find that at $x = x_D$ the maximum displacement $\delta_D = 242.76$ in. (6.166 m). The complete results are shown in Table 2.7. In the same table, the calculated values for the same quantities were obtained by assuming that $\Delta(x) = \Delta, \Delta(x) = \Delta x/L_o, \Delta(x) = \Delta[x/L_o]^{1/2}$, and $\Delta(x) = \Delta \sin(\pi x/2L_o)$, in order to compare results. These results, with a few exceptions, are in close agreement for practical purposes, which indicates that the deformation of the flexible member, for practical purposes, is not very sensitive to the variation assumed for the function $\Delta(x)$. This is verified in all the work performed in this text and in many other publications of the author. The few exceptions associated with the computation of the horizontal displacement Δ for $w =$ 10 lb in. (1,751.268 N m^{-1}), might be attributed to numerical errors that need to be investigated.

2.6 Uniform Statically Indeterminate Single Span Flexible Beam Subjected to a Vertical Concentrated Load

The statically indeterminate uniform flexible member in Fig. 2.8 is loaded with a concentrated load P at a distance L_1 from the fixed support A of the member. The right support of the member is a roller, which permits this

Table 2.7. Variation of Δ, θ_B, δ_D, x_D, and R_B with uniformly distributed load w_o for various cases of horizontal displacement function $\Delta(x)$ (1 in. = 0.0254 m, 1 lb in^{-1} = 175.1268 N m^{-1}, 1 kip = 4,448.222 N)

w_o (lb in)$^{-1}$	Δ (in.)	θ_B (deg.)	δ_D (in.)	x_D (in.)	R_B (kip)
$\Delta(x) = \Delta$					
2.0	42.80	28.56	128.806	558.05	0.7592
5.0	150.55	53.79	242.760	502.60	1.9584
10.0	274.59	70.09	303.079	440.396	4.0590
$\Delta(x) = \Delta x/L_o$					
2.0	46.35	29.75	133.390	555.98	0.7485
5.0	184.99	58.59	255.212	486.724	1.8594
10.0	357.682	78.18	337.469	395.347	3.6892
$\Delta(x) = \Delta(x/L_o)^{1/2}$					
2.0	44.65	29.20	130.896	556.300	0.7528
5.0	167.21	55.70	244.128	494.594	1.90107
10.0	315.808	74.07	321.000	415.989	3.8520
$\Delta(x) = \Delta\sin(\pi x/2L_o)$					
2.0	44.14	28.87	130.260	555.259	0.7523
5.0	161.65	53.90	241.152	491.608	1.89727
10.0	302.10	70.50	317.195	411.270	3.8410

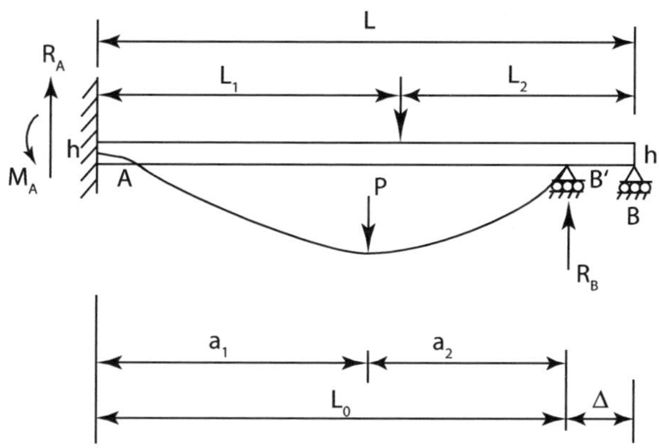

Fig. 2.8. Statically indeterminate flexible beam of uniform stiffness that is loaded with a concentrated load P

end of the beam to move horizontally. The large deformation configuration of the beam is also shown in the same figure. This is a practical case of a statically indeterminate beam subjected to large deformations. If both ends of the member were fixed, then severe axial restraints would be introduced, and the member would probably fail from large axial stresses long before the deformations become large enough to require utilization of large deflection theory.

Since the single span beam is statically indeterminate, the vertical reaction R_B at the roller support is taken as the redundant. By applying static equilibrium conditions we write the following two equations:

$$R_A + R_B = P \tag{2.80}$$

$$M_A = Pa_1 - R_B L_0 \tag{2.81}$$

The uniform cross sectional moment of inertia I of the member is

$$I = \frac{bh^3}{12} \tag{2.82}$$

where b is the constant width of the member.

The bending moment M_x at any x, where x is measured from the fixed support A, is written as follows:

$$M_x = -(Pa_1 - R_B L_0) + (P - R_B)\,x, \qquad 0 \le x \le a_1 \tag{2.83}$$

$$M_x = R_B(L_0 - x), \qquad a_1 \le x \le (a_1 + a_2) \tag{2.84}$$

where $L_o = (a_1 + a_2)$. By substituting Eqs. (2.83) and (2.84) into the Euler–Bernoulli equation, we find

$$\frac{y_1''}{\left[1 + (y_1')^2\right]^{3/2}} = \frac{1}{EI}\left[P\,(a_1 - x) - R_B(L_0 - x)\right], \qquad 0 \le x \le a_1 \tag{2.85}$$

$$\frac{y_2''}{\left[1 + (y_2')^2\right]^{3/2}} = -\frac{1}{EI}\left[R_B(L_0 - x)\right], \qquad a_1 \le x \le (a_1 + a_2) \tag{2.86}$$

By integrating Eqs. (2.85) and (2.86) once, we obtain

$$\frac{y_1'}{\left[1 + (y_1')^2\right]^{1/2}} = \frac{1}{2EI}\left[P\left(2a_1x - x^2\right) - R_B\left(2L_0x - x^2\right)\right] + C_1, \qquad 0 \le x \le a_1 \tag{2.87}$$

$$\frac{y_2'}{\left[1 + (y_2')^2\right]^{1/2}} = -\frac{1}{2EI}\left[R_B\left(2L_0x - x^2\right)\right] + C_2, \qquad a_1 \le x \le (a_1 + a_2) \tag{2.88}$$

where C_1 and C_2 are the constants of integration, which can be determined by applying the boundary conditions

$$y_1'(0) = 0 \tag{2.89}$$

$$y_1'(a_1) = y_2'(a_1) \tag{2.90}$$

By using Eqs. (2.87) and (2.88) and the above two boundary conditions, we find $C_1 = 0$ and $C_2 = Pa_1/2EI$. On this basis, Eqs. (2.87) and (2.88) may be rewritten as

$$\frac{y_1'}{\left[1+(y_1')^2\right]^{1/2}} = G_1(x), \qquad 0 \le x \le a_1 \tag{2.91}$$

$$\frac{y_2'}{\left[1+(y_2')^2\right]^{1/2}} = G_2(x), \qquad a_1 \le x \le (a_1+a_2) \tag{2.92}$$

where

$$G_1(x) = \frac{1}{2EI}\left[P\left(2a_1x - x^2\right) - R_B\left(2L_0x - x^2\right)\right] \tag{2.93}$$

$$G_2(x) = -\frac{1}{2EI}\left[R_B\left(2L_0x - x^2\right) - Pa_1^2\right] \tag{2.94}$$

By using Eqs. (2.50) and (2.51) and solving Eqs. (2.91) and (2.92) for y_1' and y_2', we can write the following equations for L, L_1, and L_2:

$$L = \int_0^{a_1} \frac{1}{\left\{1-[G_1(x)]^2\right\}^{1/2}}dx + \int_{a_1}^{(a_1+a_2)} \frac{1}{\left\{1-[G_2(x)]^2\right\}^{1/2}}dx \tag{2.95}$$

$$L_1 = \int_0^{a_1} \frac{1}{\left\{1-[G_1(x)]^2\right\}^{1/2}}dx \tag{2.96}$$

$$L_2 = \int_{a_1}^{(a_1+a_2)} \frac{1}{\left\{1-[G_2(x)]^2\right\}^{1/2}}dx \tag{2.97}$$

By observing Eqs. (2.93) and (2.94), we find that y_1' and y_2' can be determined if the horizontal displacement Δ of support B, and the reaction R_B at the same support, are known. Note that $L_o = (L - \Delta)$.

One way to determine Δ and R_B is to use a trial-and-error procedure as in the preceding section. For example, we start by assuming a value for the redundant reaction R_B. Then, since $L_o = (L - \Delta) = (a_1 + a_2)$, we initiate a trial-and-error procedure to determine the values of a_1 and a_2 which satisfy Eqs. (2.95)–(2.97). When this is done, we check if the assumed redundant reaction R_B satisfies the boundary condition of zero displacement at support B. If this condition is not satisfied, we increase or decrease the values of the assumed R_B and repeat the procedure until it is satisfied. The procedure converges fairly fast to the correct value of R_B. Note that $L_1 = a_1 + \Delta_1$, and $L_2 = a_2 + \Delta_2$, where Δ_1 and Δ_2 are the horizontal displacements of L_1 and L_2, respectively, during the large deformation configuration. Thus, when a_1 and a_2 are determined, it can be implied that Δ_1 and Δ_2 are also determined.

With known L_o, a_1, and R_B, we can determine the rotations y_1' and y_2' by using Eqs. (2.91) and (2.92), respectively. The bending moment M_x at any $0 \leq x \leq L_o$ can be obtained from Eqs. (2.83) and (2.84), and the pseudolinear analysis can be carried out as discussed in Sect. 1.8.1 and in other preceding sections of this text. The moment diagram $M_e{}'$ of the equivalent pseudolinear system can be determined using Eqs. (1.119) and (1.120) with $f(x) = g(x) = 1$. By approximating the shape of M_e with a few straight-line segments and applying statics, we obtain the pseudolinear equivalent system of constant EI which is loaded with a few concentrated vertical loads at the appropriate locations. As in preceding sections, linear analysis can be applied to the pseudolinear system to determine deflections and rotations along its length. These deflections and rotations will be closely identical to the ones at the corresponding points of the original system of length L. This correspondence is discussed in detail in Example 1.4 of Sect. 1.8.1 in the first chapter of this text, and in other parts of the book.

2.7 Flexible Uniform Cantilever Beam Under Combined Loading Conditions

In this case we consider the uniform flexible cantilever beam in Fig. 2.9 that is loaded simultaneously with a uniformly distributed vertical load w_o over its entire length, and a concentrated load P at the free end B as shown. The solution of this problem was performed by Lau [53], using the Power series method with about 12 coefficients. In this section, we will use pseudolinear analysis to solve this problem and compare the results.

By considering the deformed configuration of the member and its free end as the origin, the bending moment M_x at any $0 \leq x \leq L_o$ is

$$M_x = -\left[\frac{w_0 x}{2} x_0 + Px\right] \tag{2.98}$$

By substituting Eq. (2.98) into the Euler–Bernoulli equation, we find

$$\frac{y''}{\left[1 + (y')^2\right]^{3/2}} = \frac{w_0 x}{2EI} x_0 + \frac{Px}{EI} \tag{2.99}$$

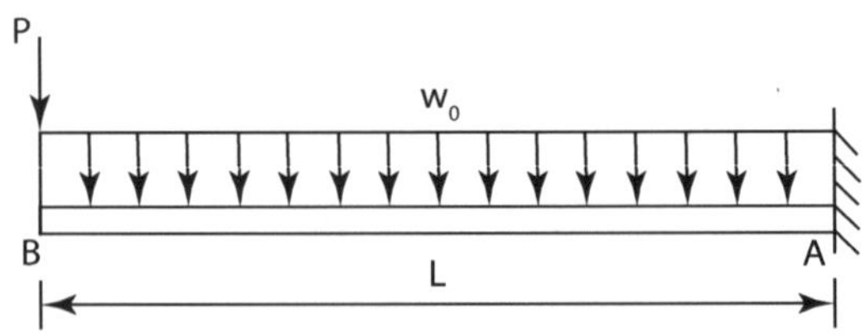

Fig. 2.9. Uniform flexible cantilever beam loaded with a uniformly distributed load w_o and a concentrated load P at the free end

By using Eq. (1.81), we can assume that $x_o = x + \Delta$, and Eq. (2.99) yields

$$\frac{y''}{\left[1+(y')^2\right]^{3/2}} = \frac{w_0 x}{2EI}(x+\Delta) + \frac{Px}{EI}$$

or, with some rearrangement,

$$\frac{y''}{\left[1+(y')^2\right]^{3/2}} = \frac{Px}{2EIL}\left[k\left(x^2+\Delta x\right)+2Lx\right] \tag{2.100}$$

where

$$k = \frac{w_0 L}{P} \tag{2.101}$$

In this case we have

$$\lambda(x) = -\frac{M}{EI} = \frac{P}{2EIL}\left[k\left(x^2+\Delta x\right)+2Lx\right] \tag{2.102}$$

$$\phi(x) = \int \lambda(x)dx = \frac{P}{2EIL}\left[k\left(\frac{x^3}{3}+\frac{x^2\Delta}{2}\right)+Lx^2\right] \tag{2.103}$$

Thus, by using Eq. (1.32), we obtain

$$\frac{y'}{\left[1+(y')^2\right]^{1/2}} = \phi(x) + C \tag{2.104}$$

where C is the constant of integration and $\phi(x)$ is given by Eq. (2.103).

We determine the constant C by using Eq. (2.104) and applying the boundary condition $y' = 0$ at $x = L_o$, where $L_o = L - \Delta$. This yields

$$\frac{P}{2EIL}\left\{k\left[\frac{(L-\Delta)^3}{3}+\frac{\Delta(L-\Delta)^2}{2}\right]+L(L-\Delta)^2\right\}+C=0$$

or

$$C = -\frac{P}{2EIL}\left\{k\left[\frac{(L-\Delta)^3}{3}+\frac{\Delta(L-\Delta)^2}{2}\right]+L(L-\Delta)^2\right\}$$

Thus,

$$\begin{aligned} G(x) = \phi(x) + C = \frac{P}{2EIL}\Bigg\{ & k\left(\frac{x^3}{3}+\frac{\Delta x^2}{2}\right)+Lx^2 \\ & -k\left[\frac{(L-\Delta)^3}{3}+\frac{\Delta(L-\Delta)^2}{2}\right]+L(L-\Delta)^2\Bigg\} \end{aligned}$$

or, after simplification,

$$G(x) = \frac{P}{12EIL}\left\{k\left(2x^3+3\Delta x^2-2L_0^3-3\Delta L_0^3\right)+6L\left(x^2-L_0^2\right)\right\} \tag{2.105}$$

By substituting Eq. (2.105) into Eq. (2.104), we obtain

$$\frac{y'}{[1+(y')^2]^{1/2}} = \frac{P}{12EIL}\left\{k\left(2x^3+3\Delta x^2-2L_0^3-3\Delta L_0^2\right)+6L\left(x^2-L_0^2\right)\right\} \tag{2.106}$$

Thus, by using Eq. (1.33), we find

$$y'(x) = \frac{G(x)}{\sqrt{1-[G(x)]^2}} \tag{2.107}$$

where G(x) is given by Eq. (2.105).

In order to determine y' from Eq. (2.107), the value of the horizontal displacement Δ of the free end of the cantilever beam must be known. This can be accomplished by using either Eq. (1.59) or (1.63). We rewrite Eq. (1.63) here for convenience.

$$L = (L-\Delta)\int_0^1 \frac{1}{\sqrt{1-[G(\xi)]^2}}d\xi \tag{2.108}$$

where

$$\xi = \frac{x}{L-\Delta} \tag{2.109}$$

$$d\xi = \frac{dx}{L-\Delta} \tag{2.110}$$

In terms of the dummy variable ξ, the function G(x) may be written as

$$G(\xi) = \frac{P}{12EIL}\left[k\left(2L_0^3\xi^3+3\Delta L_0^2\xi^2-2L_0^3-3\Delta L_0^2\right)+6LL_0^2\left(\xi^2-1\right)\right] \tag{2.111}$$

Since the horizontal displacement Δ in the above equation is not known, we can use a trial-and-error procedure, as it was done earlier, to determine it. We do this by assuming values of Δ until Eq. (2.108) or (2.76), is satisfied. With known Δ , the function G(x) at any $0 \le x \le L_o$ can be obtained from Eq. (2.105), and the values of $y'(x)$ at any $0 \le x \le L_o$ can be obtained form Eq. (2.107). With known $y'(x)$, we can carry out the pseudolinear analysis as discussed in Sect. 1.8.1 and in other parts of this text. Since in this case we have $f(x) = g(x) = 1.0$, the moment diagram M_e' of the pseudolinear system may be obtained from the equation

$$M_e' = z_e M_x \tag{2.112}$$

where M_x is given by Eq. (2.98), and

$$z_e = \left[1+(y')^2\right]^{3/2} \tag{2.113}$$

The following numerical example illustrates the procedure.

Example 2.5 The uniform flexible cantilever beam in Fig. 2.9 has a length $L = 1{,}000\,\text{in.}$ (25.4 m), $EI = 180 \times 10^3\,\text{kip in.}^2$ ($516{,}541\,\text{N m}^2$), $k = 3$, and $PL^2/EI = 2.416$. By using an equivalent pseudolinear system, determine the rotation θ_B, horizontal displacement Δ_B, and vertical displacement δ_B of the free end B of the member. Compare the results with the ones obtained by Lau [53].

Solution: By using the expression

$$\frac{PL^2}{EI} = 2.416$$

we find that $P = 0.4349\,\text{kip}$ (1,934 N). From the equation

$$k = \frac{w_0 L}{P} = 3$$

we obtain $w_o = 1.305\,\text{lb in.}^{-1}$ ($228.54\,\text{N m}^{-1}$).

By assuming that $x_o = x + \Delta$, the bending moment M_x at any $0 \le x \le L_o$, may be obtained from Eq. (2.98), yielding

$$\begin{aligned} M_x &= -\left[\frac{w_0 x}{2}(x+\Delta) + Px\right] \\ &= -\left[\frac{1.305(10)^{-3}x}{2}(x+\Delta) + 0.4349x\right] \\ &= -\left[0.6525(10)^{-3}\left(x^2 + \Delta x\right) + 0.4349x\right] \end{aligned} \tag{2.114}$$

From Eq. (2.105), we find

$$\begin{aligned} G(x) &= \frac{0.4349}{(12)(180)(10)^3(1{,}000)} \\ &\quad \times \left(6x^3 + 9\Delta x^2 - 6L_0^3 - 9\Delta L_0^2 + 6{,}000x^2 - 6{,}000L_0^2\right) \\ &= 0.201343(10)^{-9}\left(6x^3 + 9\Delta x^2 - 6L_0^3 - 9\Delta L_0^2 + 6{,}000x^2 - 6{,}000L_0^2\right) \end{aligned} \tag{2.115}$$

By applying the trial-and-error procedure stated earlier, the value of the horizontal displacement Δ of the free end of the member that satisfies Eq. (2.108) or (2.76) is 406.35 in. (10.32 m). By substituting $\Delta = 406.35$ in. (10.32 m) and $L_o = L - \Delta = 593.65$ in. (15.08 m) into Eqs. (2.114) and (2.115), we find

$$M_x = -\left[0.6525(10)^{-3}x^2 + 0.70x\right] \tag{2.116}$$

$$G(x) = 0.201343(10)^{-9}\left(6x^3 + 9,657.15x^2 - 4,658,661,864\right) \tag{2.117}$$

At $x = 0$, we find

$$G(0) = -0.937989$$

and from Eq. (2.107), we find

$$y'(0) = \frac{-0.937989}{\sqrt{1-(-0.937989)^2}} = -2.705743$$

Therefore, the rotation θ_B of the free end B of the flexible member, is

$$\begin{aligned}\theta_B = \tan^{-1} y'(0) &= 1.216782\,\text{rad}\\ &= 69.72^\circ\end{aligned}$$

The value obtained by Lau is $\theta_B = 70^\circ$, yielding a difference of 0.40%.

With known Δ we can determine the values of $y'(x)$ at any $0 \le x \le L_o$, by using Eq. (2.107). Proceeding with the pseudolinear analysis, Table 2.8 has been prepared. This table contains the values of $M_x, y'(x), z_e$, and $M_e{}'$ at the indicated values of $0 \le x \le L_o$ shown in the first column of the table. The equation used for each quantity is also shown in each column of the table. The last column, column 5, shows the values of the bending moment $M_e{}'$ of the equivalent pseudolinear system.

The moment diagram $M_e{}'$ of the pseudolinear system is shown plotted by the solid line in Fig. 2.10a. By approximating its shape with five straight-line segments, as shown by the dashed line in the same figure, and applying simple statics, we obtain the pseudolinear equivalent system shown in Fig. 2.10b.

By using the pseudolinear system in Fig. 2.10b and applying the moment-area method, we find

Table 2.8. Values of $M_x, y'(x), z_e$, and $M_e{}'$ corresponding to the assumed values of x(1 in. = 0.0254 m, 1 kip in. = 113.0 N m)

(1) x (in.)	(2) M_x (kip in.) Eq. (2.116)	(3) $y'(x)$ Eqs. (2.117) and (2.107)	(4) z_e Eq. (2.113)	(5) M'_e (kip in.) Eq. (2.112)
0	0	−2.705743	24.003072	0
100	−76.525	−2.304218	15.848363	−1,212.8
200	−166.100	−1.617333	6.875441	−1,142.0
300	−268.725	−1.069291	3.137977	−843.3
400	−384.400	−0.657814	1.714910	−659.2
500	−513.125	−0.315502	1.152969	−591.6
593.65	−767.975	0	1	−768.0

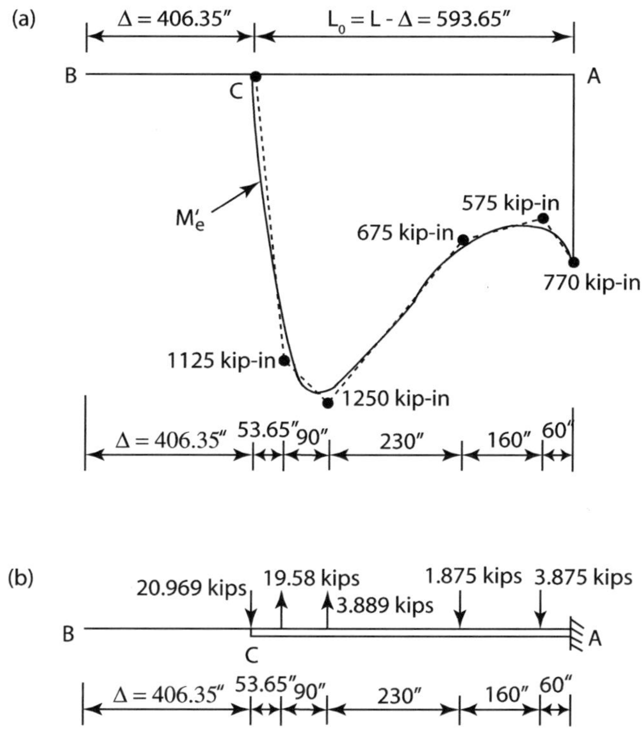

Fig. 2.10. (**a**) Moment diagram M_e' of the pseudolinear system approximated with five straight-line segments. (**b**) Pseudolinear equivalent system loaded with five concentrated loads (1 in. = 0.0254 m, 1 kip in. = 113.0 N m, 1 kip = 4, 448 N)

$$\begin{aligned}\delta_C = \delta_B &= \frac{1}{EI}\left[\frac{1}{2}(1125)(53.65)\left(\frac{2}{3}\right)(53.65) + (1125)(90)(98.65)\right.\\&\quad +\frac{1}{2}(125)(90)(113.65) + (675)(230)(258.65) + \frac{1}{2}(575)(230)(220.32)\\&\quad +(575)(160)(453.65) + \frac{1}{2}(100)(160)(426.98) + (575)(60)(563.65)\\&\quad \left.+\frac{1}{2}(195)(60)(573.65)\right]\\&= \frac{134,384,456}{EI}\end{aligned}$$

By substituting for EI, we obtain

$$\delta_C = \delta_B = \frac{134,384,456}{(180)(10)^3} = 746.58\,\text{in.}\ (18.96\,\text{m})$$

The value obtained by Lau is $\delta_B = 751$ in. (19.08 m), a difference of 0.59%.

Table 2.9. Comparison of results obtained for θ_B and δ_B using pseudolinear systems and the ones obtained by Lau (1 in. = 0.0254 m)

	pseudolinear equivalent systems			Lau's solution		
PL^2/EI	θ_B (deg.)	δ_B (in.)	Δ_B (in.)	θ_B (deg.)	δ_B (in.)	relative differnce in δ_B (%)
0.364	19.98	243.551	35.28	20.00	244.000	0.018
0.834	39.89	466.201	137.82	40.00	471.000	1.020
1.642	59.71	653.570	300.64	60.00	666.000	1.860
2.416	69.72	713.500	406.35	70.00	751.000	2.330
4.389	81.00	816.320	555.476	81.00	847.000	3.620

In Table 2.9, the values of the rotation $\theta_B = \tan^{-1}(y_B')$, horizontal displacement Δ_B and vertical displacement δ_B at the free end B of the member in Fig. 2.9 are obtained for various values of the parameter PL^2/EI, and the results are compared with the results obtained by Lau. The maximum relative difference in δ_B between pseudolinear systems and Lau's approach is 3.62% and it occurs when $PL^2/EI = 4.389$. Use of a digital computer was made for the computation of the results.

2.8 Flexible Uniform Cantilever Beam Under Complex Loading Conditions

The information in this section is very important for the practical engineer, because it demonstrates how a very complex nonlinear problem can be made very easy and able to obtain very reasonable results, by using simplified nonlinear equivalent systems as discussed in Sect. 1.8.2 of the first chapter in this book. For comparison purposes, the two methodologies to be compared here are (a) the utilization of pseudolinear equivalent systems as discussed in Sect. 1.8.1 and (b) using simplified nonlinear equivalent systems as discussed in Sect. 1.8.2.

2.8.1 Application of Equivalent Pseudolinear Systems

The problem we will use here to compare these two methodologies is the uniform flexible cantilever beam in Fig. 2.11 loaded as shown. The direct solution of this problem using other known methods of analysis as discussed in the first chapter of this text, would be either extremely difficult to obtain because the problem is nonlinear and superposition does not apply, or impossible to get, as stated in the first chapter when various other methods were compared. The use of a pseudolinear system provides an excellent and very

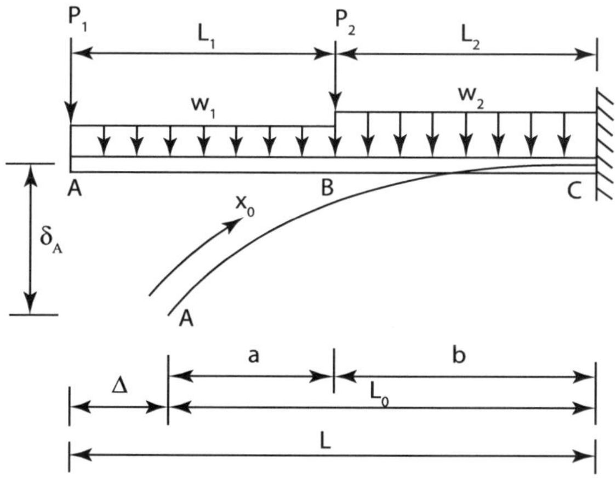

Fig. 2.11. Uniform flexible cantilever beam loaded with an arbitrary loading condition

accurate simplification of the complex nonlinear problem. However, for very complex problems, the practicing engineer could obtain first a simplified nonlinear equivalent system, which is very easy to derive, and then use it to apply pseudolinear analysis to obtain the required results. This approach would be even more important when the stiffness EI of the member is variable, as discussed in detail in Chap. 3, or when inelastic and vibration analysis is carried out as discussed in Chaps. 4 and 5.

In the deformed configuration, the horizontal projection of the length L_1 in Fig. 2.11 is represented by the length a, and the projection of the length L_2 is represented by the length b. For the given loading conditions, the expressions for the bending moments M_1 and M_2, for the intervals a and b, respectively, must be defined. In Fig. 2.11, the coordinate x_{o1} is used to represent the interval $0 \le x_{o1} \le L_1$, and x_{o2} is used for the interval $L_1 \le x_{o2} \le L$. From the deformed configuration the expressions for the bending moments M_1 and M_2 may be written as follows by using statics:

$$M_1 = \left[\frac{w_1 x}{2} x_{01} + P_1 x\right], \qquad 0 \le x \le a, \quad 0 \le x_{01} \le L_1 \tag{2.118}$$

$$M_2 = \left[\frac{w_1 L_1}{2}(2x - a) + P_1 x + \frac{w_2 x_{02}}{2}(x - a) + P_2(x - a)\right], \qquad a \le x \le (a + b), \quad L_1 \le x_{02} \le L \tag{2.119}$$

Note that $x_{o1}(x)$ and $x_{o2}(x)$ are integral equations that are defined by the expression

$$x_{01}(x) = \int_0^x \left\{1 + [y_1'(x)]^2\right\}^{1/2} dx, \qquad 0 \le x \le a, \quad 0 \le x_{01} \le L_1 \tag{2.120}$$

$$x_{02}(x) = \int_0^x \left\{1 + [y_2'(x)]^2\right\}^{1/2} dx, \qquad a \le x \le (a+b), \quad L_1 \le x_{02} \le L \tag{2.121}$$

A reasonable solution may be obtained by assuming that $x_{o1}(x)$ and $x_{o2}(x)$ are given by the expressions

$$x_{01}(x) = x + \Delta_1 \tag{2.122}$$

$$x_{02}(x) = x + \Delta_2 \tag{2.123}$$

where Δ_1 is the horizontal movement of length L_1 and Δ_2 is the horizontal movement of length L_2. This implies that $\Delta = (\Delta_1 + \Delta_2)$. On this basis, the expressions for the bending moments M_1 and M_2 are as follows:

$$M_1 = \left[\frac{w_1 x}{2}(x + \Delta_1) + P_1 x\right] \tag{2.124}$$

$$M_2 = \left[\frac{w_1 L_1}{2}(2x - a) + P_1 x + \frac{w_2 (x + \Delta_2)}{2}(x - a) + P_2 (x - a)\right] \tag{2.125}$$

By using the notation

$$k_1 = \frac{w_1 L_1}{P_1}, \quad k_2 = \frac{w_2 L_2}{P_2} \tag{2.126}$$

the expressions of M_1 and M_2 given by Eqs. (2.124) and (2.125), respectively, are written as

$$M_1 = \left[x^2\left(\frac{k_1 P_1}{2L_1}\right) + x\left(\frac{k_1 P_1 \Delta_1 + 2L_1 P_1}{2L_1}\right)\right] \tag{2.127}$$

$$M_2 = \left[x^2\left(\frac{k_2 P_2}{2L_2}\right) + x\left(\frac{2k_1 P_1 L_2 + 2L_2 P_1 - k_2 P_2 a + k_2 P_2 \Delta_2 + 2P_2 L_2}{2L_2}\right) - \left(\frac{k_1 P_1 L_2 a + k_2 \Delta_2 P_2 a + 2P_2 L_2 a}{2L_2}\right)\right] \tag{2.128}$$

Therefore, the Euler–Bernoulli equation may be written as

$$\frac{y_1''}{\left[1 + (y_1')^2\right]^{3/2}} = G_1 x^2 + G_2 x, \qquad 0 \le x \le a \tag{2.129}$$

$$\frac{y_2''}{\left[1+(y_2')^2\right]^{3/2}} = G_3x^2 + G_4x + G_5, \qquad a \le x \le (a+b) \tag{2.130}$$

where

$$G_1 = -\frac{k_1P_1}{2L_1} \tag{2.131}$$

$$G_2 = -\frac{k_1P_1\Delta_1 + 2L_1P_1}{2L_1} \tag{2.132}$$

$$G_3 = -\frac{k_2P_2}{2L_2} \tag{2.133}$$

$$G_4 = -\frac{2k_1P_1L_2 + 2L_2P_1 - k_2P_2a + k_2P_2\Delta_2 + 2P_2L_2}{2L_2} \tag{2.134}$$

$$G_5 = \frac{k_1P_1L_2a + k_2\Delta_2P_2a + 2P_2L_2a}{2L_2} \tag{2.135}$$

The integration of Eqs. (2.129) and (2.130) can be carried out as discussed in Sect. 1.4. Thus, by using Eq. (1.32), we obtain

$$\frac{y_1'}{\left[1+(y_1')^2\right]^{1/2}} = \phi_1(x) + C_1, \quad 0 \le x \le a \tag{2.136}$$

$$\frac{y_2'}{\left[1+(y_2')^2\right]^{1/2}} = \phi_2(x) + C_2, \quad a \le x \le (a+b) \tag{2.137}$$

where

$$\phi_1(x) = \int \left(G_1x^2 + G_2x\right) dx = \frac{G_1x^3}{3} + \frac{G_2x^2}{2} \tag{2.138}$$

$$\phi_2(x) = \int \left(G_3x^2 + G_4x + G_5\right) dx = \frac{G_3x^3}{3} + \frac{G_4x^2}{2} + G_5x \tag{2.139}$$

and C_1 and C_2 are the constants of integration.

The constants C_1 and C_2 are determined using the boundary condition $y_2'(L_o) = 0$, and the continuity condition $y_1'(a) = y_2'(a)$, yielding

$$C_2 = -\phi_2(L_0) = -\left(\frac{G_3x_0^3}{3} + \frac{G_4L_0^2}{2} + G_3L_0\right) \tag{2.140}$$

$$\begin{aligned} C_1 &= \phi_2(a) - \phi_1(a) - \phi_2(L_0) \\ &= \frac{a^3}{3}(G_3 - G_1) + \frac{a^2}{2}(G_4 - G_2) - bG_5 - \frac{L_0^2}{6}(2G_3L_0 - 3G_4) \end{aligned} \tag{2.141}$$

Thus, by using Eq. (1.33), we obtain

$$y_1'(x) = \frac{\phi_1(x) + C_1}{\sqrt{1 - [\phi_1(x) + C_1]^2}} \tag{2.142}$$

$$y_2'(x) = \frac{\phi_2(x) + C_2}{\sqrt{1 - [\phi_2(x) + C_2]^2}} \tag{2.143}$$

where $\Phi_1(x), \Phi_2(x), C_1$, and C_2, are given by Eqs. (2.138), (2.139), (2.141), and (2.140), respectively.

We notice, however, that Eqs. (2.142) and (2.143) contain the horizontal displacements Δ_1 and Δ_2, which can be determined from the expressions

$$L_1 = \int_0^a \left[1 + (y_1')^2\right]^{1/2} dx \tag{2.144}$$

$$L_2 = \int_a^{(a+b)} \left[1 + (y_2')^2\right]^{1/2} dx \tag{2.145}$$

by using a trial-and-error procedure as in the preceding sections of this text. With known Δ_1 and Δ_2, the values of y_1' and y_2' at any $0 \leq x \leq a$ and $a \leq x \leq (a+b)$, can be determined from Eqs. (2.142) and (2.143), respectively. With known y_1' and y_2' , the pseudolinear system, and consequently, the vertical displacements $y_1(x)$ and $y_2(x)$ of the member, can be determined as discussed earlier in this text.

The following numerical example illustrates the above methodology.

Example 2.6 For the problem in Fig. 2.11, assume that $L = 1{,}000$ in. (25.4 m), $P_1 = 0.5$ kips (2, 224 N), $P_2 = 1$ kip (4, 448 N), $EI = 180 \times 10^3$ kip in.^{2} (516, 541 N m^2), $k_1 = k_2 = 1$, and determine the values of a, b, $L_o, \Delta_1, \Delta_2, \Delta, \delta_A$, and rotation θ_A. Assume also that $L_1 = L_2 = 500$ in. (12.7 m).

Solution: The first step would be the computation of the horizontal displacements Δ_1 and Δ_2, corresponding to the lengths L_1 and L_2, respectively. This is accomplished by using Eqs. (2.144) and (2.145) and applying a trial-and-error procedure by assuming values of Δ_1 and Δ_2 until these two equations are satisfied. Assigning values for Δ_1 and Δ_2 is tantamount to assigning values for L_o, a, and b, because $L_o = a+b, L_1 = a+\Delta_1$, and $L_2 = b+\Delta_2$. Proceeding in this manner, we find out that Eqs. (2.144) and (2.145) are satisfied when $\Delta_1 = 364.452$ in. (9.257 m) and $\Delta_2 = 172.928$ in. (4.3924 m). Therefore,

$$\Delta = \Delta_1 + \Delta_2 = 537.38 \text{ in. } (13.65 \text{ m})$$
$$a = L_1 - \Delta_1 = 135.548 \text{ in. } (3.443 \text{ m})$$
$$b = L_2 - \Delta_2 = 327.072 \text{ in. } (8.3076 \text{ m})$$
$$L_0 = a + b = 462.62 \text{ in. } (11.751 \text{ m})$$

At $x = 0$, Eq. (2.142) yields $y_1'(0) = 4.185246$. Therefore, the rotation θ_A at the free end of the member is $\theta_A = \tan^{-1}[y_1'(0)] = 1.336259\,\text{rad} = 76.562^o$. At $0 \leq x \leq L_o$, the values of $y'(x)$ can be evaluated by using Eqs. (2.142) and (2.143). In this equations, Δ_1, Δ_2, a, b, and L_o, are as shown above. With known $y'(x)$, the values of z_e can be determined by using Eq. (1.120), and the moment diagram $M_e{}'$ of the pseudolinear system can be obtained using the following equation:

$$M_e{}' = z_e M_x \tag{2.146}$$

The values of M_x in Eq. (2.146) can be calculated using Eqs. (2.124) and (2.125), or by applying Eqs. (2.127) and (2.128).

The approximation of the shape of $M_e{}'$ with a few straight-line segments and then applying simple statics, yields the pseudolinear equivalent system. By using the pseudolinear system and applying the moment–area method, or any other method of linear analysis, or handbook formulas, we obtain $\delta_A = 820.375\,\text{in.}$ (20.8375 m). The reader, if he/she wishes, can practice with the mechanics of this process for better understanding.

2.8.2 Deriving Simpler Nonlinear Equivalent Systems

For complicated loading conditions such as the one discussed in Example 2.6, the solution becomes very convenient when simplified nonlinear equivalent systems, as discussed in Sect. 1.8.2, are used. The advantages in using this methodology are illustrated in the following example.

Example 2.7 For the flexible cantilever beam loaded as stated in Example 2.6, determine the rotation θ_A, horizontal displacement Δ, and vertical displacement δ_A of its free end A, by using a simplified nonlinear equivalent system. The values of EI, L, L_1, L_2, P_1, P_2, k_1, and k_2 are identical to the corresponding ones used in Example 2.6.

Solution: We will use here the procedure discussed in Sect. 1.8.2. Since $f(x) = g(x) = 1$, Eq. (1.115) shows that $M_e = M_x$ indicating that the moment diagram M_e of the simplified equivalent nonlinear system is identical to the moment diagram M_x of the initial system. Since $k_1 = k_2 = 1$, Eq. (2.126) yields

$$w_1 = \frac{k_1 P_1}{L_1} = 1.0(10)^{-3}\,\text{kips in.}^{-1}\left(175.13\,\text{kN m}^{-1}\right)$$

$$w_2 = \frac{k_2 P_2}{L_2} = 2.0(10)^{-3}\,\text{kips in.}^{-1}(350.26\,\text{kN m}^{-1})$$

By using Fig. 2.11 and the assigned values for $P_1, P_2, w_1, w_2, L_1, L_2$, and L, we plot the moment diagram $M_e = M_x$ shown by the solid line in Fig. 2.12a, using simple statics. The approximation of the shape of M_e with one straight-line AE, as shown in the same figure, leads to the simplified nonlinear equivalent

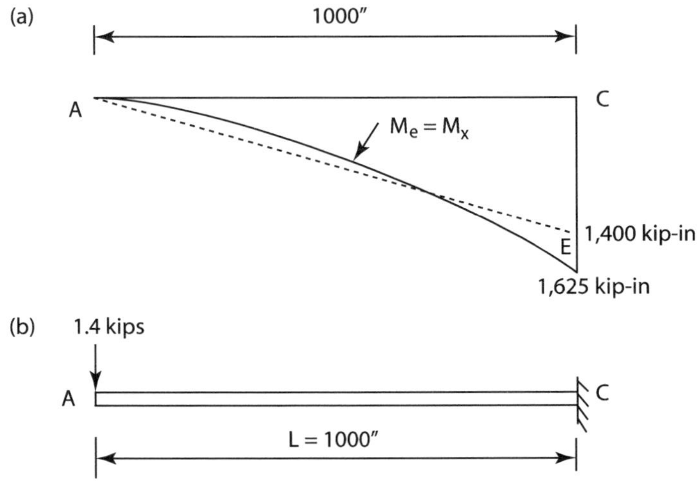

Fig. 2.12. (**a**) Moment diagram M_e of the simplified nonlinear equivalent system with its shape approximated with one straight line AE. (**b**) Simplified equivalent nonlinear system (1 in. = 0.0254 m, 1 kip = 4,448 N, 1 kip in. = 113 N m)

system shown in Fig. 2.12b. Although the approximation with one straight line looks rather crude, reasonable accuracy, however, is obtained for practical applications. The error is usually well within 5%. If better accuracy is required, then two straight-lines may be used to approximate the shape of M_e, which leads to a simplified system that is loaded with two concentrated loads. Once the M_e is plotted, then we know exactly what is best to do. Very often the approximation of M_e with one straight line gives practically exact results.

The solution of the simplified nonlinear system in Fig. 2.12b may be carried out by using either pseudolinear analysis as in Example 1.4, or by using Table 7.3 on page 516 of [85]. By using Table 7.3 of this reference, we find $\theta_A = 78.25°, \Delta = 497.89$ in. (12.65 m), and $\delta_A = 781.0$ in. (19.84 m). If we compare these values with the values obtained in Sect. 2.8.1, the difference is 2.2% for θ_A, 7.3% for Δ, and 4.8% for δ_A. Part of this difference, however, is attributed to the interpolations that need to be made in using Table 7.3 of the indicated reference. Pseudolinear analysis such as the one used in Example 1.4 should produce more accurate results.

Problems

2.1 The uniform flexible cantilever beam in Fig. 2.1, is loaded by a uniformly distributed load $w_o = 2\,\text{lb in.}^{-1}(350.2536\,\text{N m}^{-1})$ over its entire span length. By assuming that L = 1,000 in. (25.4 m), EI = $180 \times 10^3\,\text{kip in.}^2(516,541\,\text{N m}^2)$, $\Delta(x) = \text{constant} = \Delta$, and applying pseudolinear analysis, determine the rotation θ_B and the vertical displacement δ_B at its free end B.

Answer: $\theta_B = 66.30°, \delta_B = 705.72$ in. (17.9253 m).

2.2 Repeat Problem 2.1 by assuming $\Delta(x) = \Delta x/L_o$, and $w_o = 1.5$ lb in.$^{-1}$ (262.6902 Nm^{-1}). Also calculate the horizontal displacement Δ, and vertical displacement y of the member at $x = L_o/2$.
Answer: $\theta_B = 51.15°, \delta_B = 617.55$ in.(15.6858 m).

2.3 Repeat Problem 2.1 by integrating the Euler–Bernoulli equation twice, and satisfying appropriate boundary conditions for the evaluation of the constant of integration. Assume $\Delta(x) = \Delta \sin(\pi x/2L_o)$.

2.4 Repeat Problem 2.1 by determining first a simplified nonlinear equivalent system loaded with one concentrated vertical load P_e at its free end. Then use this simplified nonlinear system and apply pseudolinear analysis to determine θ_B and δ_B at its free end. Compare the results.

2.5 The uniform flexible cantilever beam in Fig. P2.5, is loaded with two concentrated loads P_1 and P_2 located as shown in the figure. Determine the horizontal displacements Δ_C and Δ_B at points C and B, respectively, the vertical displacements δ_C and δ_B, and the rotations θ_C and θ_B at the same points. To perform this task, determine first a simplified nonlinear equivalent system loaded with a load P_e at the free end, and then solve it by using pseudolinear analysis.
The modulus E = 30,000 ksi (206.84×10^9 Pa).
Answer: $\Delta_C = 28.70$ in. (0.73 m), $\delta_C = 64.47$ in.(1.6375 m), $\theta_C = 59.23°$.

2.6 The uniform simply supported beam in Fig. 2.3 is loaded with a uniformly distributed load w as shown. Determine the rotation θ_B of the end support B, the horizontal displacement Δ_A of support A, and the maximum vertical displacement δ_D, by using pseudolinear analysis with $\Delta(x) = \Delta$. Length L = 1,000 in. (25.4 m), EI = 75×10^3 kip in.^{2}(215, 224 N m^2), and w = 10 lb in.$^{-1}$(1, 751.268 N m^{-1}).
Answer: $\theta_B = 76.63°$, $\Delta_A = 440.53$ in. (11.1895 m), and $\delta_D = 367.65$ in. (9.3383 m).

2.7 Repeat Problem 2.6 by assuming that $\Delta(x) = \Delta(x/L_o)^{1/2}$ and compare the results.

2.8 Repeat Problem 2.6 by assuming that $\Delta(x) = \Delta \sin(\pi x/2L_o)$ and compare the results.

2.9 The flexible statically indeterminate uniform beam in Fig. 2.8, is loaded with a concentrated vertical load P = 3.5 kips (15, 568.78 N) at the

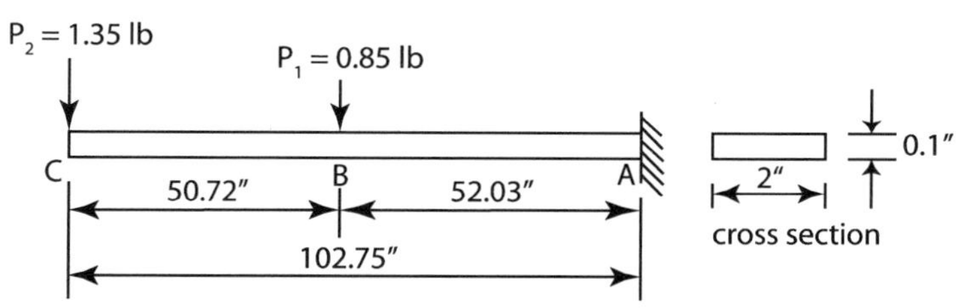

Fig. P2.5.

distance $L_1 = 600$ in. (15.24 m) from the fixed support. Determine the rotation θ_B at its simply supported end B, its maximum vertical displacement y_{max}, and the location x_D of the maximum vertical displacement, by using pseudolinear analysis. The length $L = 1{,}000$ in. (25.4 m) and the stiffness $EI = 75{,}000$ kip in.^{2} (215, 224 N m^2).

2.10 Repeat Problem 2.9 by assuming each time that $P = 1$ kip (4, 448 N), 2 kips (8.896 N), and 5 kips (22,240 N), and compare the results.

2.11 Repeat Problem 2.9 by integrating the Euler–Bernoulli nonlinear differential equation and satisfying appropriate boundary conditions to determine the constants of integration and the redundant reaction R_B. Compare the results.

2.12 Repeat Problem 2.9 by assuming that $P = 1.5$ kips (6, 672 N) and using (a) pseudolinear analysis, and (b) linear analysis, and compare the results.

2.13 The statically indeterminate uniform flexible beam in Fig. 2.7 is loaded with a uniformly distributed load $w_o = 2$ lb in.$^{-1}$ (350.254 N m^{-1}). The length $L = 1{,}000$ in. (25.4 m) and $EI = 75{,}000$ kips in.^{2} (215, 224 N m^2). Determine $\Delta, \theta_B, \delta_D, x_D$, and R_B, by using pseudolinear analysis. Assume $\Delta(x) = \text{constant} = \Delta$.
Answer: $\Delta = 42.80$ in. (1.87 m), $\theta_B = 28.56°$, $\delta_D = 128.806$ in. (3.2717 m), $R_B = 0.759$ kip (3, 376 N), and $x_D = 558.05$ in. (14.17 m).

2.14 Repeat Problem 2.13 by assuming that $\Delta(x) = \Delta x/L_o$ and $\Delta(x) = \Delta \sin(\pi x/2L_o)$, and compare the results.

2.15 The uniform flexible cantilever beam in Fig. 2.9, is acted upon by a concentrated load P at its free end B, and uniformly distributed load w_o over its entire span length, so that $w_oL = 3$ P and $PL^2/EI = 1.642$. Verify the results in Table 2.9 by using pseudolinear analysis. The length $L = 1{,}000$ in. (25.4 m), and $EI = 180{,}000$ kip in.^{2}(516, 541 N m^2).

2.16 Repeat Problem 2.15 by determining first a simplified nonlinear equivalent system loaded with an appropriate equivalent concentrated vertical load P_e at the free end. Then, use the simplified nonlinear system and apply pseudolinear analysis to compare the results in Table 2.9.

2.17 Rework the problem in Example 2.6 by assuming that $P_1 = 1$ kip (4, 448 N), $P_2 = 1.5$ kips (6, 672 N), and applying pseudolinear analysis.

3

Solution Methodologies for Variable Stiffness Flexible Beams

3.1 Introduction

The basic theories and methodologies regarding flexible member were extensively discussed and developed in the first chapter. The emphasis was placed on the method of the equivalent systems which was developed by the author and his collaborators, which simplifies a great deal the solution of the very complex flexible beam problems by introducing pseudolinear and simplified nonlinear equivalent systems of constant stiffness.

In the second chapter, many cases of statically determinate and statically indeterminate flexible beam problems with various loading conditions were examined in detail and convenient methods of analysis were developed in each case. Their analysis, however, was based on the idea that the moment of inertia I and the modulus of elasticity E, and, consequently, the stiffness EI, remain constant throughout the length of the structural member.

In this chapter, the flexible structural component is permitted to have any arbitrary variation in I and EI along its length, it is permitted to be subjected to any combination of loading conditions, but still the modulus of elasticity is assumed to remain constant. Various methodologies have been developed and very accurate and convenient simplifications for the nonlinear problem have been introduced. In Chaps. 4 and 5, the modulus of elasticity E is also permitted to vary along the length of the member, and convenient methods of analysis are introduced.

3.2 Flexible Tapered Cantilever Beam with a Concentrated Vertical Load at its Free End

The derivation of the exact integral second order nonlinear differential equation for a flexible tapered cantilever beam loaded with a concentrated vertical load P as shown in Fig. 1.6, was carried out in Example 1.2 and it is given by

Eq. (1.92). In the same example, we have also made a reasonable simplification of this differential equation by assuming that the expression for the variation of the moment of inertia I(x) is given by Eq. (1.87), which leads to a much simpler nonlinear differential equation given by Eq. (1.96). We rewrite below this equation for convenience.

$$\frac{y''}{\left[1+(y')^2\right]^{3/2}} = \frac{P(L-\Delta)^3}{EI_B} \frac{x}{[(n-1)x+(L-\Delta)]^3} \tag{3.1}$$

In Eq. (3.1), I_B is the moment of inertia at the free end B of the member which is given by Eq. (1.94), and n is the taper constant.

By following the procedure given in Sects. 1.4 and 1.6, we can integrate Eq. (3.1) to obtain the expression for the rotation $y'(x)$. By examining Eq. (3.1), we note that the function $\lambda(x)$ should be as follows:

$$\lambda(x) = \frac{P(L-\Delta)^3}{EI_B} \frac{x}{[(n-1)x+(L-\Delta)]^3} \tag{3.2}$$

The integral $\Phi(x)$ of $\lambda(x)$ may be determined by using the following general expression as our guidance:

$$\int \frac{x\,dx}{(a+bx)^m} = \frac{1}{b^2}\left[\frac{-1}{(m-2)(a+bx)^{m-2}} + \frac{a}{(m-1)(a+bx)^{m-1}}\right], \quad m \neq 1,2 \tag{3.3}$$

For our problem, $a = (L-\Delta)$, $b = (n-1)$, and $m = 3$.

On this basis, we obtain

$$\phi(x) = \frac{P(L-\Delta)^3}{EI_B}\frac{1}{(n-1)^2} \times \left\{\frac{-1}{(3-2)[(n-1)x+(L-\Delta)]^{3-2}} + \frac{(L-\Delta)}{(3-1)[(n-1)x+(L-\Delta)]^{3-1}}\right\}$$

or

$$\phi(x) = \frac{P(L-\Delta)^3}{EI_B}\frac{1}{(n-1)^2} \times \left\{\frac{-1}{[(n-1)x+(L-\Delta)]} + \frac{(L-\Delta)}{2[(n-1)x+(L-\Delta)]^2}\right\}, \quad n \neq 1 \tag{3.4}$$

Note that the taper n in Eq. (3.4) should be different than unity. It can be larger or smaller than one, but not equal to one.

By substituting Eq. (3.4) into Eq. (1.32), we obtain

$$\frac{y'}{\left[1+(y')^2\right]^{1/2}} = \phi(x) + C \tag{3.5}$$

where $\Phi(x)$ is given by Eq. (3.4) and C is the constant of integration. We determine C by applying the boundary condition that the rotation y' is zero at $x = (L - \Delta)$. On this basis, Eq. (3.5) yields

$$C = -\frac{P(L-\Delta)^3}{EI_B}\frac{1}{(n-1)^2}\left\{\frac{-1}{[(n-1)(L-\Delta)+(L-\Delta)]} + \frac{(L-\Delta)}{2[(n-1)(L-\Delta)+(L-\Delta)]^2}\right\}$$

or

$$C = -\frac{P(L-\Delta)^3}{EI_B}\frac{1}{(n-1)^2}\left\{\frac{-1}{[(L-\Delta)n]} + \frac{(L-\Delta)}{2[(L-\Delta)n]^2}\right\} \tag{3.6}$$

Thus, by solving Eq. (3.5) for $y'(x)$, we find

$$y'(x) = \frac{\phi(x) + C}{\sqrt{1 - [\phi(x) + C]^2}} \tag{3.7}$$

where $\Phi(x)$ and C are given by Eqs. (3.4) and (3.6), respectively.

Since the value of the horizontal displacement Δ in Eq. (3.4) is not known, it can be determined from the equation

$$L = \int_0^{L-\Delta}\left[1+(y')^2\right]^{1/2} dx \tag{3.8}$$

by using a trial-and-error procedure as in earlier sections of the book. That is, assume a value of Δ in Eq. (3.7) and integrate Eq. (3.8) to determine the length L. The procedure may be repeated for various values of Δ until the correct length L is obtained. With Δ known, the values of y' at $0 \le x \le (L - \Delta)$ can be computed by using Eq. (3.7). Thus, with known y', the values of any x of the moment diagram Me' of the pseudolinear system of constant stiffness EI_B of any $0 \le x \le (L - \Delta)$ can be determined by using Eqs. (1.119) and (1.120).

The following numerical example illustrates the above methodology.

Example 3.1 The tapered cantilever beam shown in Fig. 3.1a, is loaded with a vertical concentrated load P = 1kip (4,448 N) at its free end B

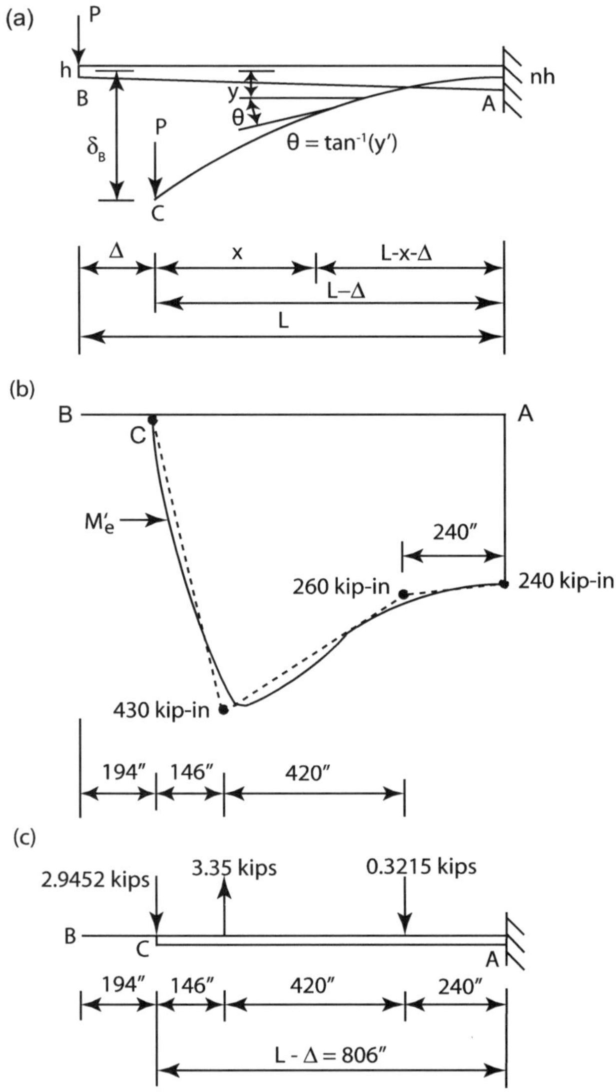

Fig. 3.1. (**a**) Original variable stiffness member. (**b**) Moment diagram M'_e of the pseudolinear system with its shape approximated with three lines. (**c**) Equivalent system of constant stiffness EI_B (1 kips in = 113N m, 1 kip = 4.448 kN, 1 in. = 0.0254 m)

as shown in the figure. The bending stiffness EI_B at the free end B is 180,000 kip in.^{2} (516.54 kN m^2), the length L = 1,000 in. (25.4 m), and the taper n = 1.5. By using a pseudolinear system, determine the vertical deflection δ_B and rotation θ_B at the free end B of the member.

Solution: By using Eq. (3.4) we find

$$\begin{aligned}\phi(\mathrm{x}) &= \frac{(1)(1{,}000-\Delta)^3}{180{,}000}\,\frac{1}{(1.5-1)^2}\left\{\frac{-1}{(1.5-1)x+(1{,}000-\Delta)}\right.\\ &\quad\left.+\frac{1{,}000-\Delta}{2\left[(1.5-1)\mathrm{x}+(1{,}000-\Delta)\right]^2}\right\}\\ &= \frac{(1{,}000-\Delta)^3}{(45)(10)^3}\left\{\frac{-1}{[0.5\mathrm{x}+(1{,}000-\Delta)]}+\frac{1{,}000-\Delta}{2\left[0.5\mathrm{x}+(1{,}000-\Delta)\right]^2}\right\}\end{aligned} \tag{3.9}$$

Also, from Eq. (3.6), we obtain

$$\mathrm{C} = \frac{(1{,}000-\Delta)^3}{(45)(10)^3}\left\{\frac{-1}{1.5(1{,}000-\Delta)}+\frac{1{,}000-\Delta}{4.5(1{,}000-\Delta)^2}\right\}$$

Thus,

$$\begin{aligned}\mathrm{Q}(\mathrm{x}) = \phi(\mathrm{x}) + \mathrm{C} = \frac{(1{,}000-\Delta)^3}{(45)(10)^3}&\left\{\frac{-1}{0.5\mathrm{x}+(1{,}000-\Delta)}\right.\\ &+\frac{1{,}000-\Delta}{2\left[0.5\mathrm{x}+(1{,}000-\Delta)\right]^2}\\ &\left.+\frac{1}{1.5(1{,}000-\Delta)}-\frac{1{,}000-\Delta}{4.5(1{,}000-\Delta)^2}\right\}\end{aligned} \tag{3.10}$$

In Eq. (3.7), the unknown horizontal displacement $\Delta_{\mathrm{B}} = \Delta$ at the free end B of the cantilever beam is calculated by applying a trial-and-error procedure. That is, we assume a value of Δ, we substitute this value into Eq. (3.7), and then perform the required integration in Eq. (3.8) to determine the length L of the member. If we find from Eq. (3.8)that the calculated value of L = 1,000 in. (25.4 m), then the correct value of Δ is assumed. If not, we repeat the procedure using new values of Δ until the correct value of L is obtained.

Let it be assumed for example that Δ = 194.0 in. (4.93 m). Then from Eq. (3.10) we find

$$\mathrm{Q}(\mathrm{x}) = \phi(\mathrm{x}) + \mathrm{C} = 11{,}635.7026\left\{\frac{-1}{(0.5\mathrm{x}+806)}+\frac{806}{2(0.5\mathrm{x}+806)^2}+0.000551\right\} \tag{3.11}$$

The integration of Eq. (3.8) to determine L, is carried out numerically by using Simpson's rule as discussed in Sect. 1.5.

To illustrate the procedure, let it be assumed that n = 10 in Eq. (1.37). This yields

$$\lambda = \frac{806-0}{10} = 80.6$$

The function f(x) is

$$f(x) = \left[1 + (y')^2\right]^{1/2} dx \tag{3.12}$$

where $y'(x)$ is given by Eq. (3.7).

The coordinate y_o of f(x) is obtained when $x = 0$. Thus, by applying Eq. (3.11), we obtain

$$\begin{aligned} Q(0) &= 11,635.7026 \left[\frac{-1}{806} + \frac{806}{2(806)} + 0.000551\right] \\ &= -0.814499 \end{aligned}$$

From Eq. (3.7), we find

$$y'(0) = \frac{-0.814499}{\sqrt{1 - (-0.814499)^2}} = -1.403909$$

and from Eq. (3.12) we obtain

$$y_0 = f(0) = \sqrt{1 + (-1.403909)^2} = 1.723648$$

For $x = 80.6$ in. (2.05 m), we find

$$\begin{aligned} Q(80.6) &= 11,635.7026 \left[\frac{-1}{40.3 + 806} + \frac{806}{2(40.3 + 806)^2} + 0.000551\right] \\ &= -0.791228 \end{aligned}$$

$$y'(80.6) = \frac{-0.791228}{\sqrt{1 - (-0.791228)^2}} = -1.293867$$

$$y_1 = f(80.6) = \sqrt{1 + (-1.293867)^2} = 1.635265$$

In a similar manner, the remaining coordinates $y_2, y_3, \ldots, y_{10}$ of f(x) can be determined.

Thus, by using the Simpson's rule given by Eq. (1.36), we find

$$\begin{aligned} L = \frac{80.6}{3} &[1.723648 + 4(1.635265) + 2(1.495411) + 4(1.375281) \\ &+ 2(1.255993) + 4(1.173764) + 2(1.101274) + 4(1.057710) \\ &+ 2(1.022682) + 4(1.006839) + 1] \\ = \frac{80.6}{3} &(37.469804) = 1{,}006.69 \text{ in. } (25.57\text{ m}) \end{aligned}$$

Since the calculated value of the length L is close to the correct one, we do not repeat the procedure and we accept $\Delta = 194$ in. (4.93 m).

With known Δ, the values of y', z_e, M_x, and M'_e at $0 \leq x \leq (L - \Delta)$, at intervals of 100 in. (2.54 m), are determined by using Eqs. (3.7), (1.120),

Table 3.1. Calculated values of the moment diagram M'_e of the pseudolinear system (1 kip in = 113.0 N m, 1 in. = 0.0254 m)

(1) x (in.)	(2) f(x)	(3) y' (rad)	(4) z_e	(5) M_x (kip in.)	(6) M_e' (kip in.)
0	1.0000	−1.4039	5.1208	0.00	0.00
100	1.0112	−1.3423	4.6897	−6.00	−27.23
200	1.2105	−1.2245	3.9514	−106.00	−346.01
300	1.4345	−1.0064	2.8557	−206.00	−410.09
400	1.6894	−0.7827	2.0598	−306.00	−374.20
500	1.9619	−0.5927	1.5708	−406.00	−325.06
600	2.2682	−0.4236	1.2809	−506.00	−285.75
700	2.6049	−0.2728	1.1137	−606.00	−259.09
800	2.9733	−0.1348	1.0274	−706.00	−243.95
806	3.3750	0	1.0000	−806.00	−238.81

Mx = −Px and (1.119), respectively. The results are shown in Table 3.1. The second column in this table, shows the values of the moment of inertia function f(x), which were obtained using Eq. (1.95) with n = 1.5. The moment diagram M'_e of the pseudolinear system of constant stiffness EI_B is shown plotted by the solid line in Fig. 3.1b. The approximation of its shape with three straight-line segments as shown by the dashed line in the same figure, leads to the very accurate pseudolinear system of constant stiffness EI_B shown in Fig. 3.1c.

The large deflections along the length of the original system can be accurately determined by using the pseudolinear system in Fig. 3.1c and applying elementary linear analysis. For example, the large deflection δ_B at the free point B in Fig. 3.1a is equal to the deflection δ_C, at point C of the pseudolinear system in Fig. 3.1c. Thus, by applying the moment–area method to the pseudolinear system in Fig. 3.1c, we find $\delta_C = 517.8$ in. (13.15 m). The value obtained by solving Eq. (1.96) using direct integration and applying Simpson's rule to carry out the required integrations, is $\delta_B = 519.85$ in. (13.2042 m). The difference is only 0.4%. In a similar manner, by using the pseudolinear system in Fig. 3.1c and applying the moment–area method, we find $y'_c = 1.3132$. Thus, the rotation θ_c at C is $\theta_c = \tan^{-1}(y_c') = 52.7°$. The solution of Eq. (1.96) yielded $\theta_B = \theta_C = 53.31°$, a difference of only 0.11%. Note that the solution of the pseudolinear system in Fig. 3.1c by using linear analysis yields the values of y' and, consequently, the rotation $\theta = \tan^{-1}(y')$.

In Table 3.2, the variation of the horizontal displacement δ_B, rotation θ_B, and vertical displacement δ_B of the free end B in Fig. 3.1a for various values of the vertical load P is shown. The variation of the same quantities for various values of the taper parameter n in Fig. 3.1a is shown in Table 3.3. In the same tables, the results obtained by using the fourth order Runge–Kutta method are shown. The last column in these tables shows the percentage difference in vertical deflection between equivalent systems and the Runge–Kutta method.

Table 3.2. Variation of Δ_B, θ_B, and δ_B for various values of the load P (1 kip = 4.448 kN, 1 in. = 0.0254 m)

	equivalent pseudolinear systems			fourth order Runge–Kutta method			
P (kip)	Δ_B (in.)	θ_B (deg.)	δ_B (in.)	Δ_B (in.)	θ_B (deg.)	δ_B (in.)	% difference
0.5	71.00	32.19	327.00	68.87	31.65	322.70	1.36
1.0	194.00	53.11	520.00	181.82	51.42	507.11	2.54
1.5	298.00	65.86	622.75	274.90	62.99	606.27	2.71
2.0	376.80	75.53	682.30	345.71	70.22	665.61	2.51
2.5	436.59	78.44	720.82	400.33	75.02	704.88	2.26
3.0	483.11	81.70	747.33	443.61	78.37	732.93	1.96
3.5	520.25	84.00	768.05	478.83	80.80	754.13	1.84
4.0	550.56	85.61	783.49	508.13	82.61	770.85	1.64

Table 3.3. Variation of Δ_B, θ_B, and δ_B for various values of the parameter n in Fig. 3.1a when P = 1 kip (1in. = 0.0254 m, 1 kip = 4.448 kN)

	equivalent pseudolinear systems			fourth order Runge–Kutta method			
n	Δ_B (in.)	θ_B (deg.)	δ_B (in.)	Δ_B (in.)	θ_B (deg.)	δ_B (in.)	% difference
1.0	414.94	71.96	732.38	414.98	71.97	732.04	0.04
1.2	314.88	64.88	649.16	304.25	63.67	641.81	1.14
1.4	229.90	57.19	562.75	216.91	55.37	550.80	2.16
1.5	194.00	53.31	519.85	181.82	51.42	507.11	2.51
1.6	162.80	49.51	478.39	151.93	47.65	465.43	2.78
1.8	113.26	42.39	401.23	105.65	40.65	389.76	2.94
2.0	78.42	36.14	334.47	73.57	34.89	325.35	2.80
2.2	54.60	30.00	278.81	51.67	29.93	272.03	2.49
2.5	32.48	24.58	214.00	31.16	24.01	209.89	1.95
3.0	14.82	17.43	142.73	14.48	17.17	140.96	1.25

The horizontal displacement Δ, of any point x along the length of the member in Fig. 3.1a, may be determined by using an expression similar to the one given by Eq. (3.8). For example, the horizontal displacement Δ at $x = L/2$, may be determined from the equation

$$\frac{L}{2} = \int_0^{(L/2)-\Delta} \left[1 + (y')^2\right]^{1/2} dx \tag{3.13}$$

using a trial-and-error procedure as it was suggested earlier. That is, assume values of Δ in Eq. (3.13), perform the integration, and repeat the procedure until the equation is satisfied. The value of Δ that satisfies Eq. (3.13) is the horizontal displacement of the member at $x = L/2$.

3.3 Doubly Tapered Flexible Cantilever Beam Subjected to a Uniformly Distributed Loading

We consider the doubly tapered flexible cantilever beam in Fig. 3.2, which is loaded with a uniformly distributed load w_o as shown. The deformed configuration of the member is also shown in the same figure. The modulus of elasticity E is constant, and the moment of inertia I_x at any $0 \leq\leq (L - \Delta)$, where Δ is the horizontal displacement of the free end B of the member, is given by the expression

$$I_x = I_B \left[1 + \frac{(n-1)}{L} x_0\right]^3 \tag{3.14}$$

where

$$I_B = \frac{bh^3}{12} \tag{3.15}$$

$$x_0 = \int_0^x \left\{1 + [y'(x)]^2\right\}^{1/2} dx \tag{3.16}$$

and b is its constant width.

The bending moment M_x at any $0 \leq x \leq (L - \Delta)$ is obtained by applying statics and it is given by the equation

$$M_x = -\frac{w_0 x}{2} x_0 \tag{3.17}$$

where x_o is given by Eq. (3.16). By using Eqs. (3.14), (3.16), and (3.17) and substituting into the Euler–Bernoulli equation, we obtain the following

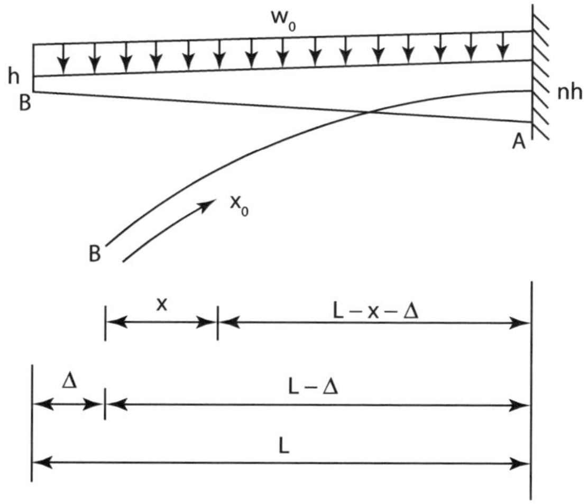

Fig. 3.2. Doubly tapered cantilever beam loaded with a uniformly distributed load w_o over its entire length

nonlinear integral differential equation, which is again difficult to solve:

$$\frac{y''}{\left[1+(y')^2\right]^{3/2}} = \frac{w_0 x}{2EI_B} \frac{\displaystyle\int_0^x \left\{1+[y'(x)]^2\right\}^{1/2} dx}{\left\{1+\frac{(n-1)}{L}\displaystyle\int_0^x \left\{1+[y'(x)]^2\right\}^{1/2} dx\right\}} \tag{3.18}$$

To simplify the solution of Eq. (3.18), we approximate the variation of the moment of inertia I_x, at any $0 \leq x \leq (L - \Delta)$, using the following equation

$$I_x = I_B \left[1 + (n-1)\frac{x}{L-\Delta}\right]^3 \tag{3.19}$$

which is the same as Eq. (1.93). If we also assume, as in preceding sections, that $x_o = (x + \Delta)$, the simplified expression for the bending moment M_x at any $0 \leq x \leq (L - \Delta)$, can be obtained by substituting for x_o in Eq. (3.17). This yields

$$M_x = -\frac{w_0 x}{2}(x + \Delta) \tag{3.20}$$

By using Eqs. (3.19) and (3.20), the Euler–Bernoulli equation yields the following simplified nonlinear differential equation:

$$\frac{y''}{\left[1+(y')^2\right]^{3/2}} = \frac{w_0 (L-\Delta)^3}{2EI_B} \frac{(x^2+\Delta x)}{[(n-1)x + (L-\Delta)]^3} \tag{3.21}$$

In this case, we note that

$$\lambda(x) = \frac{w_0 (L-\Delta)^3}{2EI_B} \frac{(x^2+\Delta x)}{[(n-1)x + (L-\Delta)]^3} \tag{3.22}$$

and the integration of $\lambda(x)$ can be carried out by using the following two standard mathematical handbook formulas [96]:

$$\int \frac{x^2 dx}{(a+bx)^3} = \frac{1}{b^3}\left[\ell n\,(a+bx) + \frac{2a}{a+bx} - \frac{a^2}{2(a+bx)^2}\right] \tag{3.23}$$

$$\int \frac{x dx}{(a+bx)^m} = \frac{1}{b^2}\left[\frac{-1}{(m-2)(a+bx)^{m-2}} + \frac{a}{(m-1)(a+bx)^{m-1}}\right], \quad m \neq 1, 2 \tag{3.24}$$

By applying Eqs. (3.23) and (3.24), we obtain

$$\begin{aligned}
\phi(x) &= \int \lambda(x)\,dx \\
&= \frac{w_0 (L-\Delta)^3}{2EI_B}\left\{\frac{1}{(n-1)^3}\left[\ell n\left((L-\Delta)+(n-1)x\right)\right.\right. \\
&\quad \left. + \frac{2(L-\Delta)}{(L-\Delta)+(n-1)x} - \frac{(L-\Delta)^2}{2\left[(L-\Delta)+(n-1)x\right]^2}\right] \\
&\quad \left. + \frac{\Delta}{(n-1)^2}\left[-\frac{1}{(L-\Delta)+(n-1)x} + \frac{(L-\Delta)}{2\left[(L-\Delta)+(n-1)x\right]^2}\right]\right\}
\end{aligned} \tag{3.25}$$

Thus, by using Eq. (1.32), we obtain

$$\frac{y'}{\left[1+(y')^2\right]^{1/2}} = \phi(x) + C \tag{3.26}$$

where $\Phi(x)$ is as shown by Eq. (3.25) and C is the constant of integration.

By using Eq. (3.26) and applying the boundary condition of $y' = 0$ at $x = (L - \Delta)$, we determine C, and it is as follows:

$$\begin{aligned}
C &= -\frac{w_0 (L-\Delta)^3}{2EI_B}\left\{\frac{1}{(n-1)^3}\left[\ell n\left((L-\Delta)+(n-1)(L-\Delta)\right)\right.\right. \\
&\quad \left. + \frac{2(L-\Delta)}{(L-\Delta)-(n-1)(L-\Delta)} - \frac{(L-\Delta)^2}{2\left[(L-\Delta)+(n-1)(L-\Delta)\right]^2}\right] \\
&\quad \left. + \frac{\Delta}{(n-1)^2}\left[\frac{1}{(L-\Delta)+(n-1)(L-\Delta)} + \frac{(L-\Delta)}{2\left[(L-\Delta)+(n-1)(L-\Delta)\right]^2}\right]\right\} \\
&= -\frac{w_0 (L-\Delta)^3}{2EI_B}\left\{\frac{1}{(n-1)^3}\left[\ell n\left((L-\Delta)n + \frac{4n-1}{2n^2}\right)\right] + \frac{\Delta}{(n-1)^2}\left[\frac{1-2n}{2n^2(L-\Delta)}\right]\right\}
\end{aligned} \tag{3.27}$$

Therefore,

$$G(x) = \phi(x) + C \tag{3.28}$$

and, by using Eq. (1.33),

$$y'(x) = \frac{G(x)}{\sqrt{1-\left[G(x)\right]^2}} \tag{3.29}$$

where $G(x)$ is given by Eq. (3.28), and $\Phi(x)$ and C by Eqs. (3.25) and (3.27), respectively.

At the free end of the flexible beam, the unknown horizontal displacement Δ can be determined, as in preceding section, by using the equation

$$L = \int_0^{L_0} \left[1+(y')^2\right]^{1/2} dx \tag{3.30}$$

and applying a trial-and-error procedure. When Δ is evaluated, the values of $y'(x)$ at any $0 \leq x \leq L_o$ may be determined by using Eq. (3.29), where $G(x) = \Phi(x) + C$, in this equation, can be obtained by using Eqs. (3.25) and (3.27). With known $y'(x)$, the pseudolinear analysis can be carried out as discussed in Sect. 1.8.1, and in preceding sections of this text. The following numerical example illustrates the application of the above methodology.

Example 3.2 For the doubly tapered flexible cantilever beam in Fig. 3.2, assume that $L = 1{,}000$ in. (25.4 m), $EI_B = 180 \times 10^3$ kip in.^{2} (516,541 N m^2), $w_o = 5$ lb in.$^{-1}$(875.63 N m^{-1}), taper $n = 1.5$, and determine the horizontal displacement Δ_B, vertical displacement δ_B, and rotation θ_B, at the free end B of the beam. Compare the results by assuming (a) $\Delta(x) = \text{constant} = \Delta$, and (b) $\Delta(x) = \Delta x/L_o$.

Solution: Consider first the case where $\Delta(x) = \text{constant} = \Delta$.

By assuming that $\Delta = 363.0$ in. (9.22 m) and substituting into Eq. (3.30), we find that this equation is satisfied, indicating that Δ is correct. By using Eqs. (3.25) and (3.27), making the appropriate substitutions for L, n, w_o, and EI_B, and then substituting into Eq. (3.28), we obtain

$$\begin{aligned} G(x) = {} & \frac{(5)(637)^3}{(2)(180)(10)^3(10)^3}\left\{\frac{1}{0.125}\left[\ell n(637+0.5x) + \frac{2(637)}{637+.05x}\right.\right. \\ & \left.\frac{(637)^2}{2(637+0.5x)^2}\right] + \frac{363}{0.25}\left[-\frac{1}{637+0.5x} + \frac{637}{2(637+0.5x)^2}\right] \\ & \left. -\frac{1}{0.125}\left[\ell n(955.5) + 1.111111\right] + 1.013082\right\} \qquad (3.31) \\ = {} & 3.589929\left\{\frac{1}{0.125}\left[\ell n(637+0.5x) + \frac{1274}{637+0.5x} - \frac{(637)^2}{2(637+0.5x)^2}\right]\right. \\ & \left. +1452\left[\frac{637}{2(637+0.5x)^2} - \frac{1}{637+0.5x}\right] - 62.773684\right\} \end{aligned}$$

At $x = 0$, Eq. (3.31) yields

$$\begin{aligned} G(0) &= 3.589929\left\{\tfrac{1}{0.125}\left[\ell n(375) + 2 - 0.5\right] - 63.913402\right\} \\ &= 3.589929\{63.654157 - 63.913402\} \\ &= -0.930671 \end{aligned}$$

and from Eq. (3.29), we obtain

$$y'(0) = \frac{-0.930671}{\sqrt{1-(-0.930671)^2}} = -2.543811$$

Thus, the rotation θ_B, at the free end B, is

$$\theta_B = \tan^{-1}[y'(0)] = 68.54°$$

Since Δ is known, we can use Eq. (3.29) to determine $y'(x)$ at any $0 \le x \le L_o$, and proceed with the pseudolinear analysis as in the preceding examples of this text. For this purpose, Table 3.4 is prepared which includes the values of $f(x)$, M_x, $y'(x)$, z_e, and M_e' for the assumed values of $0 \le x \le L_o$. The equations used for this purpose are shown in the table.

By using the values of M_e' in the last column of Table 3.4, the moment diagram of the pseudolinear system of constant stiffness EI_B is shown plotted by the solid line in Fig. 3.3a. The approximation of its shape with five straight-line segments is also shown in the same figure. By using it and applying simple statics, we derive the pseudolinear system shown in Fig. 3.3b, which has a constant stiffness EI_B throughout its length.

The vertical displacement Δ_C at the free end C of the pseudolinear system, is equal to the vertical displacement Δ_B at the free end B of the original system in Fig. 3.2. By using the pseudolinear system in Fig. 3.3b and applying the moment–area method, the vertical displacement δ_C is equal to the first moment about point C of the M_e'/EI_B area between A and C. See Fig. 3.3a. This yields

$$\begin{aligned}\delta_C = \delta_B = {} & \frac{1}{EI_B}\left[\frac{1}{2}(1{,}000)(57)\left(\frac{2}{3}\right)(57) + (1{,}000)(80)(97) + \frac{1}{2}(250)(80)(110.333)\right. \\ & + (700)(160)(217) + \frac{1}{2}(550)(160)(190.333) + (450)(200)(397) \\ & \left. + \frac{1}{2}(250)(200)(363.667) + (450)(140)(567) + \frac{1}{2}(20)(140)(590.333)\right] \\ = {} & \frac{123{,}994{,}123}{EI_B}\end{aligned}$$

Table 3.4. Calculated values of $f(x)$, Mx, $y'(x)$, z_e and M_e', corresponding to the assumed values of x (1 in. = 0.0254 m, 1 kip in. = 113.0 N m)

(1) x (in.)	(2) f(x) Eq. (3.19)	(3) M_x (kip in.) Eq. (3.20)	(4) $y'(x)$ Eq. (3.29)	(5) z_e Eq. (1.120)	(6) M_e' (kip in.) Eq. (1.119)
0	1	0	−2.543811	20.4205477	0
50	1.1224	−51.625	−2.421856	17.988596	−827.4
100	1.2544	−115.750	−2.128371	13.004187	−1,200.0
200	1.5488	−281.500	−1.480507	5.702612	−1,036.5
300	1.8858	−497.250	−0.992784	2.797977	−737.8
400	2.2686	−763.000	−0.637330	1.667499	−560.8
500	2.6999	−1,078.850	−0.352679	1.192260	−476.4
600	3.1827	−1,444.500	−0.095180	1.013620	−460.0
637	3.3780	−1,592.500	0	1.000000	−471.9

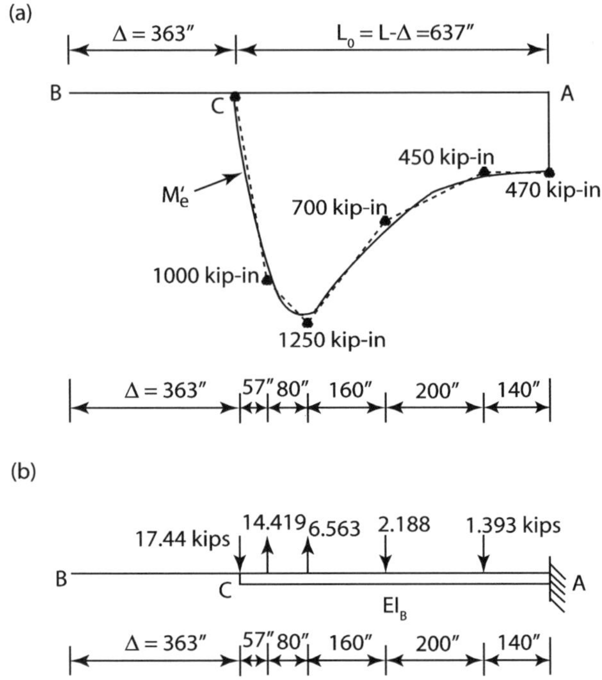

Fig. 3.3. (**a**) Moment diagram $M_e{}'$ of the pseudolinear system approximated with five straight-line segments. (**b**) Pseudolinear equivalent system of constant stiffness EI_B (1 in. = 0.0254 m, 1 kip in. = 113.0 N m, 1 kip = 4,448 N)

By substituting for EI_B, we find

$$\delta_C = \delta_B = \frac{123,994,123}{(180)(10)^3} = 688.86\,\text{in.}\ (17.50\text{m})$$

Table 3.5 has been also prepared which provides the values of Δ_B, θ_B, and δ_B, at the free end B of the member, for various values of the distributed load w_o, and assuming $x_o = x + \Delta$. In the same table, the values obtained using the fourth order Runge–Kutta method are shown.

Assume now that $\Delta(x) = \Delta x / L_o$. This yields

$$x_0 = x + \frac{\Delta x}{L_0}$$

In this case we have

$$M_x = -\frac{w_0 x}{2} x_0 = -\frac{w_0 x}{2}\left(x + \frac{\Delta x}{L_0}\right) \tag{3.32}$$

and the Euler–Bernoulli nonlinear differential equation yields

Table 3.5. Variation of Δ_B, θ_B, and δ_B for various values of the distributed load w_o(1 in. = 0.0254 m, 1 lb in.$^{-1}$ = 175.1268M m^{-1})

	equivalent pseudolinear systems			fourth order Runge–Kutta method			
w_o (lb in.$^{-1}$)	Δ_B (in.)	θ_B (deg.)	δ_B (in.)	Δ_B (in)	θ_B (deg.)	δ_B (in.)	% difference in δ_B
2	121.20	37.99	435.41	118.27	36.66	432.74	0.61
5	363.42	68.38	691.35	354.17	63.80	697.14	−0.83
10	557.08	84.08	799.36	557.53	79.35	819.88	−2.50
15	645.08	88.13	837.99	654.68	84.79	862.43	−2.83

$$\frac{y''}{\left[1+(y')^2\right]^{3/2}} = \frac{w_0\,(L-\Delta)^2}{2EI_B}\,\frac{\left[(L-\Delta)\,x^2+\Delta x^2\right]}{\left[(L-\Delta)+(n-1)\,x\right]^3} \tag{3.33}$$

Equation (3.33) may be solved in exactly the same way as Eq. (3.21), and the required integrations may be carried out by using Eq. (3.23).

By using the same values for w_o, EI_B, L, and taper n, and proceeding with the solution, we obtain $\Delta_B = 314.78$in. (7.955m), $\theta_B = 59.86°$, and $\delta_B = 666.03$in.(16.917m). These results are closely identical to the exact ones, because the only approximation made in the computation of θ_B is the one for x_o, which is proven to be an excellent approximation. The same applies for Δ_B, because the trial-and-error procedure can be repeated until we get as close as we wish to the exact value of Δ_B. For δ_B, we introduced, in addition, the approximation of the $M_e{}'$ with straight-line segments, which is also proven to be very accurate.

3.4 Solution of the Problem in the Preceding Section by Using a Simplified Nonlinear Equivalent System

The solution of the problem in Fig. 3.2 can be made very convenient for practical applications by using a simplified nonlinear equivalent system of uniform stiffness as discussed in Sect. 1.8.2. The following numerical example illustrates the application of the theory.

Example 3.3 Solve the flexible doubly tapered cantilever beam of Example 3.2 by using a simplified nonlinear equivalent system of constant stiffness $EI_B = 180 \times 10^3$ kip in.^{2} (516, 541 N m^2). Its length L = 1,000 in. (25.4m), w_o = 5 lb in.$^{-1}$ (875.63 N m^{-1}), and the taper n = 1.5, and compare the results. The modulus of elasticity is constant, indicating that the function g(x) = 1.

Solution: The variation of the moment of inertia I_x at any $0 \leq x \leq L_o$, is given by the following equation:

$$\begin{aligned} I_x &= I_B \left[1 + \frac{(n-1)}{L} x_0\right]^3 \\ &= I_B \left[1 + 0.5\frac{x_0}{L}\right]^3, \quad 0 \leq x_0 \leq L \end{aligned} \tag{3.34}$$

where

$$I_B = \frac{bh^3}{12} \tag{3.35}$$

$$f(x) = \left[1 + 0.5\frac{x_0}{L}\right]^3 \tag{3.36}$$

The bending moment M_x at any $0 \leq x_o \leq L$, is

$$M_x = -\frac{w_0 x_0^2}{2} = -2.5x_0^2 \tag{3.37}$$

Table 3.6 contains the values of f(x), M_x, and M_e, for the assumed values of x_o. The equations used for this purpose are shown in the same table. The values of the bending moment M_e of the simplified nonlinear equivalent system of constant stiffness EI_B are shown in the last column of the table. They are exact values, and the exact M_e is shown plotted by the solid line in Fig. 3.4a. The approximation of M_e with one straight line BD, as shown in the same figure, yields the simplified nonlinear equivalent system of constant stiffness EI_B, shown in Fig. 3.4b.

The only approximation involved in the derivation of the simplified nonlinear system is the approximation of M_e with one straight line BD.

Table 3.6. Values of f(x), M_x, and M_e, corresponding to the assumed values of x_o(1 in. = 0.0254 m, 1 kip in. = 113.0 N m)

(1) x_o (in.)	(2) f(x) Eq. (3.36)	(3) M_x (kip in.) Eq. (3.37)	(4) $M_e = M_x/f(x)$ (kip in.) Eq. (1.115) or (1.136)
0	1	0	0
100	1.1576	−25.000	−21.60
200	1.3310	−100.000	−75.13
300	1.5209	−225.000	−147.94
400	1.7280	−400.000	−231.48
500	1.9531	−625.000	−320.00
600	2.1970	−900.000	−409.65
700	2.4604	−1,225.000	−497.89
800	2.7440	−1,600.000	−583.09
900	3.0486	−2,025.000	−664.24
1,000	3.3750	−2,500.000	−740.74

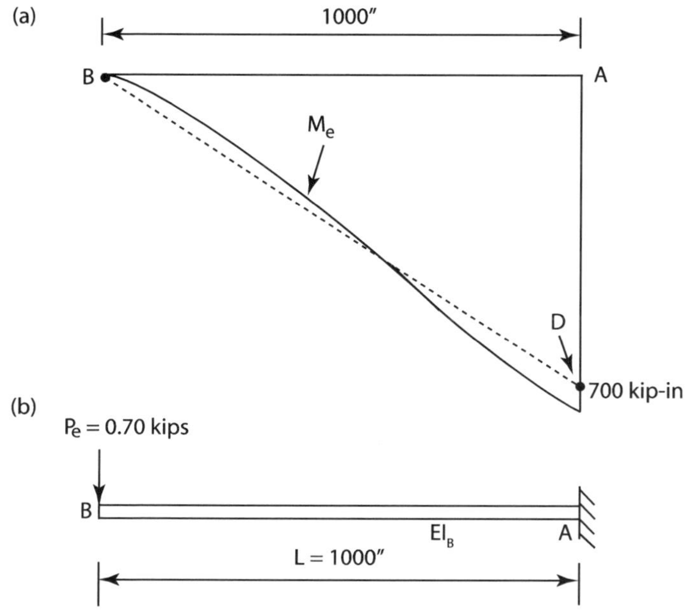

Fig. 3.4. (**a**) Moment diagram M_e of the simplified nonlinear equivalent system approximated with one straight line BD. (**b**) Simplified nonlinear equivalent system of constant stiffness EI_B (1 in. = 0.0254 m, 1 kip in. = 113.0 N m, 1 kip = 4,448 N)

This type of approximation is proven in many ways to provide accurate results, particularly for practical applications. Thus, the very complicated problem in Fig. 3.2 is now represented by a very simple nonlinear problem where solutions are available. We can solve the simplified system in Fig. 3.4b by using either pseudolinear analysis as discussed in Sect. 1.8.1, or by using Table 7.3, page 516 of [85]. By using the table in this reference, we find $\theta_B = 63.4°$, $\Delta_B = 320.66$ in. (8.145 m), and $\delta_B = 662.55$ in. (16.829 m). These results are in close agreement with the ones obtained in Example 3.2 using pseudolinear analysis and $x_o = x + \Delta x/L_o$. This shows that for very complicated nonlinear problems, the use of simplified nonlinear equivalent systems provide reasonable results for practical applications, and at the same time, they save a great deal of time, effort, and, consequently, money.

3.5 Flexible Tapered Simply Supported Beam with Uniform Load

A rectangular tapered flexible simply supported beam loaded with a distributed load w over its entire span length, and its large deformation configuration, are shown in Fig. 3.5. The moment of inertia I_x of the member at any

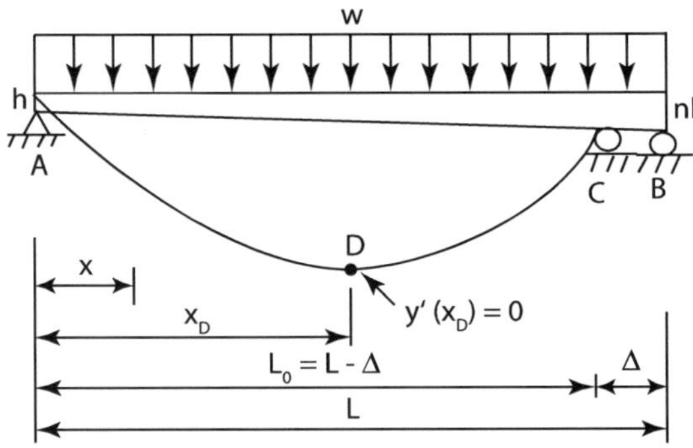

Fig. 3.5. Tapered simply supported flexible beam loaded as shown

$0 \leq x \leq L$ may be expressed as

$$\begin{aligned} I_x &= I_A \left\{1 + \frac{(n-1)}{L} x_0\right\}^3 \\ &= \frac{bh^3}{12}\left\{1 + \frac{n-1}{L}\int_0^x \left[1 + (y'(x))^2\right]^{1/2} dx\right\}^3 \end{aligned} \tag{3.38}$$

where the expression for the moment of inertia I_A at the end A of the member, and the expression for x_o, are as follows

$$I_A = \frac{bh^3}{12} \tag{3.39}$$

$$x_0 = \int_0^x \left[1 + (y'(x))^2\right]^{1/2} dx \tag{3.40}$$

The letter b in Eq. (3.39) denotes the constant width of the member.

By applying simple statics, the bending moment at any $0 \leq x \leq L_o$ is as follows

$$M_x = \frac{wLx}{2} - \frac{wxx_0}{2} \tag{3.41}$$

or, by substituting for x_o, we find

$$M_x = \frac{wLx}{2} - \frac{wx}{2}\int_0^x \left[1 + (y'(x))^2\right]^{1/2} dx \tag{3.42}$$

By substituting into the Euler–Bernoulli equation, we obtain

$$\frac{y''}{\left[1 + (y')^2\right]^{3/2}} = -\frac{wx}{2EI_A}\left\{\frac{L - \int_0^x \left[1 + (y'(x))^2\right]^{1/2} dx}{\left\{1 + \frac{n-1}{L}\int_0^x \left[1 + (y'(x))^2\right]^{1/2} dx\right\}^3}\right\} \tag{3.43}$$

As stated many times earlier in this text, Eq. (3.43) is a nonlinear integral differential equation and it is extremely difficult to solve. The modulus of elasticity E in Eq. (3.43) is assumed to be constant. In later parts of this text, the modulus E will also be permitted to vary along the length of the member.

We can simplify the complexity of this problem by assuming that $x_o = x + \Delta$, where Δ is the horizontal displacement of the end B of the member in Fig. 3.5. However, anyone of the Eqs. (1.81)–(1.84) could be used for this purpose without complicating the methodology. The one selected here is the simplest one. With this in mind, the expressions for the bending moment M_x and the moment of inertia I_x at any $0 \leq x \leq L_o$, are written as follows:

$$M_x = \frac{wLx}{2} - \frac{wx}{2}x_0 = \frac{w(L-\Delta)x}{2} - \frac{wx^2}{2} \tag{3.44}$$

$$I_x = \frac{bh^3}{12}\left[1 + (n-1)\frac{x}{L-\Delta}\right]^3 = I_A f(x) \tag{3.45}$$

where

$$I_A = \frac{bh^3}{12} \tag{3.46}$$

$$f(x) = \left[1 + (n-1)\frac{x}{L-\Delta}\right]^3 \tag{3.47}$$

and b is the constant width of the member. By substituting Eqs.(3.44), and (3.45), into the Euler–Bernoulli equation, we find

$$\frac{y''}{\left[1 + (y')^2\right]^{3/2}} = \frac{w(L-\Delta)^3}{2EI_A}\left\{\frac{-(L-\Delta)x + x^2}{[(n-1)x + (L-\Delta)]^3}\right\} \tag{3.48}$$

which is a simpler differential equation to solve compared to the one given by Eq. (3.43).

The integration of Eq. (3.48) for the computation of $y'(x)$, can be carried out as suggested in Sect. 1.4. On this basis, we have

$$\lambda(x) = \frac{w(L-\Delta)^3}{2EI_A}\left\{\frac{x^2 - (L-\Delta)x}{[(n-1)x + (L-\Delta)]^3}\right\} \tag{3.49}$$

The integration of $\lambda(x)$ can be performed by using the two handbook mathematical expressions given by Eqs. (3.23) and (3.24). This yields

$$\begin{aligned}\phi(x) = \int \lambda(x)\,dx = \frac{w(L-\Delta)^3}{2EI_A}&\left\{\frac{1}{(n-1)^3}\left[\ell n\left((L-\Delta) + (n-1)x\right)\right.\right.\\ &+ \left.\frac{2(L-\Delta)}{(L-\Delta)+(n-1)x} - \frac{(L-\Delta)^2}{2\left[(L-\Delta)+(n-1)x\right]^2}\right]\\ &+\left.\frac{\Delta}{(n-1)^2}\left[-\frac{1}{(L-\Delta)+(n-1)x} + \frac{(L-\Delta)}{2\left[(L-\Delta)+(n-1)x\right]^2}\right]\right\}\end{aligned} \tag{3.50}$$

Therefore, by using Eq. (1.32), we have

$$\frac{y'}{\left[1+(y')^2\right]^{1/2}} = \phi(x) + C \tag{3.51}$$

where C is the constant of integration and $\Phi(x)$ is given by Eq. (3.50).

We can determine C by applying the boundary condition $y'(x_D) = 0$, where x_D defines the point D in Fig. 3.5 where the deflection is maximum. By using Eq. (3.51) and applying this boundary condition, we obtain

$$\begin{aligned} C = -\frac{w(L-\Delta)^3}{2EI_A}\Bigg\{&\frac{1}{(n-1)^3}\Bigg[\ell n\left((L-\Delta)+(n-1)x_D\right) \\ &+\frac{2(L-\Delta)}{(L-\Delta)+(n-1)x_D}-\frac{(L-\Delta)^2}{2\left[(L-\Delta)+(n-1)x_D\right]^2}\Bigg] \\ &+\frac{\Delta}{(n-1)^2}\left[-\frac{1}{(L-\Delta)+(n-1)x_D}+\frac{(L-\Delta)}{2\left[(L-\Delta)+(n-1)x_D\right]^2}\right]\Bigg\} \end{aligned} \tag{3.52}$$

Thus, by using Eq. (1.33), the expression for $y'(x)$, is

$$y'(x) = \frac{G(x)}{\sqrt{1-[G(x)]^2}} \tag{3.53}$$

where

$$G(x) = \phi(x) + C \tag{3.54}$$

and $\Phi(x)$ and C are given by Eqs. (3.50) and (3.52), respectively.

The unknown horizontal displacement Δ in Eq. (3.53) can be determined in the usual way by applying a trial-and-error procedure. We can do this by using Eq. (3.53) and assuming a value for x_D, where $x = x_D$ defines the position where the maximum vertical displacement occurs. Then from Eq. (1.59), by using the value of the assumed x_D, we determine Δ, and, consequently, $L_o = (L - \Delta)$, by applying a trial-and-error procedure using assumed values for Δ until Eq. (1.59) is satisfied. We can facilitate this procedure by using Simpson's rule. When we find the value of Δ that satisfies Eq. (1.59) for the assumed value of x_D, we check to find out if the zero vertical displacement condition at the end C or the end A of the member is satisfied. If not, the procedure is repeated with new values for x_D until this condition is satisfied. The procedure converges rather fast to the required value of Δ, and, consequently, to that of L_o.

When x_D and L_o or Δ are determined, the values of y' at any $0 \le x \le L_o$ may be obtained from Eq. (3.53). The large deflection y at any $0 \le x \le L_o$ may be determined either by (a) integrating Eq. (3.53) once and satisfying one displacement boundary condition for the constant of integration, or, (b)

Table 3.7. Variation of the horizontal displacement Δ_B, Rotation θ_A, and vertical displacement δ_D for various values of the distributed load w (1 in. = 0.0254 m, 1 lb in.$^{-1}$ = 175.1268 N m^{-1})

	equivalent pseudolinear systems			fourth order Runge–Kutta method			
w (lb in.$^{-1}$)	Δ_B (in.)	θ_A (deg.)	δ_D (in.)	Δ_B (in.)	θ_A (deg.)	δ_D (in.)	% difference in δ_D
2	62.99	33.05	155.48	67.36	32.15	145.17	6.63
5	194.21	57.73	262.04	219.35	58.14	251.59	3.99
10	326.59	73.49	326.98	368.78	74.79	315.15	3.61
15	403.10	80.17	358.55	452.27	81.41	342.81	4.39
20	454.32	83.60	377.10	504.30	84.68	359.31	4.71

by applying pseudolinear analysis as explained earlier. Both procedures yield accurate results, but the pseudolinear analysis would be the easier to apply.

The above procedure is used to solve the problem in Fig. 3.5 by assuming that $EI_A = 75 \times 10^3$ kip in.^{2} (215,224 N m^2), L = 1,000 in. (25.4 m), taper n = 1.5, and using various values for the uniformly distributed vertical load w. The results are given in Table 3.7, where Δ_B is the horizontal displacement of the end B of the member, θ_A is its rotation at the end A, and δ_D is the maximum vertical displacement occurring at $x = x_D$. For comparison purposes, the table also includes the results obtained by using the fourth order Runge–Kutta method. We see in the last column of this table that the maximum difference in the deflection δ_D, by comparing these two methods, is 6.63%, and occurs when w = 2 lb in.$^{-1}$(350.236 N m^{-1}). It reduces to 3.61% when w = 10 lb in.$^{-1}$ (1,751.18 N m^{-1}).

3.6 Flexible Tapered Simply Supported Beam Carrying a Trapezoidal Load

The flexible doubly tapered simply supported beam in Fig. 3.6, is loaded with a vertical distributed trapezoidal loading as shown in the same figure. Its large deformation configuration is also shown in the same figure. By applying simple statics, we obtain the following expression for the bending moment M_x at any $0 \le x \le L_o$:

$$M_x = \frac{w_1}{6L_0}\left[x^3(1-m) + 3mL_0x^2 - L_0^2x(1+2m)\right] \tag{3.55}$$

The variation of its moment of inertia I_x at any $0 \le x \le L_o$ is given by the equation

$$I_x = I_B f(x) \tag{3.56}$$

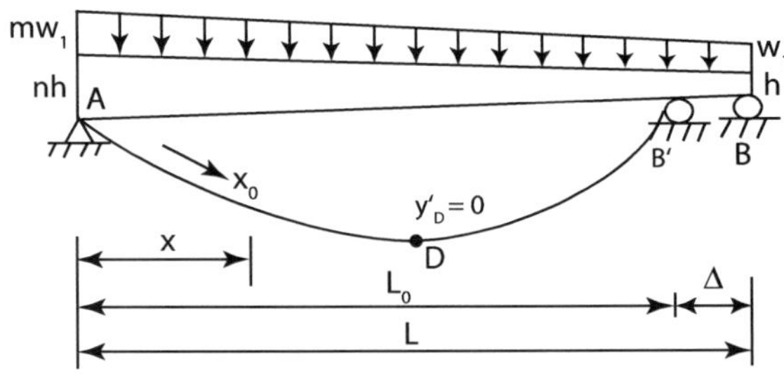

Fig. 3.6. Doubly tapered flexible simply supported beam loaded with a trapezoidal load

where

$$I_B = \frac{bh^3}{12} \tag{3.57}$$

$$f(x) = \left[\frac{L_0 + (n-1)(L_0 - x)}{L_0}\right]^3 \tag{3.58}$$

and b is the constant width of the member.

Since M_x and I_x are functions of the large deformation, their derivation given by Eqs. (3.55) and (3.56), is based on the assumption that $\Delta(x) =$ constant $= \Delta$ and, consequently, $x_o = x + \Delta$. However, anyone of the Eqs. (1.81)–(1.84) could be used for this purpose, but what we have used here is very reasonable for practical applications.

By substituting Eqs. (3.55) and (3.56) into the Euler–Bernoulli equation, we obtain

$$\frac{y''}{\left[1 + (y')^2\right]^{3/2}} = \frac{w_1 L_0^2}{6EI_B}\left\{\frac{x^3(1-m) + 3mL_0x^2 - L_0^2x(1+2m)}{\left[L_0 + (n-1)(L_0 - x)\right]^3}\right\} \tag{3.59}$$

The integration of Eq. (3.59) to obtain the values of $y'(x)$ at any $0 \le x \le L_o$ can be carried out by following the procedure discussed in Sect. 1.4 and in other sections of this text. Thus, by integrating both sides of Eq. (3.59) once and applying the boundary condition $y'(x_D) = 0$ for the computation of the constant of integration C, where x_D is the distance from the support A to the point D where the rotation is zero (also position of maximum vertical displacement), we find the following expression for $y'(x)$ at any $0 \le x \le L_o$:

$$y'(x) \frac{Q(x)}{\left\{1 - [Q(x)]^2\right\}^{1/2}} \tag{3.60}$$

where

$$
\begin{aligned}
Q(x) = {} & \frac{w_1 L_0^2}{12EI_B (n-1)^4 \left[L_0 + (n-1)(L_0 - x)\right]^2} \Big\{ L_0^3 n (n-1)^2 (1+2m) \\
& - L_0^2 x \left[2(n-1)^3 (1+2m) + 6n(n-1)(n-m)\right] \\
& + L_0 x^2 \left[9m(n-1)^3 + 9n(n-1)^2 (1-m)\right] - 2x^3 (n-1)^3 (1-m) \\
& - 6L_0 \left[L_0 + (n-1)(L_0 - x)\right]^2 \ell n \left[L_0 + (n-1)(L_0 - x)\right](n-m) \Big\} + C
\end{aligned}
\tag{3.61}
$$

and

$$
\begin{aligned}
C = {} & \frac{w_1 L_0^2}{12EI_B (n-1)^4 \left[L_0 + (n-1)(L_0 - x_D)\right]^2} \Big\{ L_0^3 n (n-1)^2 (1+2m) \\
& - L_0^2 x_D \left[2(n-1)^3 (1+2m) + 6n(n-1)(n-m)\right] \\
& + L_0 x_D^2 \left[9m(n-1)^3 + 9n(n-1)^2 (1-m)\right] - 2x_D^3 (n-1)^3 (1-m) \\
& - 6L_0 \left[L_0 + (n-1)(L_0 - x_D)\right]^2 \ell n \left[L_0 + (n-1)(L_0 - x_D)\right](n-m) \Big\}
\end{aligned}
\tag{3.62}
$$

Note that n in the above equations is the taper parameter and m is the loading parameter, as shown in Fig. 3.6.

The values of x_D and L_o in Eqs. (3.60)–(3.62) are not known, but they can be determined by applying a trial-and-error procedure as it was done in the preceding section. That is, we assume first a value for x_D and we substitute it into the equation

$$
L = \int_0^{L-\Delta} \left[1 + (y')^2\right]^{1/2} dx \tag{3.63}
$$

where y' in Eq. (3.63) is given by Eq. (3.60).

The next step is to use Eq. (3.63) and assume values of Δ and perform the required integration, preferably by using Simpson's rule, until the equation is satisfied. When this is done and Δ, or $L_o = L - \Delta$, is determined, we need to check and see if the vertical displacement condition at one of the end supports of the member is satisfied. If it is not, the above procedure is repeated with new values for x_D until this condition is satisfied. When this is completed and the values of x_D and L_o are obtained, the values of $y'(x)$ at any $0 \leq x \leq L_o$ can be obtained by using Eq. (3.60).

The large vertical displacement y may now be determined by using one of the two ways (a) by integrating Eq. (3.60) once, and applying the condition of zero deflection at one of the end supports to determine the constant of integration and (b) by using pseudolinear analysis as discussed in the preceding sections. Both procedures yield accurate results, but the pseudolinear analysis would be the easier to use.

Table 3.8. Variation of Δ, θ_A, θ_B, δ_D, and x_D for various values of w_1 and for $m = n = 2$ (1 in. = 0.0254 m, 1 kip in.$^{-1}$ = 175,126.8 N m^{-1})

w_1 (kips in.$^{-1}$)	Δ (in.)	θ_A (deg.)	θ_B (deg.)	δ_D (in.)	x_D (in.)
0.0025	77.86	27.14	39.26	170.69	517.36
0.0050	181.20	41.63	59.38	252.10	468.63
0.0075	257.65	49.95	69.89	293.60	431.55
0.0100	313.85	55.49	76.10	318.60	403.42
0.0125	356.91	59.54	80.05	335.83	381.31

Table 3.9. Variation of Δ, θ_A, θ_B, δ_D, and x_D for various values of the taper parameter n and with $m = 2$ and $w_1 = 0.005$ kips in.$^{-1}$ (1 in. = 0.0254 m, 1 kip in.$^{-1}$ = 175,126.8 N m^{-1})

n	Δ (in.)	θ_A (deg.)	θ_B (deg.)	δ_D (in.)	x_D (in.)
1.0	338.56	73.52	71.08	350.49	298.62
1.5	267.63	55.67	65.59	301.55	398.60
2.0	181.20	41.63	59.38	252.10	468.63
3.0	77.62	24.11	44.51	167.16	553.67

Numerical results are obtained here by assuming that $m = n = 2, w_1 = 0.005$ kip in.$^{-1}$ (875.634 N m^{-1}), $L = 1{,}000$ in. (25.4 m), and $EI_B = 75{,}000$ kip in.^{2} (215,224 N m^2). The application of the above methodology using a pseudolinear system yields $\theta_A = 41.63°$, $\Delta = 181.20$ in. (4.6025 m), and $y_{max} = \delta_D = 252.10$ in. (6.4033 m). In Table 3.8 the variation of θ_A, θ_B, Δ, and δ_D, for various values of the load w_1 and $m = n = 2$ is shown. The variation of the same quantities for various values of the load parameter n, when $m = 2$ and $w_1 = 0.005$ kip in.$^{-1}$ (875.634 N m^{-1}), is shown in Table 3.9. The values of the distance x_D from the end support A to the point of zero rotation are also shown in the last column of these two tables.

3.7 Using an Alternate Approach to Derive a Simpler Equivalent Nonlinear System of Constant Stiffness

The procedure and methodology discussed in Sect. 1.8.2 concerning the derivation of simplified nonlinear equivalent systems of constant stiffness, is further simplified in this section in order to facilitate even more the solution of complex nonlinear problems. This procedure is particularly useful for cases where the nonlinear flexible member involves complex loading conditions and complicated variations in their stiffness along the length of the member. Such complicated conditions are often encountered in practical engineering problems, and their frequency of occurrence is increasing.

In order to proceed with the discussion, we consider the Euler–Bernoulli nonlinear differential equation which is as follows:

$$\frac{y''}{\left[1+(y')^2\right]^{3/2}} = -\frac{1}{E_1 I_1}\,\frac{M}{g(x)\,f(x)} \tag{3.64}$$

where the function g(x) represents the variation of the modulus of elasticity E_x, with reference value E_1 and the function f(x) represents the variation of the moment of inertia I_x with reference value I_1.

We rewrite Eq. (3.64) as shown below.

$$\begin{aligned}\frac{y''}{\left[1+(y')^2\right]^{3/2}} &= -\frac{1}{E_1 I_1}\,\frac{M}{g(x)\,f(x)} \\ &= -\frac{M_e}{E_1 I_1}\end{aligned} \tag{3.65}$$

where

$$M_e = \frac{M_x}{f(x)\,g(x)} \tag{3.66}$$

is the expression for the moment M_e of the simplified nonlinear equivalent system of constant stiffness E_1I_1. It should be noted here that Eq. (3.66) is identical to Eqs. (1.115) and (1.136).

An accurate and convenient alternate nonlinear equivalent system of constant stiffness E_1I_1 can be easily obtained here by using Eq. (3.66) and plotting the moment diagram M_e of the equivalent nonlinear system in terms of the length $L_o = (L - \Delta)$. In Sect. 1.8.2, as well as in other sections of the text, we have plotted M_e in terms of the length L of the initial member.

3.7.1 Application to Cantilever Flexible Beam Problems

In order to illustrate the procedure, we consider the tapered cantilever beam in Fig. 3.7a, and we assume that $EI_B = 180 \times 10^3$ kip in.^{2} (516.54×10^3 N m^2), and P = 1 kip (4.448 kN). The moment of inertia I_x and bending moment M_x at any $0 \le x \le L_o$, are as shown below.

$$I_x = \frac{bh^3}{12}\left[\frac{L_O + 0.5x}{L_O}\right]^3 = I_B f(x) \tag{3.67}$$

$$M_x = -Px = -x \tag{3.68}$$

where

$$I_B = \frac{bh^3}{12} \tag{3.69}$$

$$f(x) = \left[\frac{L_O + 0.5x}{L_O}\right]^3 \tag{3.70}$$

and b is the constant width of the beam. By using Eqs. (3.68) and (3.70), and letting $g(x) = 1$, Eq. (3.66) yields

$$M_e = -\frac{x}{\left[\frac{L_O+0.5x}{L_O}\right]^3} = -\frac{x}{\left[1+0.5\frac{x}{L_O}\right]^3} \tag{3.71}$$

By using Eq. (3.71), we determine the values of M_e in terms of L_o, as shown in Table 3.10. Note in this table that both x and M_e are in terms of L_o. The plot of M_e is shown by the solid line in Fig. 3.7b. The approximation of M_e with one straight line CF, as shown in the same figure, leads to the nonlinear system of constant stiffness EI_B shown in Fig. 3.7c. If the M_e diagram in Fig. 3.7b is approximated by the two straight lines CD and DE shown by the dashed line in the same figure, then we obtain the constant stiffness nonlinear equivalent system shown in Fig. 3.7d, which is loaded with two concentrated loads as shown in the same figure.

To compare methodologies, the same problem is used to derive the simplified nonlinear equivalent systems shown in Fig. 3.8c and d, by using the method discussed in Sect. 1.8.2. It can be observed that the procedure followed in this section, and the procedure used to obtain the results in Fig. 3.8, are very similar. The main difference is that the M_e diagram in Fig. 3.7b is plotted in terms of $L_o = (L - \Delta)$ along the length L_o. For both methods, the one-line approximation of M_e yields identical nonlinear equivalent systems as shown by the results in Fig. 3.8d and c. The two-line approximation, however, yields some important differences. In Fig. 3.8c, the location of the equivalent loads P_1 and P_2 is given along the length L of the member, and their position along the length L_o during the large deformation of the member, as well as L_o, must be determined. In Fig. 3.7d, where the alternate approach is used, the position P_1 and P_2 along the length L_o has been determined with L_1, L_2, and L_o as the unknowns. Both methods yield accurate results, but in the alternate approach, since the location of P_1 and P_2 along L_o is known, it was only

Table 3.10. Values of M_e, vs. x in terms of L_o (1 in. = 0.0254 m, 1 kip in. = 113.0 N m)

x (in.)	M_e (kip in.)
0	0
$0.1L_o$	$0.086384L_o$
$0.2L_o$	$0.150263L_o$
$0.3L_o$	$0.197255L_o$
$0.4L_o$	$0.231481L_o$
$0.5L_o$	$0.256000L_o$
$0.6L_o$	$0.273100L_o$
$0.7L_o$	$0.284509L_o$
$0.8L_o$	$0.291545L_o$
$0.9L_o$	$0.295215L_o$
$1.0L_o$	$0.296296L_o$

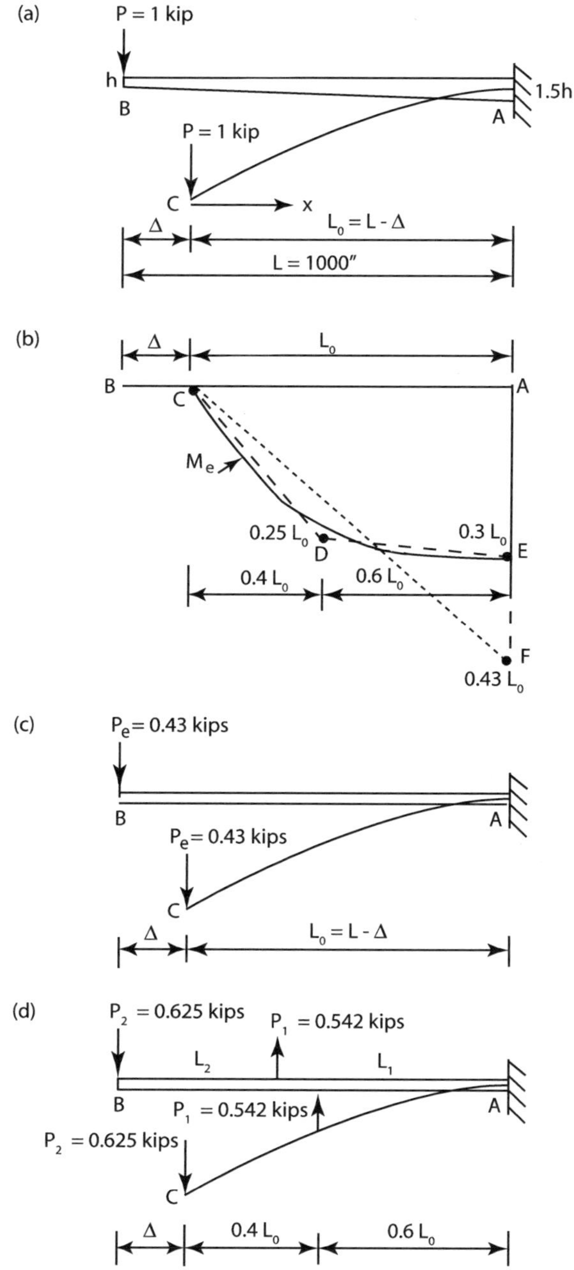

Fig. 3.7. (**a**) Original variable stiffness member. (**b**) M_e diagram of the equivalent nonlinear system of constant stiffness EI_B. (**c**) Constant stiffness equivalent nonlinear system loaded with one concentrated load at the free end. (**d**) Constant stiffness equivalent nonlinear system loaded with two concentrated loads (1 in. = 0.0254 m, 1 kip = 4.448 kN)

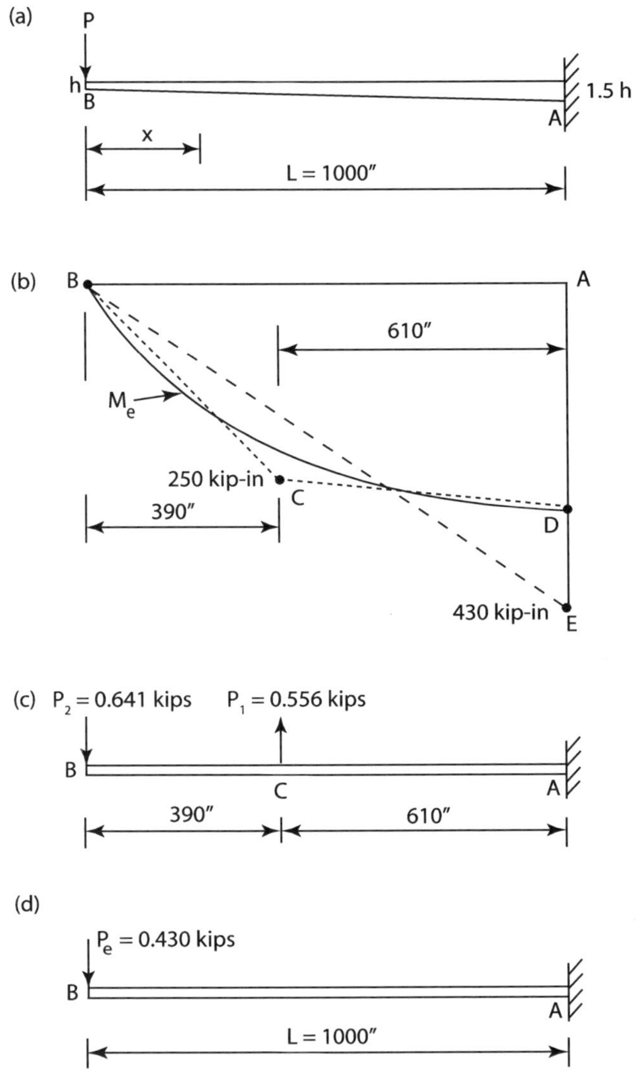

Fig. 3.8. (**a**) Original variable stiffness member. (**b**) M_e moment diagram of the equivalent nonlinear system following the method of Sect. 1.8.2. (**c**) Constant stiffness equivalent nonlinear system loaded with two concentrated loads. (**d**) Constant stiffness equivalent nonlinear system loaded with one concentrated load (1 in. = 0.0254 m, 1 kip = 4.448 kN)

necessary to determine L_o in order to compute the values of y' along L_o. The approach used for Fig. 3.8c, will require in addition to determine the location of P_1 and P_2 along L_o in order to obtain y'. The alternate approach would be even more advantageous when more than two straight lines are used to approximate the shape of the M_e diagram. The length L_o and, consequently,

Δ for the system in Fig. 3.7d may be determined from the equation

$$L = \int_0^{0.4L_0} \left[1 + (y')^2\right]^{1/2} dx + \int_{0.4L_0}^{L_0} \left[1 + (y')^2\right]^{1/2} dx \tag{3.72}$$

by a trial-and-error procedure. That is, assume values of Δ and repeat the procedure until Eq. (3.72) is satisfied, as discussed in earlier sections of this chapter and text.

The simplified nonlinear equivalent systems in Fig. 3.8c and d or the ones in Fig. 3.8d and c, may be solved using pseudolinear analysis as discussed earlier, or, if preferred, by applying other methods of a nonlinear analysis suitable for such types of problems.

For comparison purposes, the problem in Fig. 3.7c, which is the same as the one in Fig. 3.8d, is solved by using elliptic integrals. This solution yielded $\delta_B = 532.78$ in. (13.58 m), $\theta_B = 49.87°$, and $\Delta_B = 189.23$ in. (4.81 m) at the free end B of the member. In this solution we assumed that $P = 1$ kip (4.448 kN), $EI_B = 180 \times 10^3$ kip in.^{2} (516.54×10^3 Nm^2), $n = 1.5$, and $L =$ 1,000 in. (25.4 m). The solution of the original problems was also obtained in Sect. 3.2 by using pseudolinear analysis. The results obtained are $\delta_B =$ 519.85 in. (13.20 m), $\theta_B = 52.7°$, and $\Delta_B = 194$ in.(4.93 m). For practical purposes, we conclude that very reliable results are obtained by approximating the shape of M_e in Fig. 3.7b, or Fig. 3.8b, by only one straight-line segment. However, the shape of the M_e diagram may be approximated by as many straight-line segments as desired.

It should be pointed out, however, that the elliptic integral method does not apply to members where their stiffness E_xI_x, or their moment of inertia I_x, are not constant along their length. Also, elliptic integrals cannot be used when a member is subjected to distributed loadings. The utilization of the elliptic integral method was made possible here when the initial variable stiffness member in Fig. 3.7a was transformed into an equivalent simplified nonlinear system of constant stiffness EI_B as shown in Fig. 3.7c or Fig. 3.8d.

3.7.2 Application to Flexible Simply Supported Beam Problems

Consider now the variable stiffness simply supported flexible beam loaded as shown in Fig. 3.9a. From its deformed configuration, at any distance x from the end A, $0 \le x \le (L-\Delta)$, we write the following expressions for the moment of inertia I_x, the bending moment M_x, and the moment of inertia function f(x):

$$I_x = \frac{bh^3}{12}\left[\frac{(n-1)(L-\Delta-x)+(L-\Delta)}{(L-\Delta)}\right]^3 \tag{3.73}$$

$$M_x = \frac{wLx}{2} - \frac{wxx_0}{2} = \frac{w(L-\Delta)x}{2} - \frac{wx^2}{2} \tag{3.74}$$

$$f(x) = \left[\frac{(n-1)(L-\Delta-x)+(L-\Delta)}{(L-\Delta)}\right]^3 \tag{3.75}$$

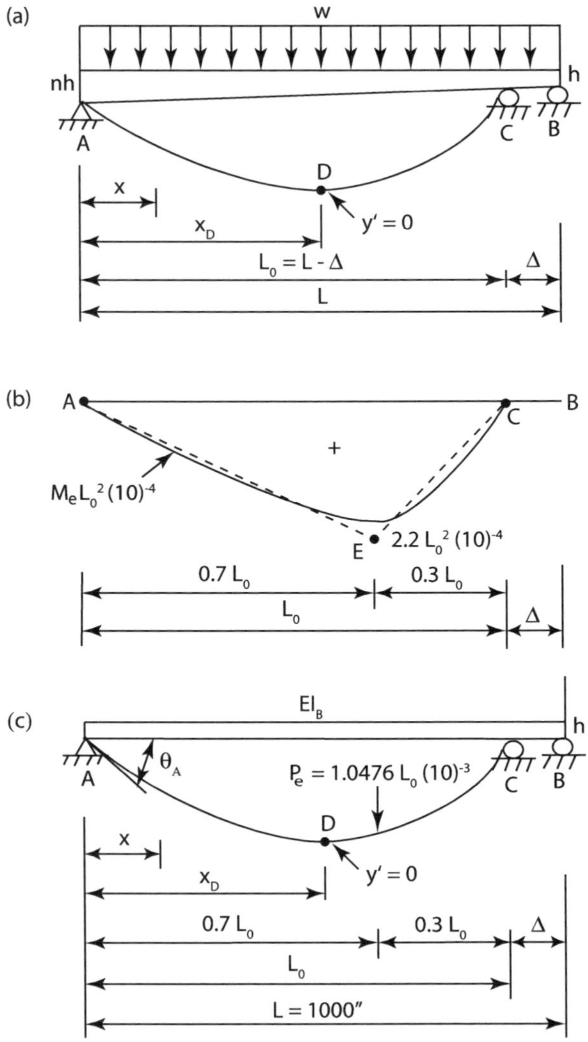

Fig. 3.9. (**a**) Tapered simply supported beam loaded with a uniformly distributed load w as shown. (**b**) Moment diagram M_e of the simplified nonlinear equivalent system of constant stiffness with its shape approximated with two straight lines. (**c**) Constant stiffness equivalent simplified nonlinear system loaded with one concentrated load (1 in. = 0.0254 m)

Note that x_o in Eq. (3.74) was taken to be equal to $(x + \Delta)$.

If it is assumed that $n = 1.5$ and $L_o = (x - \Delta)$, then Eqs. (3.75) and (3.74) yield

$$f(x) = \left[\frac{1.5L_0 - 0.5x}{L_0}\right]^3 \tag{3.76}$$

$$M_x = \frac{wL_0x - wx^2}{2} \tag{3.77}$$

Thus, by substituting into Eq. (3.66) and assuming $g(x) = 1$, we find

$$M_e = \frac{M_x}{f(x)} = \frac{wx(L_0 - x)}{2\left(1.5 - 0.5\frac{x}{L_0}\right)} \tag{3.78}$$

By using Eq. (3.78) with $w = 0.0025\,\text{kip in.}^{-1}$ ($437.2\,\text{N m}^{-1}$), the values of M_e for various values of x, in terms of L_0^2, are shown in Table 3.11. The plot of M_e is shown by the solid line in Fig. 3.9b. The approximation of its shape with two straight line segments AE and EC leads to the simplified nonlinear equivalent system of constant stiffness EI_B shown in Fig. 3.9c. The solution of the constant stiffness nonlinear system in Fig. 3.9c is simpler and it yields very accurate results.

This solution can be obtained by writing first the expressions for the moment due to P_e for the intervals , $0 \le x \le 0.7L_o$, and $0.7L_o \le x \le L_o$. By utilizing the Euler–Bernoulli equation and the indicated moment expressions for the two intervals, the two nonlinear differential equations for these two intervals of x may be written. By integrating each equation once, the expression of y' for each interval may be derived. The associated two constants of integration and the distance x_D to the point D of maximum deflection can be determined by utilizing the boundary conditions of zero deflection at the end A of the beam, $y_D{}'$ at $x = x_D$ being zero, and the rotation at $x = 0.7\,L_o$ for portions $0.7\,L_o$, and $0.3\,L_o$ being equal. The distance L_o can be determined from the equation

$$L = \int_0^{0.7L_0} \left[1 + (y')^2\right]^{1/2} dx + \int_{0.7L_0}^{L_0} \left[1 + (y')^2\right]^{1/2} dx \tag{3.79}$$

by a trial-and-error procedure as discussed earlier. That is, we assume values of L_o in Eq. (3.79) and perform the required integrations until the equation is satisfied. Simpson's rule may be used to carry out the integrations.

Table 3.11. Values of M_e vs. x in terms of L_o (1 in. = 0.0254 m, 1 kip in. = 113.0 N m)

xL_o (in.)	$M_eL_0^2\,(10)^{-4}$ (kip in.)
0	0
0.10	0.3690
0.20	0.7289
0.30	1.0669
0.40	1.3655
0.50	1.6000
0.60	1.7361
0.70	1.7260
0.80	1.5026
0.90	0.9718
1.00	0

When L_o is determined, the values of y' at any $0 \leq x \leq L_o$ can be determined from the equations of y' for portions 0.7 L_o and 0.3 L_o. With known y', the pseudolinear analysis may be carried out as discussed in Sect. 1.8.1.

The following example illustrates the application of the methodology.

Example 3.4 By using a simplified nonlinear constant stiffness EI_B equivalent system loaded with one vertical concentrated load P_e, determine the rotation θ_A at support A, the maximum vertical displacement δ_D, and the horizontal displacement Δ, for the tapered simply supported flexible beam loaded as shown in Fig. 3.9a. Assume that $EI_B = 75{,}000$ kip in.^{2} ($215{,}224\,\mathrm{N\,m^2}$), $L =$ 1,000 in. (25.4 m), $w = 0.0025$ kip in.$^{-1}$ ($437.8\,\mathrm{N\,m^{-1}}$), and taper $n = 1.5$.

Solution: With taper $n = 1.5$, the values of the moment of inertia function $f(x)$, at any $0 \leq x \leq L_o$, can be obtained from Eq. (3.76), the actual moment M_x at the same interval can be determined from Eq. (3.77), and the moment M_e of the simplified nonlinear equivalent system of constant stiffness EI_B can be obtained by using Eq. (3.78). With $w = 0.0025$ kip in.$^{-1}$ ($437.8\,\mathrm{N\,m^{-1}}$), the values of M_e at any $0 \leq x \leq L_o$ are calculated using Eq. (3.78) and they are shown in Table 3.11. The M_e diagram is plotted as shown in Fig. 3.9b. By approximating this diagram with two straight line segments AE and EC, as shown in the figure, and applying simple statics, we determine the simplified nonlinear equivalent system in Fig. 3.9c, which is loaded with a concentrated vertical load $P_e = 1.0476 L_o\ (10)^{-3}$. This equivalent nonlinear system has a constant stiffness EI_B throughout its length.

In this manner, the initial complex variable stiffness member in Fig. 3.9a has been reduced to a simply supported beam of constant stiffness EI_B, which is loaded with the equivalent load P_e at the indicated location. This problem is simpler to solve, and the methodology discussed in Sect. 2.4 may be used for its solution. It is extremely important at this point to understand the advantages associated with the utilization of simplified nonlinear equivalent systems. Even if the initial member in Fig. 3.9a is assumed to have any arbitrary variation in depth along its length, and also any combinations and variations in applied loading, a reasonably accurate simplified nonlinear equivalent system such as the one in Fig. 3.9c can be always obtained. The method of the simplified nonlinear equivalent systems is specifically developed for a convenient and very accurate solution of such difficult, large deformation problems.

By using the equivalent system and applying statics, the expressions for the bending moments M_1 and M_2 for portions a and b, respectively, are

$$M_1 = \frac{P_e b x}{L_0} = 0.3 P_e x, \quad 0 \leq x \leq a \tag{3.80}$$

$$M_2 = \frac{P_e b x}{L_0} - P_e (x - a) = 0.7 P_e (L_0 - x), \quad a \leq x \leq L_0 \tag{3.81}$$

where

$$P_e = 1.0476 L_0 \left(10\right)^{-3} \tag{3.82}$$

Note also that $L_o = (a + b)$.

By substituting into the Euler–Bernoulli nonlinear differential equation, we obtain the following two equations:

$$\frac{y_1''}{\left[1 + (y_1')^2\right]^{3/2}} = -\frac{M_1}{EI_B} = -\frac{0.3P_e x}{EI_B}, \quad 0 \le x \le a \tag{3.83}$$

$$\frac{y_2''}{\left[1 + (y_2')^2\right]^{3/2}} = -\frac{M_2}{EI_B} = -\frac{0.7P_e\left(L_0 - x\right)}{EI_B}, \quad a \le x \le L_0 \tag{3.84}$$

By integrating each of the Eqs. (2.83) and (2.84) once, as it was done in preceding sections, and in Sect. 2.4, particularly, we find

$$\frac{y_1'}{\left[1 + (y_1')^2\right]^{1/2}} = -\frac{0.15P_e x^2}{EI_B} + C_1, \quad 0 \le x \le a \tag{3.85}$$

$$\frac{y_2'}{\left[1 + (y_2')^2\right]^{1/2}} = -\frac{0.7P_e}{EI_B}\left(\frac{x^2}{2} - L_o x\right) + C_2, \quad a \le x \le L_0 \tag{3.86}$$

where C_1 and C_2 are the constants of integration.

We can determine C_1 and C_2 by applying the boundary conditions $y_1'(0) = y_A{}'$ and $y_1'(a) = y_2'(a)$, where $y_A{}'$ is the value of $y'(x)$ at the left end A of the member. By applying these two boundary conditions, we obtain

$$C_1 = \frac{y_A{}'}{\left[1 + (y_A{}')^2\right]^{1/2}} \tag{3.87}$$

$$C_2 = \frac{P_e a^2}{2EI_B} + \frac{y_A{}'}{\left[1 + (y_A{}')^2\right]^{1/2}} \tag{3.88}$$

Thus, by applying Eq. (1.33), we can write y_1' and y_2' as follows:

$$y_1' = \frac{G_1(x)}{\sqrt{1 - \left[G_1(x)\right]^2}}, \quad 0 \le x \le a \tag{3.89}$$

$$y_2' = \frac{G_2(x)}{\sqrt{1 - \left[G_2(x)\right]^2}}, \quad a \le x \le L_0 \tag{3.90}$$

where

$$G_1(x) = -\frac{0.15P_e x^2}{2EI_B} + \frac{y_A'}{\left[1 + (y_A')^2\right]^{1/2}} \tag{3.91}$$

$$G_2(x) = \frac{0.7P_e}{2EI_B}\left(\frac{x^2}{2} - L_0 x\right) + \frac{P_e a^2}{2EI_B} + \frac{y_A'}{\left[1 + (y_A')^2\right]^{1/2}}, \quad a \le x \le L_0 \tag{3.92}$$

At $x = x_D$, we have $y_1'(x = x_D) = 0$, and Eq. (3.89) yields

$$\frac{G_1(x_D)}{\sqrt{1 - [G_1(x_D)]^2}} = 0 \tag{3.93}$$

Solving Eq. (3.93) for $G_1(x_D)$, we find

$$G_1(x_D) = -\frac{0.15P_e {x_D}^2}{2EI_B} + \frac{y_A'}{\left[1 + (y_A')^2\right]^{1/2}} \tag{3.94}$$

We can apply now a trial-and-error procedure to determine the unknowns y_A' and L_o in the above equation and, consequently, the distance x_D from support A.

For example, we can start the trial-and-error procedure by assuming first that $y_A' = 0.57735$, corresponding to a $\theta_A = 30°$. Then, by assigning various values for L_o and performing the required integrations in Eq. (3.79), we find that $L_o = 909.05$ in. (23.09 m) satisfies this equation. Also, by integrating Eq. (3.89) once, we verify that the zero vertical displacement boundary condition at the support A is satisfied, which means that the assumed value $y_A' = 0.57735$ is correct.

Since we know y_A' and L_o, we can obtain the rotations y_1' and y_2' at any $0 \le x \le a$ and $a \le x \le L_o$ using Eqs. (3.89) and (3.90), respectively. From here and on, the pseudolinear analysis can be carried out as discussed in Sect. 1.8.2 and 2.4, as well as in many other parts of this text. If we use Eq. (3.93), we find that $x_D = 512.4$ in. (13.015 m). At this position the vertical displacement δ_D is maximum. Thus, by using the pseudolinear system, which is derived from the pseudolinear analysis and applying linear analysis, moment–area method, or handbook formulas, we find that $\delta_D = 186.31$ in. (4.73 m). For experience purposes, the reader can carry out the pseudolinear analysis using the simplified nonlinear system in Fig. 3.9c and verify this value of δ_D.

Problems

3.1 Solve the cantilever flexible beam problem in Example 3.1 by assuming that the vertical concentrated load $P = 1.5$ kips (6,672N).

3.2 Repeat the problem in Example 3.1 by assuming that the taper n = 2 and compare the results.

3.3 A tapered cantilever beam is loaded as shown in Fig. P3.3. By applying pseudolinear analysis, determine the vertical displacements δ_B and δ_C, at points B and C, respectively. Note the deformation configuration shown in the figure. The stiffness $EI_C = 180{,}000$ kip in.^{2} (516, 541 N m^2). Answer: $\delta_B = 238.36$ in. (6.0543 m), $\delta_C = 746.41$ in.(18.9588 m).

3.4 The stepped cantilever flexible beam in Fig. P3.4, is loaded as shown by the uniformly distributed load $w_o = 5.0$ lb in^{-1} (875.634 N m^{-1}). By assuming

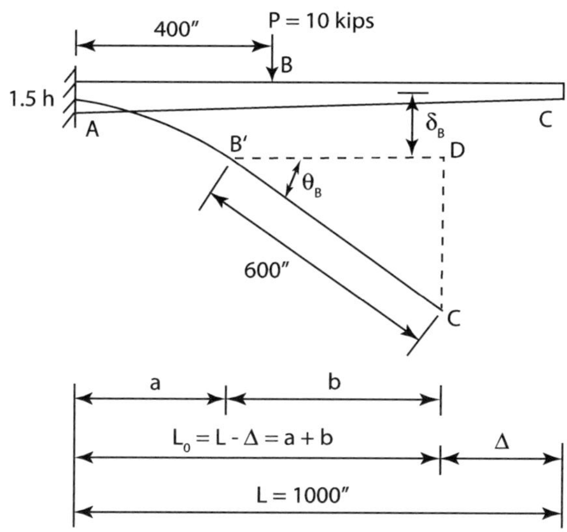

Fig. P3.3.

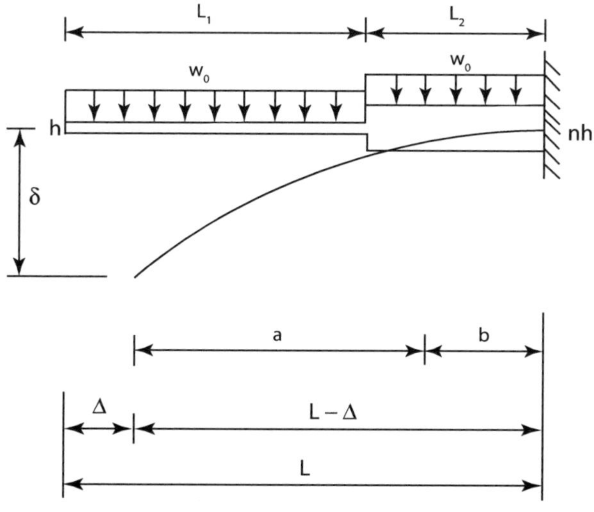

Fig. P3.4.

that L = 1,000 in. (25.4 m), $L_1 = L_2 = 500$ in. (12.7 m), n = 2, and EI_1 at the free end equal to 180,000 kip in.^{2} (516,541 N m^2), determine the horizontal displacement Δ, the rotation θ, and the vertical displacement δ, at the free end.
Answer: $\Delta = 155.25$(3.94 m), $\theta = 49.84°$, and $\delta = 460.89$ in. (11.71 m).

3.5 The doubly tapered cantilever beam in Fig. 3.2 is loaded with a uniformly distributed load $w_o = 10.0$ lb in^{-1}. (1,751.268 N m^{-1}) as shown. The stiffness $EI_B = 180{,}000$ kip in.^{2} (516,541 N m^2), length L = 1,000 in.(25.4 m), and taper n = 1.5. Determine the horizontal displacement Δ_B, rotation θ_B, and the vertical displacement δ_B, at the free end B of the beam, by using pseudolinear analysis and $\Delta(x) = \Delta$.
Answer: $\Delta_B = 557.08$ (74.1498 m), $\theta_B = 84.08°$, and $\delta_B = 799.36$ in. (20.3037 m).

3.6 Solve Problem 3.5 for $w_o = 2.0$ lb in.$^{-1}$ (350.2536 N m^{-1}), and $\Delta(\chi) = \Delta$.
Answer: $\Delta_B = 121.20$ in. (3.0785 m), $\theta_B = 37.99°$, and $\delta_B = 435.41$ in. (11.0594 m).

3.7 Solve Problem 3.5 by assuming that $\Delta(x) = \Delta(x/L_o)$ and compare the results.

3.8 Solve Problem 3.5 by using a simplified equivalent nonlinear system of constant stiffness EI_B that is loaded with a concentrated equivalent load P_e at the free end. Compare the results.

3.9 For the flexible tapered cantilever beam in Problem 3.5, determine the horizontal and vertical displacements, as well as the rotation, at $x = L/2$.

3.10 For the tapered cantilever beam loaded as shown in Fig. P3.10, determine Δ, θ_B, and δ_B, at the free end B by using a simplified nonlinear equivalent system of constant stiffness $EI_B = 180{,}000$ kip in.^{2} (516,541 N m^2), and loaded with an equivalent concentrated load P_e at the free end. The beam is loaded as shown in the figure.
Answer: $\Delta = 312.11$ (8.69 m), $\theta_B = 65.46°$, and $\delta_B = 679.78$ in. (17.27 m).

3.11 Solve Problem 3.10, by integrating the Euler–Bernoulli equation twice, and determining the two constants of integration by applying appropriate boundary conditions. Compare the results. Assume $\Delta(x) = \Delta$.

3.12 The cantilever beam in Fig. P3.12 is loaded with a trapezoidal distributed loading as shown. By using pseudolinear analysis, determine Δ_B, θ_B, and

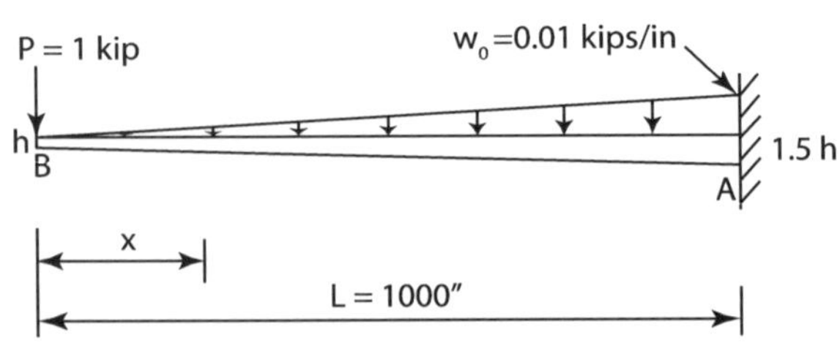

Fig. P3.10.

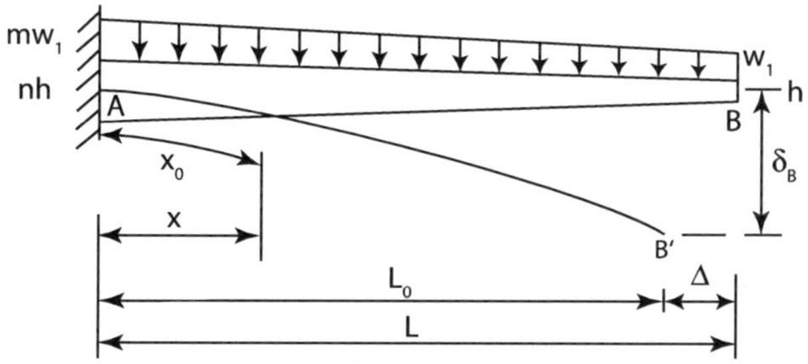

Fig. P3.12.

δ_B, at the free end of the member by assuming $\Delta(x) = \Delta$, m = 2, and n = 1.5. The stiffness EI_B = 180,000 kip in.^{2} (516,541 N m^2), L = 1,000 in. (25.4 m), and w_1 = 0.005 kips in.$^{-1}$ (87.6 kN m^{-1}).

Answer: Δ_B = 284.72 (7.2319 m), θ_B = 56.3°, and δ_B = 642.41 in. (16.3172 m).

3.13 Solve Problem 3.12 by assuming $\Delta(x) = \Delta\sin(\pi x/2L_o)$, and compare the results.

3.14 Solve Problem 3.12, by integrating the Euler–Bernoulli equation twice, and determining the two constants of integration by applying appropriate boundary conditions. Compare the results.

3.15 The flexible tapered simply supported beam in Fig. 3.5, is loaded with a uniformly distributed load w = 5.0 lb in.$^{-1}$ (875.634 N m^{-1}) as shown in the figure. By applying pseudolinear analysis and assuming $\Delta(x)$ = constant = Δ, determine the horizontal displacement Δ_B, of the end support B, the rotation θ_A of the end support A, and the maximum vertical displacement δ_D. The stiffness $EI_A = 75\times10^3$ kip in.^{2} (215, 224 N m^2), L = 1,000 in.(25.4 m), and taper n = 1.5.

Answer: Δ_B = 194.21(4.93 m), θ_A = 57.73°, and δ_D = 262.04 in. (6.656 m)

3.16 Solve Problem 3.15 by assuming that $\Delta(x) = \Delta(x/L_o)$, and compare the results.

3.17 Repeat the solution of Problem 3.15 by utilizing a simplified nonlinear equivalent system of constant stiffness EI_A and loaded with a vertical concentrated load P_e. Compare the results.

3.18 Solve Problem 3.4 by using a simplified nonlinear equivalent system of constant stiffness EI loaded with an equivalent concentrated load P_e at the free end. Compare the results.

3.19 The statically indeterminate beam in Fig. P3.19 is loaded with a distributed trapezoidal loading w_o = 0.0125 kips in.$^{-1}$(2, 189.085N m^{-1}). Determine Δ, θ_A, θ_B, x_D, and maximum vertical displacement y_{max}, by using pseudolinear analysis. The length L = 1,000 in. (25.4 m), EI_B = 75,000 kip in.^{2} (215, 224 N m^2), and m = n = 2. Assume Δ_x = constant = Δ. Follow the procedure discussed in Sects. 2.5 and 2.6.

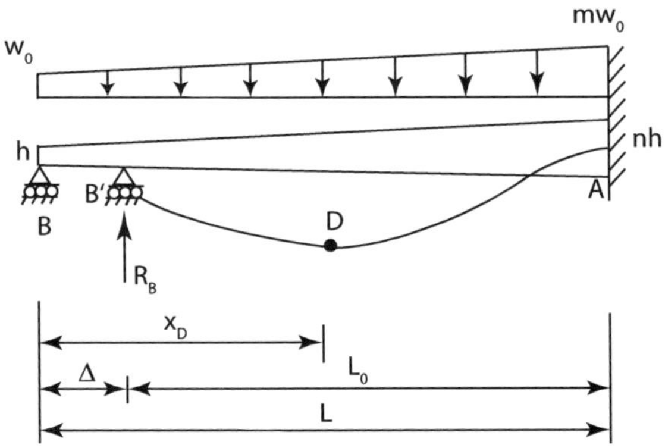

Fig. P3.19.

Answer: $\Delta = 166.61$ in. (4.232 m), $\theta_A = 0$, $\theta_B = 65.87°$, $x_D = 559.79$ in. (14.219 m), $y_{max} = 234.01$ in. (5.944 m).

3.20 Solve Problem 3.19 with $n = 1.5$ and $w_o = 0.01$ kips in.$^{-1}$ (1,751.268 N m^{-1}).
Answer: $\Delta = 219.28$ in. (5.57 m), $\theta_B = 69.72°$, $x_D = 504.34$ in. (12.81 m), $y_{max} = 270.56$ in. (6.872 m).

4

Inelastic Analysis of Structural Components

4.1 Introduction

In the preceding three chapters, the nonlinear analysis of a flexible member was carried out by assuming that the material of the member is not permitted to be stressed beyond its elastic limit and, therefore, Hooke's law would be applicable. The deformations, however, can be either large or small. This covers a great deal of problems, particularly when the member is flexible, and they deserve the special attention.

In many practical situations, structures that are subjected to blast and earthquake, for example, or structures where weight control is important, are often permitted to be stressed beyond the elastic limit of their material. In such situations, it is extremely important to know how structural members react to loading conditions, which cause the material to be stressed well beyond its elastic limit, and all the way to failure. Because of the unavailability of analytical closed-form solutions, the structural engineer is not always sure as to how far into the inelastic range his structure should be permitted to go and still be safe.

Although numerical methods such as the finite element method and the boundary element method are widely used, it is generally agreed that analytical work based on closed form solutions is definitely needed in order to establish the reliability of the results produced by numerical methods. The method of the equivalent systems, developed by the author and his collaborators, in combination with Timoshenko's reduced modulus concept, can be effectively used for inelastic analysis and provide many of the needed answers. The method is general and provides convenient solutions to many complicated structural engineering problems.

In this chapter, the material of the member will be permitted to be stressed well beyond its elastic limit and all the way to failure. Such a condition will cause the modulus E to vary along the length of the member and it must be taken into consideration for a reasonable solution to the practical problem. Therefore, in this chapter, both the modulus E and moment of inertia I, are permitted to vary along the length of the member.

In the analysis in this chapter, the deformation of the member is assumed to be small, there are no axial restraints acting on the member, and the member is statically determinate. This covers a wide range of practical problems. However, the inelastic analysis of statically indeterminate problems, members with axial restraints, and flexible members subjected to large deformations, can be found in Chapt. 6 of this text, the author's work in [2,3,5,6,84,97], as well as in other references at the end of this text.

4.2 Theoretical Aspects of Inelastic Analysis

When the material of a structural component is stressed beyond its elastic limit, the stress will seize to be proportional to the strain, Hooke's law will no longer apply, and at cross sections along the length of the member the modulus of elasticity E_x will not be the same. Since the deformations of structural members are dependant on their modulus of elasticity E_x, such modulus must be accurately and reliably determined, in order to be able to have an accurate solution to the problem, at least for practical applications.

A reasonable solution would be to determine a reduced modulus E_r by applying Timoshenko's [2,3,6,97] method. With the help of this method we can accurately establish the variation of the reduced modulus E_r at cross sections along the length of the member, and, consequently, the variation of the modulus of elasticity E_x along the length of a member. At cross sections where the material is elastic, the reduced modulus E_r will be equal to the elastic modulus E. Thus, with the elastic modulus E as a reference value, the modulus function g(x), which represents the variation of E_x along the length of the member, can be established.

In the analysis included in this section, the member will be permitted to be stressed well beyond the elastic limit of its material, its moment of inertia I_x can vary in any arbitrary manner along its length, and its modulus of elasticity E_x will be also variable. We also assure here that the deformations of the structural component are small and conform with small-deflection theory, and no axial forces or axial strains are applied to the member. Such types of restraints are discussed in detail in Chapt. 6. A great deal of practical problems can be handled with such limitations and deserve the special attention.

The discussion in this section includes (a) the theory and concept of the reduced modulus E_r and (b) the application of the method of the equivalent systems for the inelastic analysis of structural components.

4.2.1 The Theory and Concept of the Reduced Modulus E_r

The computation of the reduced modulus E_r is initiated by considering a portion of a beam of rectangular cross section as shown in Fig. 4.1b, where r is the radius of curvature of the neutral surface produced by the bending moment M. The stress–strain curve representing the mechanical properties of the beam's material is shown in Fig. 4.1a. In Fig. 4.1b the quantities h_1 and h_2

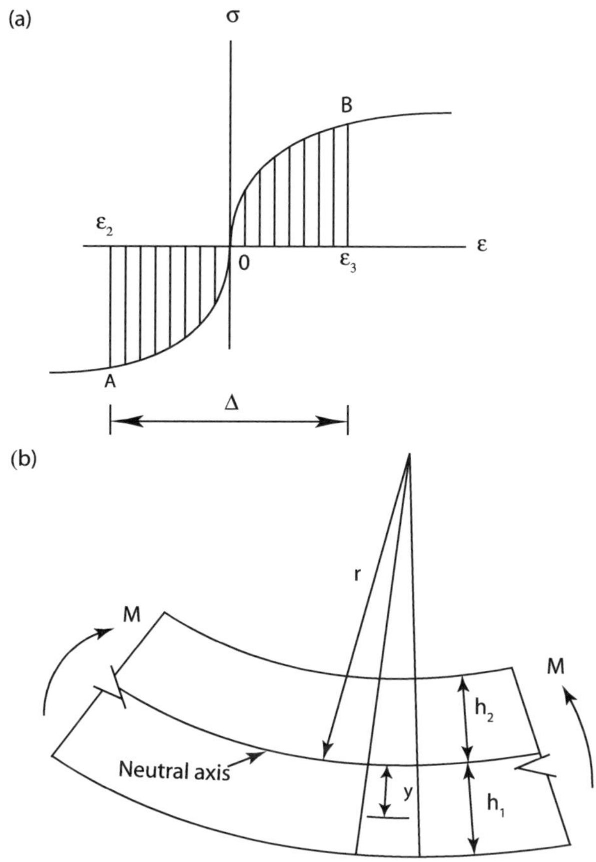

Fig. 4.1. (**a**) Stress–strain curve. (**b**) Portion of a member with rectangular cross section

denote the distances from the neutral axis to the lower and upper surfaces of the member, respectively. If ε_1 and ε_2 in Fig. 4.1a represent the unit elongations of the extreme fibers, and using the symbol Δ to denote the sum of the absolute values of the strains ε_1 and ε_2, we can write the following expressions by using basic mechanics:

$$\varepsilon_1 = \frac{h_1}{r} \tag{4.1}$$

$$\varepsilon_2 = \frac{h_2}{r} \tag{4.2}$$

We also have

$$\varepsilon = \frac{y}{r} \tag{4.3}$$

where ε is the unit elongation of a fiber at a distance y from the neutral axis.

Note that the symbol Δ above, was also used in the preceding three chapters to denote the horizontal displacement of flexible members, and it should not be confused with the discussion that takes place in this chapter.

The position of the neutral axis and the radius of curvature r can be determined from the following two equations of statics:

$$b\int_{-h_2}^{h_1} \sigma\, dy = 0 \tag{4.4}$$

$$b\int_{-h_2}^{h_1} \sigma y\, dy = M \tag{4.5}$$

where b is the width of the member. From Eq. (4.3) we can write the following two expressions:

$$y = \varepsilon r \tag{4.6}$$

$$dy = r\, d\varepsilon \tag{4.7}$$

By substituting Eq. (4.7) into Eq. (4.4), we obtain

$$br\int_{\varepsilon_2}^{\varepsilon_1} \sigma\, d\varepsilon = 0 \tag{4.8}$$

which indicates that the position of the neutral axis, since $r \neq 0$, requires to satisfy the identity

$$\int_{\varepsilon_2}^{\varepsilon_1} \sigma\, d\varepsilon = 0 \tag{4.9}$$

In order to determine the position of the neutral axis we use the curve AOB in Fig. 4.1a, where Δ is the sum of the absolute values of ε_1 and ε_2. That is, we have

$$\Delta = \varepsilon_1 + \varepsilon_2 = \frac{h_1}{r} + \frac{h_2}{r} = \frac{h}{r}\ \text{(a)}, \qquad r = \frac{h}{\Delta}\ \text{(b)} \tag{4.10}$$

where h is the total depth at the considered cross section of the member. It should be noted, however, that h can vary along the length of the member. In order to satisfy Eq. (4.9), we define Δ, and consequently ε_1 and ε_2, so that the two shaded areas in Fig. 4.1a are equal. This procedure yields the values of ε_1, ε_2, and Δ.

From Eqs. (4.1) and (4.2), we obtain

$$\frac{\varepsilon_1}{\varepsilon_2} = \frac{\frac{h_1}{r}}{\frac{h_2}{r}}$$

or

$$\frac{h_1}{h_2} = \left|\frac{\varepsilon_1}{\varepsilon_2}\right| \tag{4.11}$$

Equation (4.11) defines the position of the neutral axis.

From Eq. (4.3) we note that the strain ε is proportional to the distance y from the neutral axis. We also note that the curve AOB in Fig. 4.1a represents the bending stress distribution along the depth h of the member, if we substitute h for Δ. Note that Eq. (4.11) defines the position of the neutral axis. This conclusion should be very reasonable for practical applications.

The radius of curvature r may be determined by substituting Eqs. (4.6) and (4.7) into Eq. (4.5). This procedure yields

$$br^2 \int_{\varepsilon_2}^{\varepsilon_1} \sigma\varepsilon \, d\varepsilon = M \tag{4.12}$$

or, by using Eq. (4.10) and rearranging, we can write the following equation:

$$\frac{bh^3}{12}\frac{1}{r}\frac{12}{\Delta^3}\int_{\varepsilon_2}^{\varepsilon_1} \sigma\varepsilon \, d\varepsilon = M \tag{4.13}$$

For the elastic range, the curvature 1/r that is produced by the bending of the member, can be obtained from the equation

$$\frac{EI}{r} = M \tag{4.14}$$

where $I = bh^3/12$ is the cross-sectional moment of inertia of the rectangular member, and E is the elastic modulus. If the proportional limit of the material is exceeded, the curvature produced by the bending moment M may be determined from the expression

$$\frac{E_r I}{r} = M \tag{4.15}$$

where E_r is the reduced modulus which is defined, see Eqs. (4.13) and (4.14), by the expression

$$E_r = \frac{12}{\Delta^3}\int_{\varepsilon_2}^{\varepsilon_1} \sigma\varepsilon \, d\varepsilon \tag{4.16}$$

The integral in Eq. (4.16) represents the first moment of the shaded area in Fig. (4.1a) with respect to the origin 0. The units of E_r are force per unit area, since the ordinates of the curve in Fig. 4.1a represent stresses and the abscissas represent strain, which are exactly the same quantities as the ones for E.

A trial-and-error procedure may be used to determine E_r for any specific values of h and I at a given cross section of a member. Since Δ is unknown, a trial-and-error procedure may be initiated by assuming values of Δ and using the curve in Fig. 4.1a to determine the extreme elongations ε_1 and ε_2 for each assumed value of Δ. For the assumed Δ, the reduced modulus E_r can be obtained by using Eq. (4.16), and the curvature r can be determined from Eq. (4.10b) Thus, with known E_r and r, Eq. (4.15) may be used to determine the required bending Moment M_{req}. The M_{req} at any cross section of the member, must be equal to the actual bending moment M_x at the corresponding cross section of the member that can be obtained by using statics. If this is not so, the procedure may be repeated with new values of Δ until this is satisfied. This is a valid and correct assumption when the structural component is statically determinate.

When the member is statically indeterminate, the bending moment at a given cross section depends on both the external load and the distribution of the inelastic bending moment throughout the length of the member. The inelastic analysis of statically indeterminate members is discussed in detail in Chapter 6 and in [2, 3, 6]. In this section, we consider only statically determinate problems where both I and E can vary along the length of the structural members and where the deformations of the member are small. Many practical structural problems fall into this category.

When the trial-and-error procedure is finished and we have the correct values of Δ and M_{req}, we use the expression

$$E_r = Eg\,(x) \tag{4.17}$$

to determine the modulus function g(x) as follows:

$$g\,(x) = \frac{E_r}{E} \tag{4.18}$$

Note that the elastic modulus E in Eqs. (4.17) and (4.18) is used as the reference value.

When both functions f(x) and g(x) are known, the inelastic analysis of members of either uniform or variable cross section along their length can be performed by using the method of the equivalent systems as discussed in Sect. 1.8.3.

The procedure for the computation of the function g(x) is general, and it can be applied to beams with other types of cross sections. The following example illustrates the application of the methodology.

Example 4.1 The tapered cantilever beam in Fig. 4.2 is loaded with a uniformly distributed load $w = 1,600\,\text{lb in.}^{-1}$.$(280,203\,\text{N m}^{-1})$. The material of the member is monel, with a stress–strain curve as shown in Fig. 4.3a. The depth $h = 8$ in. (0.2032 m), width $b = 6$ in. (0.1524 m), taper $n = 1.5$, and length $L = 120$ in. (3.048 m). In practice, the yield strength of monel is taken equal to 50,000 psi (344,750 kPa) at 70°F, with a modulus of elasticity $E = 26 \times 10^6$psi $(179.27 \times 10^6$ kPa). Perform an inelastic analysis of the steel member and determine the reduced modulus E_r, the modulus function g(x), and the required bending moment M_{req} at cross section along the length of the member. Perform the analysis by using a two-line approximation of the stress–strain curve of monel. To be somewhat more general regarding the application of the theory, both E_x and I_x are permitted here to be variable.

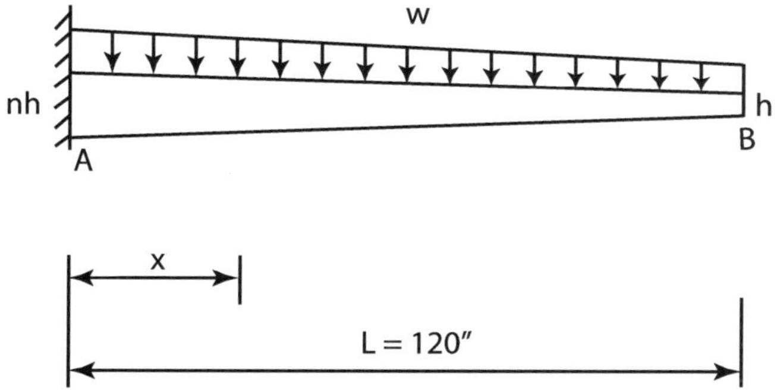

Fig. 4.2. Tapered cantilever beam loaded with a uniformly distributed load w (1 in. = 0.0254 m)

Solution: The moment of inertia I_x at any x measured from the fixed end A of the member, is given by the following expression:

$$I_x = \frac{bh^3}{12}\left[\frac{(n-1)(L-x)+L}{L}\right]^3 \tag{4.19}$$

If the taper n = 1, then the cantilever beam has a uniform cross section along its length. By using the moment of inertia $I_B = bh^3/12$ at the free end B as the reference value, the function f(x) that represents the variation of I_x along the length of the member is

$$f(x) = \left[\frac{(n-1)(L-x)+L}{L}\right]^3 \tag{4.20}$$

For n = 1.5 Eq. (4.20) yields

$$f(x) = \left[\frac{1.5\,L - 0.5\,x}{L}\right]^3 \tag{4.21}$$

The evaluation of the modulus function g(x) will be carried out by calculating the reduced modulus E_r at cross sections along the length of the member and then, by using Eq. (4.18), where E is the elastic modulus, we obtain the variation of g(x). To simplify the procedure, it would be reasonable and accurate to approximate the stress–strain curve of monel in Fig. 4.3a with two, three, or more straight-line segments, depending upon practical design requirements. A two-line or three-line approximation is usually sufficient [2,3,6] for the practical applications.

Tables 4.1 and 4.2 provide the values of the modulus E and the corresponding bending stresses σ for the two-line, three-line, and six-line approximation of the stress–strain curve of monel. For convenience, the methodology here is illustrated by using the two-line approximation of the stress–strain curve of monel, which yields $E_1 = 26\times10^6$ psi(179.27×10^6 kPa),

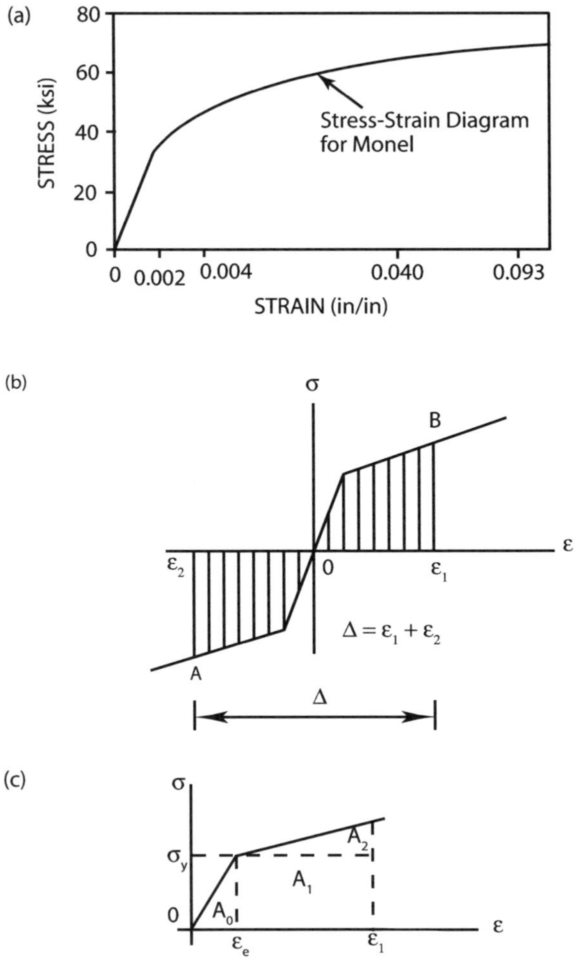

Fig. 4.3. (**a**) Stress–strain curve of monel. (**b**) Bilinear approximation of the stress–strain curve. (**c**) Area under the stress–strain curve (1 in. = 0.0254 m, 1 psi = 6.895 kPa)

Table 4.1. Values of E for Two-Line, Three-Line, and Six-Line approximations of the stress–strain curve of monel (1 psi = 6.895 kPa)

modulus E (psi)	two-line approximation	three-line approximation	six-line approximation
E_1	26×10^6	22×10^6	30×10^6
E_2	53×10^3	504×10^3	15×10^6
E_3	–	125×10^3	364×10^3
E_4	–	–	400×10^3
E_5	–	–	244×10^3
E_6	–	–	220×10^3

Table 4.2. Values of σ for Two-Line, Three-Line, and Six-Line approximations of the stress–strain curve of monel (1 psi = 6.895 kPa)

stress σ (psi)	two-line approximation	three-line approximation	six-line approximation
σ_1	50×10^3	48×10^3	30×10^3
σ_2	–	59×10^3	42×10^3
σ_3	–	–	50×10^3
σ_4	–	–	58×10^3
σ_5	–	–	60.2×10^3

$E_2 = 53 \times 10^3$ psi(365, 435 kPa), and yield stress $\sigma_y = 50,000$ psi(344, 750 kPa). For monel material, the two-line approximation should provide reasonable results. The three- and six-line approximations are discussed later in this text and in [2, 3, 6].

The schematic representation of the two-line approximation of the stress–strain curve of monel is shown in Fig. 4.3b and c. The compressive yield stress of monel is assumed to be equal to its tensile yield strength, which implies that we have a symmetrical stress–strain curve. This is not mandatory for the application of the methodology, but it follows typical engineering practices for steels. In Fig. 4.3b, if we substitute h for Δ, the curve AOB represents the bending stress distribution along the depth h of the member.

We apply now the theory discussed in this section to calculate E_r, which involves a trail-and-error procedure. That is, at any given cross section located at a distance x from the fixed support A in Fig. 4.2, which has a specific value of moment of inertia I and depth h, we assume values of Δ. Note that because of symmetry, we have $\Delta = (\varepsilon_1 + \varepsilon_2) = 2\varepsilon$, since $\varepsilon_1 = \varepsilon_2 = \varepsilon$. For each assumed value of Δ, we find from the curve AOB in Fig. 4.3b or c the values of ε_1 and ε_2, which in this case, because of symmetry, we have $\varepsilon_1 = \varepsilon_2 = \Delta/2$.

The next step would be to use Eq. (4.16) and calculate the reduced modulus E_r, which is associated with the assumed value of Δ. Note that the integral in Eq. (4.16) represents the first moment of the area under the curve AOB in Fig. 4.3b about the origin 0. The associated areas of this curve are shown in Fig. 4.3c. Note the symmetry involved. Then, with known E_r, we use Eq. (4.18) to calculate at the considered cross section the value of the modulus function g(x). In this equation the symbol E in the denominator represents the elastic modulus and, consequently, the reference value of g(x) is E.

With known E_r, the inelastic (or required) bending moment M_{req} at the given cross section is determined by using Eq. (4.15). The radius of curvature r in this equation is determined by using Eq. (4.10b), since h and Δ are known. Since the member is statically determinate, the calculated M_{req} should be equal to the bending moment M_x at the given cross section which is produced by the applied load w, which can be determined by applying elastic static analysis. If, for the assumed value of Δ, the required bending moment M_{req} is not equal to M_x, then the procedure is repeated with new values of Δ until this condition is satisfied.

Table 4.3. Summary of reduced modulus of elasticity E_r, and required moment M_{req} corresponding to the assumed values of x (1 in. = 0.0254 m, 1 psi = 6.895 kPa, 1 lb = 4.448 N)

x (in.)	h_x (in.)	I_x (in.^{4})	Δ (in. in.$^{-1}$)	Strain ε (in. in.$^{-1}$)	E_r (psi)	$M_{req} = M_x$ (in. lb)
0	12.0	864.00	$19.4790(10)^{-2}$	$\Delta/2$	$0.8214(10)^6$	$11.52(10)^6$
3	11.9	842.58	$9.5345(10)^{-2}$	$\Delta/2$	$1.6222(10)^6$	$10.95(10)^6$
6	11.8	821.52	$2.1996(10)^{-2}$	$\Delta/2$	$6.7892(10)^6$	$10.40(10)^6$
9	11.7	800.81	$1.0846(10)^{-2}$	$\Delta/2$	$13.2760(10)^6$	$9.86(10)^6$
12	11.6	780.45	$7.9961(10)^{-3}$	$\Delta/2$	$17.3300(10)^6$	$9.32(10)^6$
15	11.5	760.44	$6.6463(10)^{-3}$	$\Delta/2$	$20.0620(10)^6$	$8.82(10)^6$
18	11.4	740.77	$5.7961(10)^{-3}$	$\Delta/2$	$22.0890(10)^6$	$8.32(10)^6$
21	11.3	721.45	$5.1962(10)^{-3}$	$\Delta/2$	$23.6000(10)^6$	$7.83(10)^6$
24	11.2	702.46	$4.7462(10)^{-3}$	$\Delta/2$	$24.6890(10)^6$	$7.35(10)^6$
27	11.1	683.82	$4.3962(10)^{-3}$	$\Delta/2$	$25.4160(10)^6$	$6.88(10)^6$
30	11.0	665.50	$4.1462(10)^{-3}$	$\Delta/2$	$25.8010(10)^6$	$6.47(10)^6$
33	10.9	647.51	$3.8962(10)^{-3}$	$\Delta/2$	$25.9940(10)^6$	$6.02(10)^6$
36	10.8	629.86	$3.7227(10)^{-3}$	$\Delta/2$	$26.0000(10)^6$	$5.64(10)^6$

Table 4.3 summarizes the values of $h_x, I_x, \Delta, \varepsilon, E_r$, and $M_{req} = M_x$ at various cross sections x along the length of the member. Note that the stress at locations beyond 36 in. (0.9144 m) is below the yield strength of the material, and, consequently, $E_r = E$ in these locations.

For illustrative purposes, we verify the preceding methodology by using the cross section at the fixed end of the tapered cantilever beam in Fig. 4.2. This is represented by x = 0 in Table 4.3. From this table, the value of Δ at this section is 19.479×10^{-2}, or, $\varepsilon_1 = \varepsilon_2 = \varepsilon = 9.7395 \times 10^{-2}$. With these values of Δ and ε, we can define completely the bilinear curve shown in Fig. 4.3c. By observing Table 4.2, we note that $\sigma_y = \sigma_1 = 50 \times 10^3$ psi(344, 750 kPa), and the elastic strain ε_e is,

$$\varepsilon_e = \frac{\sigma_y}{E} = \frac{50,000}{(26)(10)^6} = 0.1923 \times 10^{-2}$$

We reproduced Fig. 4.3c as shown in Fig. 4.4. The bending stress σ_A at point A in this figure, is

$$\begin{aligned}\sigma_A &= (50)(10)^3 + \left[(9.7395)(10)^{-2} - (0.1923)(10)^{-2}\right](53)(10)^3 \\ &= 55,060\,\text{psi}\,\left(379,639 \times 10^3\,\text{kPa}\right)\end{aligned}$$

Thus, by using Fig. 4.4 and realizing that the other side of the σ axis is symmetrical, the first moment about the origin 0 of the associated areas is as follows:

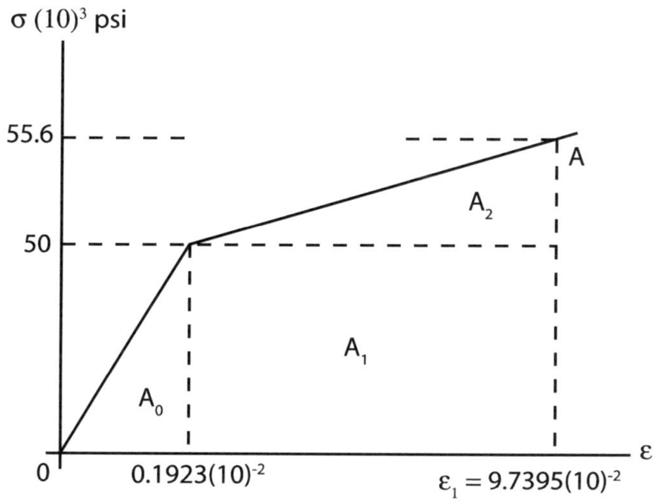

Fig. 4.4. Bilinear stress–strain curve (1 psi = 6, 895 Pa)

$$\begin{aligned}\int_{\varepsilon_2}^{\varepsilon_1} \sigma\varepsilon \, d\varepsilon = 2 \Big[& \frac{1}{2} (0.1923) (10)^{-2} (50) (10)^3 (0.1282) (10)^{-2} \\ & + (50) (10)^3 (9.5472) (10)^{-2} (4.9659) (10)^{-2} \\ & + \frac{1}{2} (5.06) (10)^3 (9.5472) (10)^{-2} (6.5571) (10)^{-2} \Big] \\ = {} & 505,904.2533 \times 10^3\end{aligned}$$

This is the value of the integral in Eq. (4.16). By substituting Δ, and the preceding calculated value of the integral, into Eq. (4.16), we find

$$\begin{aligned}E_r &= \frac{12}{[19.479 \times 10^{-2}]^3} \left(505,904.2533 \times 10^{-3}\right) \\ &= 0.8214 \times 10^6 \, \text{psi} (5,633,444 \, \text{kPa})\end{aligned}$$

which is identical to the value shown in Table 4.3. From Eq. (4.10b), we find

$$r = \frac{h}{\Delta} = \frac{12}{(19.479)(10)^{-2}} = 61.605 \, \text{in.} \, (1.565 \, \text{m})$$

and from Eq. (4.15), the value of the required moment M_{req} is

$$\begin{aligned}M_{req} &= \frac{E_r I_x}{r} = \frac{(0.8214)(10)^6}{61.605} (864) \\ &= 11.52 \times 10^6 \, \text{in. lb} \left(1.3018 \times 10^6 \, \text{N m}\right)\end{aligned}$$

which agrees with the results obtained for x = 0 in Table 4.3.

Table 4.4. Values of f(x), g(x), E_r, M_x, and M_e for inelastic analysis (1 in. = 0.0254 m, 1 psi = 6.895 kPa, 1 in. lb = 0.1130 N m)

(1) x (in.)	(2) f(x)	(3) $E_r \times 10^6$ (psi)	(4) g(x)	(5) M_x (in. lb)	(6) $M_e = M_x/f(x)g(x)$ (in.lb)
0	3.38	0.821	0.0316	$11.520(10)^6$	$108.040(10)^6$
6	3.21	6.789	0.2611	$10.397(10)^6$	$12.407(10)^6$
12	3.05	17.330	0.6665	$9.323(10)^6$	$4.588(10)^6$
18	2.89	22.089	0.8496	$8.319(10)^6$	$3.384(10)^6$
24	2.74	24.689	0.9496	$7.349(10)^6$	$2.821(10)^6$
30	2.60	25.801	0.9924	$6.472(10)^6$	$2.509(10)^6$
36	2.46	26.00	1.0000	$5.645(10)^6$	$2.294(10)^6$
42	2.33	26.00	1.0000	$4.867(10)^6$	$2.092(10)^6$
48	2.20	26.00	1.0000	$4.147(10)^6$	$1.888(10)^6$
54	2.07	26.00	1.0000	$3.485(10)^6$	$1.681(10)^6$
60	1.95	26.00	1.0000	$2.880(10)^6$	$1.475(10)^6$
66	1.84	26.00	1.0000	$2.333(10)^6$	$1.269(10)^6$
72	1.73	26.00	1.0000	$1.843(10)^6$	$1.067(10)^6$
78	1.62	26.00	1.0000	$1.411(10)^6$	$0.870(10)^6$
84	1.52	26.00	1.0000	$1.037(10)^6$	$0.682(10)^6$
90	1.42	26.00	1.0000	$0.720(10)^6$	$0.506(10)^6$
96	1.33	26.00	1.0000	$0.461(10)^6$	$0.346(10)^6$
102	1.24	26.00	1.0000	$0.259(10)^6$	$0.209(10)^6$
108	1.16	26.00	1.0000	$0.115(10)^6$	$0.100(10)^6$
114	1.08	26.00	1.0000	$0.029(10)^6$	$0.027(10)^6$
120	1.00	26.00	1.0000	0	0

Table 4.4 was also prepared which provides the values of f(x), g(x), E_r, M_x, and the moment M_e of the equivalent system of constant stiffness EI_B at various locations of x. The reference value for the function f(x) is I_B, which is the value of the moment of inertia at the free end B of the member in Fig. 4.2. The application of the method of the equivalent systems to determine the inelastic deflections of members where their material is stressed beyond their elastic limit, is discussed in the following subsection of this chapter.

The values of f(x) in Table 4.4 are obtained by using Eq. (4.21), E_r is computed as shown above and in Table 4.3, g(x) is calculated by using Eq. (4.18), Fig. 4.2 and applying simple statics, and M_e is determined by using equation

$$M_e = \frac{M_x}{f(x)\,g(x)} \tag{4.22}$$

as discussed in Sect. 1.8.

By observing Table 4.4, we note that for $0 \le x \le 30$ the function g(x) is smaller than one and, consequently, the member behaves inelastically. On the other hand, for $36 \le x \le 120$ the function g(x) is equal to unity, which means that the member behaves elastically.

4.2.2 Application of the Method of the Equivalent Systems for Inelastic Analysis

The theory and method of the equivalent systems and its application to various types of flexible and nonflexible structural components was discussed in detail in the preceding three chapters of this text. In these chapters, the moment of inertia was permitted to vary in any arbitrary manner along the length of the structural member, and various loading conditions and load combinations were examined. However, in these three chapters, the modulus of elasticity E remained constant along the length of the member and, consequently, the modulus function g(x) was equal to one.

In this chapter, the material of the member is permitted to be stressed well beyond its elastic limit, and all the way to failure, a situation that makes the modulus E to vary along the length of the structural member. Therefore, under these conditions, both the moment of inertia function f(x) and the modulus function g(x), are permitted to vary along the length of the member. However, the deformations in this section will be kept small. The subject of large inelastic deformations is discussed in Chapter 6 and in [2, 3, 6, 84].

The following example illustrates the application of the method.

Example 4.2 For the tapered cantilever beam loaded as shown in Example 4.1, determine an equivalent system of constant stiffness $E_1 I_B$, where E_1 is the elastic modulus of its material and I_B is the moment of inertia at the free end B of the member. Then, use the equivalent system to determine the inelastic deflection and rotation at the free end.

Solution: The values of f(x), g(x), and static bending moment M_x, were calculated in Example 4.1 at various locations x along the length of the member and they are shown in Table 4.4. The last column of the same table contains the values of the moment diagram M_e of the equivalent system of constant stiffness $E_1 I_B$. They are calculated by using Eq. (4.22). For example, at $x = 0$, we have

$$f(x) = (1.5)^3 = 3.375$$

$$E_r = 0.8214 \times 10^6 \text{psi}\,(5,663,444\,\text{kPa})$$

$$g(x) = \frac{E_r}{E} = \frac{(0.8214)(10)^6}{(26)(10)^6} = 0.0316$$

$$M_x = (1,600)(120)(60) = 11.52 \times 10^6\ \text{in. lb}\ \left(1.302 \times 10^6\ \text{N m}\right)$$

$$M_e = \frac{M_x}{f(x)\,g(x)} = \frac{(11.52)(10)^6}{(3.375)(0.0316)} = 108.02 \times 10^6\ \text{in. lb}\ \left(12.206 \times 10^6\ \text{N m}\right)$$

The elastic modulus $E_1 = E$ of monel, which is used as the reference value for the function g(x) is 26×10^6 psi(179.27×10^6 kPa).

By using the values of M_e in the last column of Table 4.4, the moment diagram of the equivalent system of constant stiffness $E_1 I_B$, is shown plotted

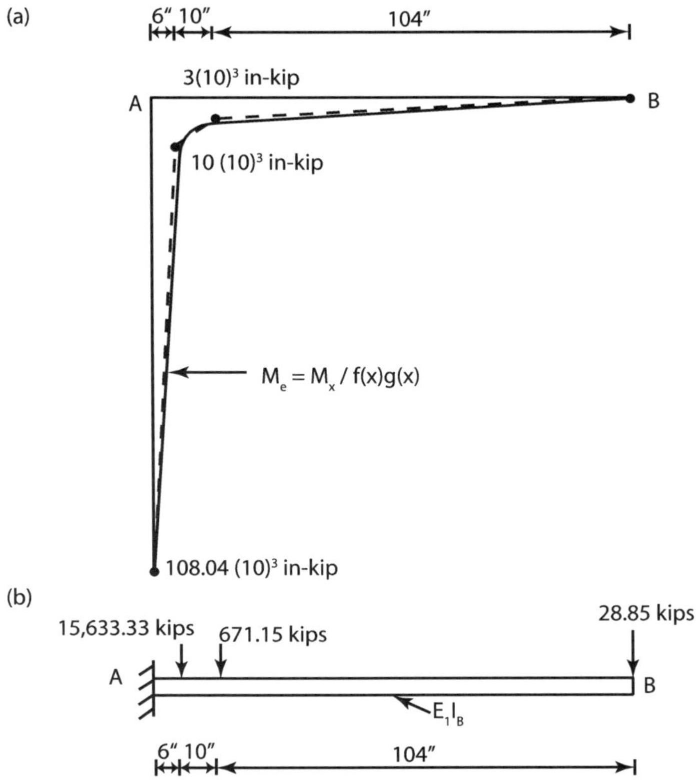

Fig. 4.5. (**a**) M_e diagram with its shape approximated by three straight lines. (**b**) Equivalent system of constant stiffness E_1I_B (1 in. = 0.0254 m, 1 kip = 4.448 kN, 1 in. kip = 112.9848 N m)

by the solid line in Fig. 4.5a. By approximating the shape of this diagram with three straight-line segments, as shown by the dashed line in the same figure, and applying simple statics, we obtain the constant stiffness equivalent system loaded as shown in Fig. 4.5b.

The deflection and rotation at any point x along the length of the equivalent system will be closely identical to the ones at the corresponding point of the original system in Fig. 4.2. However, the task in solving directly the original inelastic system is extremely difficult. The equivalent system in Fig. 4.5b can be easily solved by using handbook formulas or linear methods of elementary mechanics.

If we divide the approximated M_e diagram in Fig. 4.5a by the stiffness E_1I_B and apply the moment–area method, the vertical deflection δ_B at the free end B, is equal to the first moment of the M_e/E_1I_B area between points A and B, taken about B. That is

$$\delta_B = \frac{(10)^6}{E_1 I_B}\left[\frac{1}{2}(3)(104)\left(\frac{2}{3}\right)(104) + (3)(16)(112) + (7)(6)(117) + \frac{1}{2}(7)(10)(110.6667) + \frac{1}{2}(98.04)(6)(118)\right]$$

$$= \frac{(10)^6}{E_1 I_B}[59,685.4933]$$

or, by substituting for E_1 and I_B, we find

$$\delta_B = \frac{(10)^6}{(26)(10)^6(256)}[59,685.4933]$$
$$= 8.967\,\text{in.}\,(0.2278\,\text{m})$$

In accordance with the moment–area method, the rotation θ_B at the free end B would be equal to the $M_e/E_1 I_B$ area between points A and B. This yields

$$\theta_B = \frac{(10)^6}{E_1 I_B}\left[\frac{1}{2}(3)(104) + (3)(16) + (7)(6) + \frac{1}{2}(7)(10) + \frac{1}{2}(98.04)(6)\right]$$
$$= \frac{(10)^6}{E_1 I_B}[575.12]$$

By substituting for $E_1 I_B$, we obtain

$$\theta_B = \frac{(10)^6}{(26)(10)^6(256)}[575.12]$$
$$= 0.086406\,\text{rad}$$
$$= 4.951^\circ$$

In a similar manner, the deflections and rotations at any other point of the structural member can be determined.

By retaining the two-line approximation of the stress–strain curve of monel, the inelastic analysis of the member in Fig. 4.2 was repeated with depth $h = 8$ in.(0.2032 m), and assuming that the taper $n = 1.5, 1.75$, and 2.0. For each value of the taper n, the distributed load w was permitted to increase until near failure. The results of these three cases of n are shown plotted in Fig. 4.6.

The starting point P in these curves represents the transition from elastic to inelastic behavior. Note that as the value of n increases, additional load w is needed to reach this transition stage. We also note that when a certain value of the load w is reached, any further increase in w is associated with large increases in deflection, indicating that the ultimate capacity of the member to resist load is reached. For example, for $n = 1.75$, the curve shows that the ultimate load w_u should be about 2,000 lb in.$^{-1}$(350,253.6 N m^{-1}. More discussion on this subject is given later in this chapter and in [2,3,84].

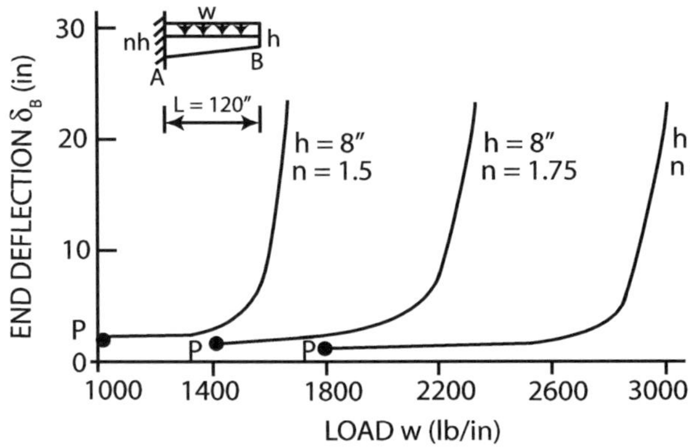

Fig. 4.6. Load–deflection curves for the two-line approximation of the stress–strain curve of monel with n = 1.5, 1.75, and 2 (1 in. = 0.0254 m, 1 lb = 4.448 N, 1 lb in.$^{-1}$ = 175.1268 N m^{-1})

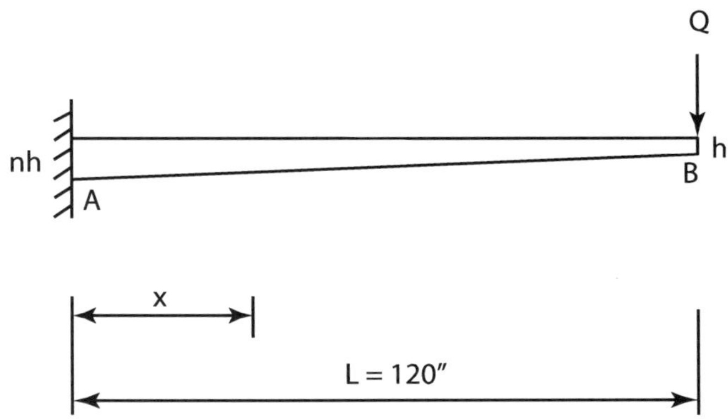

Fig. 4.7. Tapered cantilever beam loaded with a concentrated load Q at its free end (1 in. = 0.0254 m)

Example 4.3 The tapered cantilever beam in Fig. 4.7 is loaded at its free end B with a vertical concentrated load Q. By assuming that Q = 135 kips (600, 510 N), h = 8 in. (0.2032 m), width b = 6 in. (0.1524 m), taper n = 1.75, and L = 120 in.(3.048 m), perform the inelastic analysis of the structural member by using the three-line approximation of the stress–strain curve of monel. Also using an equivalent system of constant stiffness E_1I_B, determine the rotation θ_B and the vertical deflection δ_B at the free end B.

Solution: We start the analysis using the three-line approximation of the stress–strain curve of monel. Tables 4.1 and 4.2 show, respectively, the values of E_1, E_2, and E_3, and the ones for the stresses σ_1 and σ_2, corresponding to this approximation. Then, by following the procedure discussed earlier regarding the evaluation of the reduced modulus E_r at cross sections along the length

Table 4.5. Values of M_x, M_{req}, h_x, f(x), Δ, E_r, and g(x) at intervals of 6.0 in. Along the length of the tapered cantilever beam (1 in. = 0.0254 m, 1 in. lb = 0.113 N m, 1 psi = 6,895 Pa)

(1) x (in.)	(2) $M_x \times 10^6$ (in. lb)	(3) $M_{req} \times 10^6$ (in. lb)	(4) h_x (in.)	(5) f(x)	(6) $\Delta \times 10^{-3}$	(7) $E_r \times 10^6$ (psi)	(8) $g(x) = E_r/E_1$
0	16.200	16.199	14.00	5.36	49.563	3.3351	0.1516
6	15.390	15.388	13.70	5.02	46.963	3.4915	0.1587
12	14.580	14.578	13.40	4.70	40.313	3.6977	0.1681
18	13.770	13.770	13.10	4.39	40.313	3.9807	0.1809
24	12.960	12.959	12.80	4.10	36.063	4.3866	0.1994
30	12.150	12.148	12.50	3.81	31.163	4.9896	0.2268
36	11.340	11.340	12.20	3.55	25.813	5.9029	0.2683
42	10.530	10.529	11.90	3.29	20.213	7.3568	0.3344
48	9.720	9.719	11.60	3.05	15.163	9.5269	0.4330
54	8.910	8.907	11.30	2.82	11.364	12.2770	0.5581
60	8.100	8.095	11.00	2.60	8.864	15.0960	0.6862
66	7.290	7.280	10.70	2.39	7.214	17.6300	0.8014
72	6.480	6.461	10.40	2.20	6.064	19.7020	0.8956
78	5.670	5.668	10.10	2.01	5.264	21.1111	0.9596
84	4.860	4.853	9.80	1.84	4.614	21.9070	0.9958
90	4.050	4.050	9.50	1.67	4.080	22.0000	1.0000
96	3.240	3.240	9.20	1.52	3.480	22.0000	1.0000
102	2.430	2.430	8.90	1.38	2.789	22.0000	1.0000
108	1.620	1.620	8.60	1.24	1.991	22.0000	1.0000
114	0.810	0.810	8.30	1.12	1.069	22.0000	1.0000
120	0	0	8.00	1.00	0	22.0000	1.0000

of the member, as well as other quantities, such as g(x), f(x), M_{req}, M_e, etc., we prepare Tables 4.5 and 4.6.

The variation of the modulus function g(x) is shown in column 8 of Table 4.5. At the values of $0 \leq x \leq 84$, the member behaves inelastically, while at values of $90 \leq x \leq 120$, the member remains elastic and its modulus of elasticity is constant and equal to $E_1 = 22 \times 10^6$ psi(151.70×10^6 kPa). Consequently, g(x) = 1.0, and I_B is the moment of inertia I_B at the free end B, which is equal to 256 in.^{4}(106.555×10^{-6} m^4). Thus, at the free end B, we have f(x) = 1.0. The required moment M_{req} is given in column 3 of Table 4.5, and column 5 of Table 4.6 provides the values of the bending moment $M_e = M_x/f(x)g(x)$ of the equivalent system of constant stiffness E_1I_B.

For illustration purposes, we verify the results obtained by using the cross section at the fixed end of the tapered cantilever beam. At this section, the value obtained for Δ from the trail-and-error procedure is 49.563×10^{-3}, or, because of symmetry, $\varepsilon_1 = \varepsilon_2 = \Delta/2 = 24.7815 \times 10^{-3}$. The tri-linear stress–strain curve approximation is shown in Fig. 4.8a, and the values of the stresses and the strain corresponding to the cross section at the fixed end where x = 0, are shown in Fig. 4.8b. Because of symmetry, only the right-hand side is shown.

Table 4.6. Values of f(x), g(x), $M_{req} = M_x$, and M_e along the length of the member (1 in. = 0.0254 m, 1in. lb = 0.113 N m)

(1) x (in.)	(2) f(x)	(3) g(x)	(4) $M_{req} = M_x \times 10^6$ (in. lb)	(5) $M_e \times 10^6$ (in. lb)
0	5.36	0.1516	16.199	19.939
6	5.02	0.1587	15.388	19.306
12	4.70	0.1681	14.578	18.457
18	4.39	0.1809	13.770	17.332
24	4.10	0.1994	12.959	15.868
30	3.81	0.2268	12.148	14.041
36	3.55	0.2683	11.340	11.916
42	3.29	0.3344	10.529	9.567
48	3.05	0.4330	9.719	7.362
54	2.82	0.5581	8.907	5.664
60	2.60	0.6862	8.095	4.538
66	2.39	0.8014	7.280	3.797
72	2.20	0.8956	6.461	3.284
78	2.01	0.9596	5.668	2.935
84	1.84	0.9958	4.853	2.651
90	1.67	1.0000	4.050	2.419
96	1.52	1.0000	3.240	2.130
102	1.38	1.0000	2.430	1.765
108	1.24	1.0000	1.620	1.304
114	1.12	1.0000	0.810	0.725
120	1.00	1.0000	0	0

From Fig. 4.8b, by taking the first moment of the areas between ε_1 and 0, about the origin 0 and multiplying by 2 because of symmetry, we evaluate the following integral:

$$
\begin{aligned}
\int_{\varepsilon_2}^{\varepsilon_1} \sigma\varepsilon \, d\varepsilon = 2 \Bigg[& \frac{1}{2}(48)(10)^3(2.1818)(10)^{-3}(1.4545)(10)^{-3} \\
& + (48)(10)^3(21.8)(10)^{-3}(13.0818)(10)^{-3} \\
& + \frac{1}{2}(21.8)(10)^{-3}(11)(10)^3(16.7151)(10)^{-3} \\
& + (0.7995)(10)^{-3}(59)(10)^3(24.3817)(10)^{-3} \\
& + \frac{1}{2}(0.7797)(10)^{-3}(0.09996)(10)^3(24.5149)(10)^{-3} \Bigg] \\
= {} & 33,842.8854\,(10)^{-3}
\end{aligned}
$$

Thus, from Eq. (4.16), we find

$$
\begin{aligned}
E_r &= \frac{12}{\left[(49.563)(10)^{-3}\right]^3}(33,842.8854)(10)^{-3} \\
&= 3.3356\,(10)^6 \text{ psi } \left(22.999 \times 10^6 \text{ kPa}\right)
\end{aligned}
$$

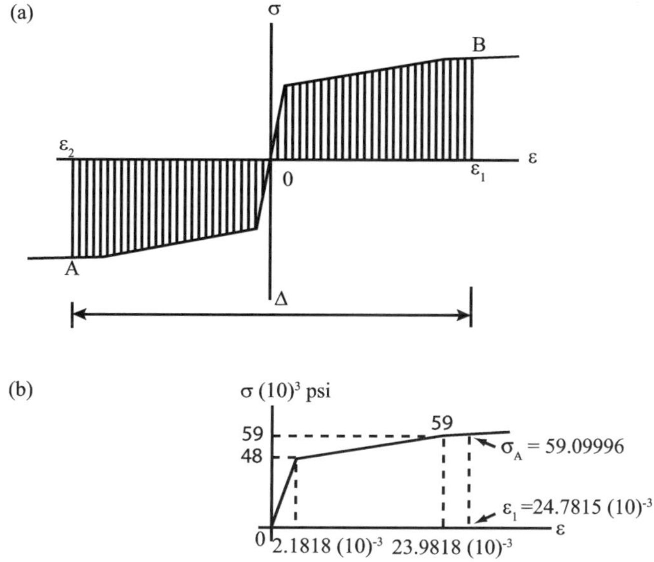

Fig. 4.8. (**a**) Tri-linear stress–strain curve approximation of monel. (**b**) Areas under the stress–strain curve at the fixed end of the member

This value of E_r is identical to the one shown in the seventh column of Table 4.5 for $x = 0$.

From Eq. (4.10), we find

$$r = \frac{h}{\Delta} = \frac{14}{(49.563)\,(10)^{-3}}$$
$$= 282.4677\,\text{in.}\ (7.1747\,\text{m})$$

Therefore, from Eq. (4.15), we find that the required bending moment M_{req} is

$$M_{req} = \frac{E_r I_x}{r}$$
$$= \frac{3.3356\,(10)^6}{282.4677}\left[\frac{(6)\,(14)^3}{12}\right]$$
$$= 16.2016\,(10)^6\ \text{in. lb}\ \left(1.8308\,(10)^6\ \text{N m}\right)$$

This is identical to the value of M_{req} shown in the third column of Table 4.5 for $x = 0$, and also identical to the value of M_x, at $x = 0$, in the second column of this table, which is obtained using static analysis.

The reader may verify the data given in Tables 4.5 and 4.6 by investigating additional cross sections along the length of the structural member.

From Table 4.6, since f(x), and $M_{req} = M_x$, are known, we can determine the bending moment M_e of the equivalent system of constant stiffness $E_1 I_B$

at the indicated values of x, by applying Eq. (4.22). These values of M_e are shown in column 5 of Table 4.6. Note that

$$I_1 = I_B = \frac{(6)(8)^3}{12} = 256\,\text{in.}^4 \left(107 \times 10^{-6}\,\text{m}^4\right)$$

By using the values of M_e in the last column of Table 4.6, we plot the bending moment diagram M_e of the equivalent system of constant stiffness E_1I_B, shown by the solid line in Fig. 4.9a. The approximation of its shape with three straight-line segments leads to the constant stiffness equivalent system shown in Fig. 4.9b.

The deflections and rotations at any point along the length of the equivalent system are closely identical to the corresponding ones of the initial system in Fig. 4.7. As an example, we will use the equivalent system in Fig. 4.9b and we will apply the moment–area method to determine the vertical deflection δ_B and rotation θ_B at the free end B. The deflection δ_B is equal to the first moment of the approximated M_e area between points A and B in Fig. 4.9a, divided by E_1I_B, and then taken about point B. That is

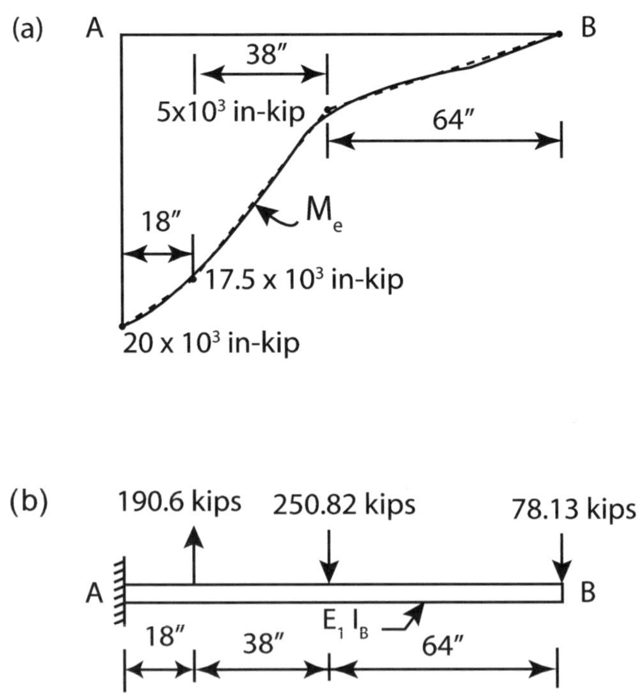

Fig. 4.9. (**a**) Approximation of the M_e diagram with three straight-line segments. (**b**) Equivalent system of constant stiffness E_1I_B (1 in. = 0.0254 m, 1 in. kip = 113 N m, 1 kip = 4.448 kN)

$$\delta_B = \frac{1}{E_1 I_B}\left[\frac{1}{2}(64)(5)(10)^3\left(\frac{2}{3}\right)(64) + (56)(5)(10)^3(92)\right.$$
$$+\frac{1}{2}(2.5)(10)^3(38)(89.333)$$
$$\left. + (18)(12.5)(10)^3(111) + \frac{1}{2}(18)(2.5)(10)^3(114)\right]$$
$$= \frac{1}{E_1 I_B}[6{,}826.6667 + 25{,}760.0000 + 21{,}216.6588 + 24{,}975.0000$$
$$+2{,}565.0000](10)^3$$
$$= \frac{1}{E_1 I_B}[81{,}343.3255](10)^3$$

or, by substituting for E_1 and I_B, we find

$$\delta_B = \frac{1}{(22)(10)^3(256)}[81,343.3255](10)^3$$
$$= 14.44\,\text{in.}\ (0.3669\,\text{m})$$

Proceeding with the moment–area method, the rotation θ_B at the free end B, is equal to the M_e/E_1I_B area between points A and B in Fig. 4.9a. By using the approximated M_e area, we find

$$\theta_B = \frac{1}{E_1 I_B}\left[\frac{1}{2}(64)(5)(10)^3 + (56)(5)(10)^3 + \frac{1}{2}(12.5)(10)^3(38)\right.$$
$$\left. + (18)(12.5)(10)^3 + \frac{1}{2}(18)(2.5)(10)^3\right]$$
$$= \frac{1}{E_1 I_B}[160.00 + 280.00 + 237.50 + 225.00 + 22.50](10)^3$$
$$= \frac{1}{E_1 I_B}[925.00](10)^3$$

By substituting for $E_1 I_B$, we find

$$\theta_B = \frac{1}{(22)(10)^3(256)}[925.00](10)^3$$
$$= 0.164240\,\text{rad}$$
$$= 9.41^\circ$$

The inelastic analysis of the variable stiffness cantilever beam is again performed here by varying the concentrated load Q, and considering the beam cases with taper $n = 1.25, 1.5, 1.75$, and 2.0. The three-line and the six-line approximations of the stress–strain curve of monel were used for this purpose. Figure 4.10 shows a comparison of the results obtained by using the three-line

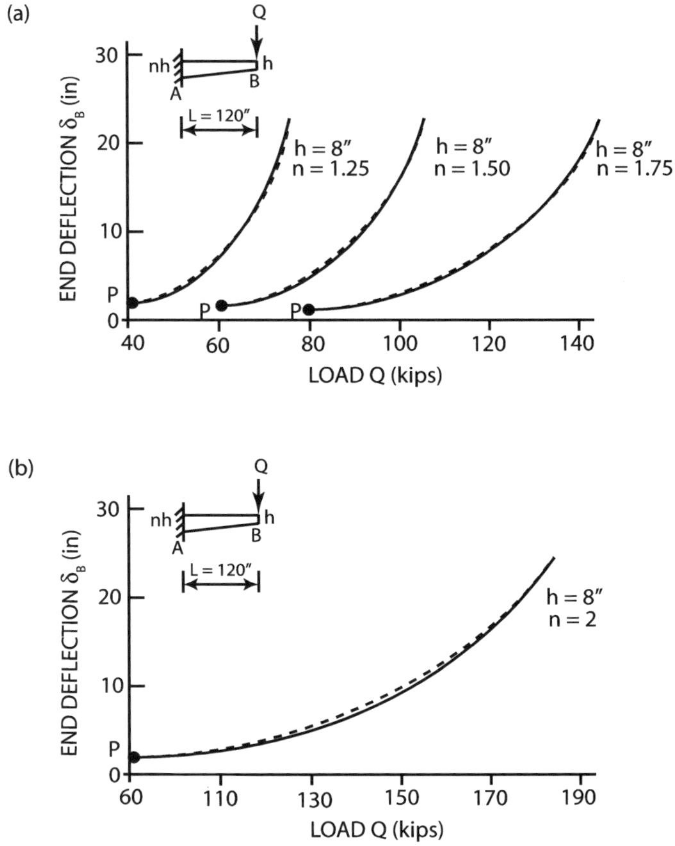

Fig. 4.10. (**a**) Comparison of the results for the three-line and the six-line approximations of the stress–strain curve of monel with n = 1.25, 1.50, and 1.75. (**b**) Comparison of the results for the three-line and the six-line approximations of the stress–strain curve of monel with n = 2.0(1 in. = 0.0254 m, 1 kip = 4.448 kN)

approximation of the stress–strain curve of monel and the six-line approximation of the same curve. The three-line approximation is represented by the dashed line, while the solid line is used for the six-line approximation. For all practical purposes, the results for both cases are considered to be identical. This shows that for monel, a three-line approximation of its stress–strain curve should be sufficiently accurate for practical analysis.

The two-line approximation of the stress–strain curve of monel, as discussed earlier, also provides reasonable results for practical applications.

Figure 4.11 shows the variation of the modulus function g(x) along the length of the tapered cantilever beam for various values of the vertical concentrated load Q and with taper n = 2. We note that for Q = 90 kips (400.32 kN)

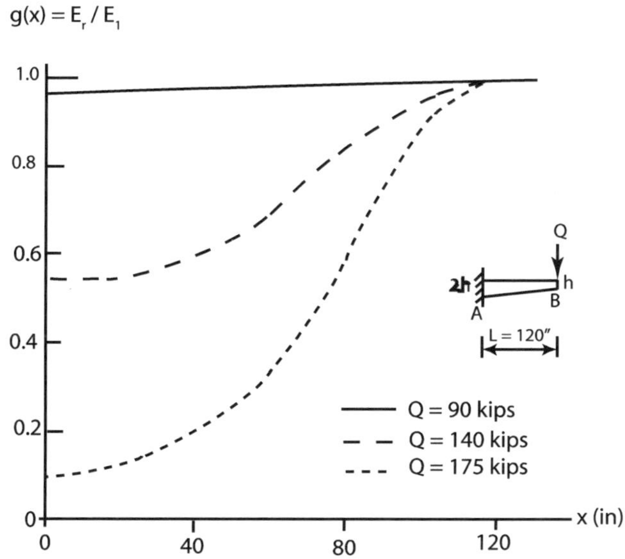

Fig. 4.11. Curves representing the variation of the function g(x) with increasing Q and n = 2.0(1 in. = 0.0254 m, 1 kip = 4.448 kN)

the function g(x) is nearly constant, but for Q = 175 kips (778.40 kN), the member becomes very inelastic. Note also that g(x) is plotted with the elastic modulus E_1 as the reference point.

4.3 Inelastic Analysis of Simply Supported Beams

The inelastic analysis of simply supported beams is investigated in this section. The procedure is illustrated in the following example by considering a simply supported beam of variable cross section along its length. In this case, both functions f(x) and g(x) will be variable, a situation that makes the problem more general. However, by using equivalent systems of constant stiffness EI, such variations in f(x) are easily incorporated in the solution of the problem, as observed in preceding chapters of this text.

Example 4.4 The tapered simply supported beam in Fig. 4.12a, is loaded by a uniformly distributed load $w = 2,500\,\text{lb in.}^{-1}(437,817\,\text{N m}^{-1})$ as shown. At support B the depth $h_B = 8\,\text{in.}(0.2032\,\text{m})$, and it is 12 in. (0.3048 m), at support A. The width b is constant and equal to 6 in. (0.1524 m), and the length L = 120 in. (3.048 m). The material of the member is mild steel, and its stress–strain curve is as shown by the solid line in Fig. 4.12b. By using the four-line approximation of the stress–strain curve, perform an inelastic analysis to determine the quantities $E_r, g(x), M_{req} = M_x$, and $M_e = M_x/f(x)g(x)$, at

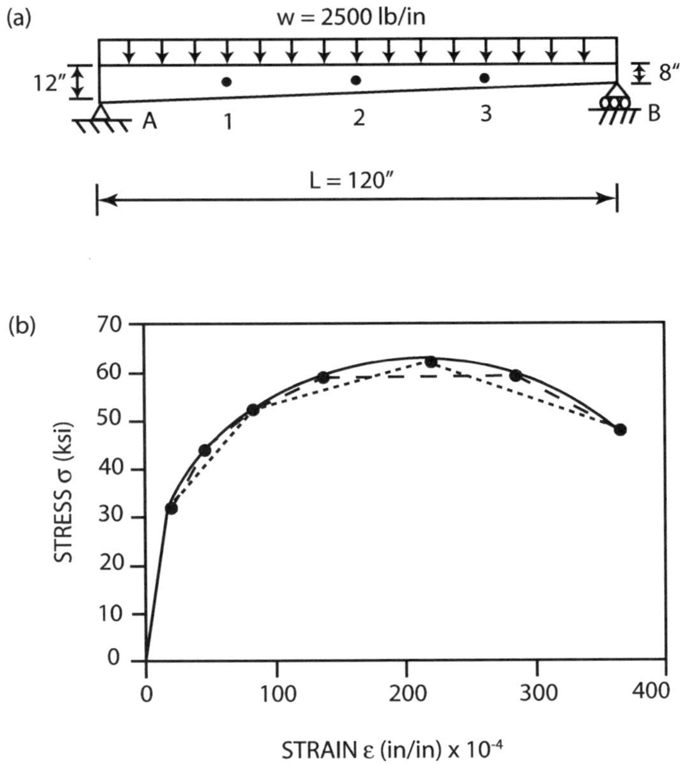

Fig. 4.12. (**a**) Tapered simply supported beam loaded with a uniformly distributed load w. (**b**) Stress–strain curve of the mild steel with its shape approximated by four and six straight lines (1 in. = 0.0254 m, 1 lb in.$^{-1}$ = 175.1268 N m^{-1}, 1 ksi = 6,895 kPa)

points along the length of the member. Also, by using an equivalent system of constant stiffness EI_B, where E is the elastic modulus and I_B is the moment of inertia at the end B, determine the vertical deflections and rotations along the length of the member.

Solution: The stress–strain curve of the mild steel is approximated with four and six straight-line segments, as shown by the dashed lines in Fig. 4.12b. For each stress–strain approximation, the corresponding values of the modulus E and the bending stress σ are shown in Table 4.7.

The inelastic analysis discussed in the preceding section is again applied here to solve the problem in Fig. 4.12a. In this case, however, since the depth is variable, the moment of inertia function f(x) will vary along the length of the member. The moment of inertia $I_B = 256.0\,\text{in.}^4 (106.555 \times 10^{-6}\,\text{m}^4)$ is taken as the reference value. By using the four-line approximation of the stress–strain curve of the mild steel and performing the inelastic analysis for the member,

Table 4.7. Values of E and σ for a four-line and a six-line approximation of the stress–strain curve of the mild steel (1 psi = 6.895 kPa)

(1)	(2) four-line approximation	(3) six-line approximation
modulus E (psi)		
E_1	94.366×10^5	94.366×10^5
E_2	28.289×10^4	40.351×10^4
E_3	3.785×10^4	21.053×10^4
E_4	-7.971×10^4	3.785×10^4
E_5		-1.942×10^4
		-11.561×10^4
stress σ (psi)		
σ_1	33.5×10^3	33.5×10^3
σ_2	55.0×10^3	45.0×10^3
σ_3	61.0×10^3	55.0×10^3
σ_4		61.0×10^3
σ_5		60.0×10^3

we find that the values of $f(x), \Delta, E_r, g(x), M_{req} = M_x$, and $M_e = M_x/f(x)g(x)$ are shown in Table 4.8. The values of the moment M_e of the equivalent system of constant stiffness E_1I_B, where $E_1 = 94.366 \times 10^6$ psi(650.6306×10^5 kPa) and $I_B = 256.0\,\text{in.}^4(106.555 \times 10^{-6}\,\text{m}^4)$ are shown in column 8 of the same table.

The results in Table 4.8 were obtained with computer assistance at intervals of 6 in. (0.1524 m) starting from support A where $x = 0$. The procedure is discussed in detail in the preceding section of the chapter. At each cross section, Δ is determined by a trial-and-error procedure. This is thoroughly discussed in Example 4.1. The calculated values of Δ at each cross section of the member are shown in column 4 of Table 4.8.

With known Δ, we can use Eq. (4.16) to calculate the reduced modulus E_r, then Eq. (4.18) to calculate $g(x)$, and so on. We will check and verify the results for the beam cross section located at $x = 60$ in. (1.524 m). At this cross section, we note from Table 4.8 that $\Delta = 12.25 \times 10^{-3}$. Then, because of symmetry, we have $\varepsilon_1 = \varepsilon_2 = \Delta/2 = 6.125 \times 10^{-3}$.

Next, by using the four-line approximation of the stress–strain curve of the mild steel in Fig. 4.12b, we plot the stress–strain curve at $x = 60$ in. (1.524 m) shown in Fig. 4.13. Because of symmetry we plot only the right-hand side of the diagram. Then we calculate the following integral by taking the first moment of the area $0A\varepsilon_1$ about point 0, having first calculated the bending stress σ_A corresponding to the strain ε_1. On this basis, we find

$$\begin{aligned}\sigma_A &= 33.5\,(10)^3 + \left[(6.125 - 3.55)\,(10)^{-3}\,(0.28289)\,(10)^6\right] \\ &= [33.5 + 0.7284]\,(10)^3 \\ &= 34.2284\,(10)^3 \text{ psi } \left(236.0048 \times 10^3 \text{ kPa}\right)\end{aligned}$$

Table 4.8. Values of f(x), E_r, g(x), M_{req}, and M_e for inelastic analysis (1 in. = 0.0254 m, 1 in. lb = 0.1130 N m, 1 psi = 6.895 kPa)

(1) x (in.)	(2) h_x (in.)	(3) f(x)	(4) $\Delta \times 10^{-3}$ (in. in.$^{-1}$)	(5) $E_r \times 10^6$ (psi)	(6) g(x)	(7) $(M_{req} = M_x)(10)^6$ (in. lb)	(8) $M_e = M_x/f(x)g(x)$ (in. lb)
0	12.00	3.38	0	9.4366	1.0000	0	0
6	11.80	3.21	1.3014	9.4366	1.0000	−0.8550	$-0.2664(10)^6$
12	11.60	3.05	2.5516	9.4366	1.0000	−1.6200	$-0.5314(10)^6$
18	11.40	2.89	3.7427	9.4366	1.0000	−2.2950	$-0.7931(10)^6$
24	11.20	2.74	4.8660	9.4366	1.0000	−2.8800	$-1.0496(10)^6$
30	11.00	2.60	5.9116	9.4366	1.0000	−3.3750	$-1.2983(10)^6$
36	10.80	2.46	6.8685	9.4366	1.0000	−3.7800	$-1.5364(10)^6$
42	10.60	2.33	7.8000	9.3293	0.9886	−4.0950	$-1.7776(10)^6$
48	10.40	2.20	8.9999	8.8677	0.9397	−4.3200	$-2.0906(10)^6$
54	10.20	2.07	10.5000	8.1524	0.8639	−4.4550	$-3.3982(10)^6$
60	10.00	1.95	12.2500	7.3499	0.7789	−4.5000	$-2.9593(10)^6$
66	9.80	1.84	13.8000	6.7239	0.7125	−4.4550	$-2.4018(10)^6$
72	9.60	1.73	14.4500	6.4865	0.6874	−4.3200	$-3.6362(10)^6$
78	9.40	1.62	13.7500	6.7428	0.7145	−4.0950	$-3.5337(10)^6$
84	9.20	1.52	11.9000	7.5030	0.7951	−3.7800	$-3.1247(10)^6$
90	9.00	1.42	9.8500	8.4660	0.8971	−3.3750	$-2.6439(10)^6$
96	8.80	1.33	8.0500	9.2529	0.9805	−2.8800	$-2.2099(10)^6$
102	8.60	1.24	6.5766	9.4366	1.0000	−2.2950	$-1.8474(10)^6$
108	8.40	1.16	4.8660	9.4366	1.0000	−1.620	$-1.3994(10)^6$
114	8.20	1.08	2.6950	9.4366	1.0000	−0.8550	$0.7940(10)^6$
120	8.00	1.00	0	9.4366	1.0000	0	0

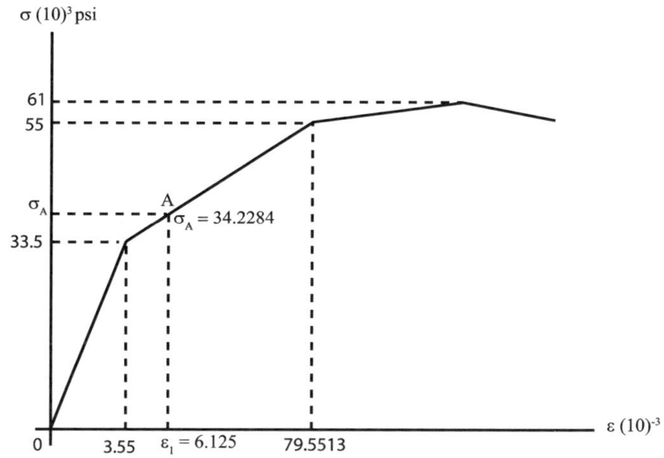

Fig. 4.13. Stress–strain diagram at x = 60 in. using the four-line approximation of the mild steel (1 in. = 0.0254 m, 1 psi = 6.895 kPa)

$$\int_{\varepsilon_2}^{\varepsilon_1} \sigma\varepsilon \, d\varepsilon = 2\left\{\frac{1}{2}(33.5)(10)^3(3.55)(10)^{-3}(2.3667)(10)^{-3} + (33.5)(10)^3(2.575)(10)^{-3}(4.8375)(10)^{-3} + \frac{1}{2}(2.575)(10)^{-3}(0.7284)(10)^3(5.2667)(10)^{-3}\right\}$$
$$= 2\{140.7299 + 417.2948 + 4.9392\}(10)^{-3}$$
$$= 1,125.9278(10)^{-3}$$

Thus,

$$E_r = \frac{12}{\Delta^3}\int_{\varepsilon_2}^{\varepsilon_1} \sigma\varepsilon \, d\varepsilon = \frac{12}{(0.01225)^3}(1,125.9278)(10)^{-3}$$
$$= 7.3499(10)^6 \text{ psi } \left(50.6776 \times 10^6 \text{ kPa}\right)$$

This value of E_r checks the value shown at x = 60 in. (1.524 m) in column 5 of Table 4.8.

From Eq. (4.10b), we find

$$r = \frac{h}{\Delta} = \frac{10}{(12.25)(10)^{-3}} = 816.3265 \text{ in. } (20.7347 \text{ m})$$

and from Eq. (4.15), we find that the value of the required moment M_{req} is

$$M_{req} = \frac{E_r I_x}{r} = \frac{(7.3499)(10)^6\left[\frac{(6)(10)^3}{12}\right]}{816.32650}$$
$$= 4.50(10)^6 \text{ in. lb } \left(0.5087 \times 10^6 \text{ N m}\right)$$

which is identical to the value given in column 7 of Table 4.8 at x = 60 in. (1.524 m).

The value of the modulus function g(x) can be determined by using Eq. (4.18), which yields

$$g(x) = \frac{E_r}{E} = \frac{7.3499\,(10)^6}{9.4366\,(10)^6}$$
$$= 0.7789$$

This value is identical to the one shown in the table for x = 60 in. (1.524 m).

The values of the moment diagram $M_e = M_x/f(x)g(x)$ of the equivalent system of constant stiffness E_1I_B, are shown in the last column of Table 4.8. The M_e diagram is shown plotted by the solid line in Fig. 4.14a. By approximating the shape of M_e with four straight lines, as shown by the dashed lines in the same figure, and applying simple statics, as in preceding sections and

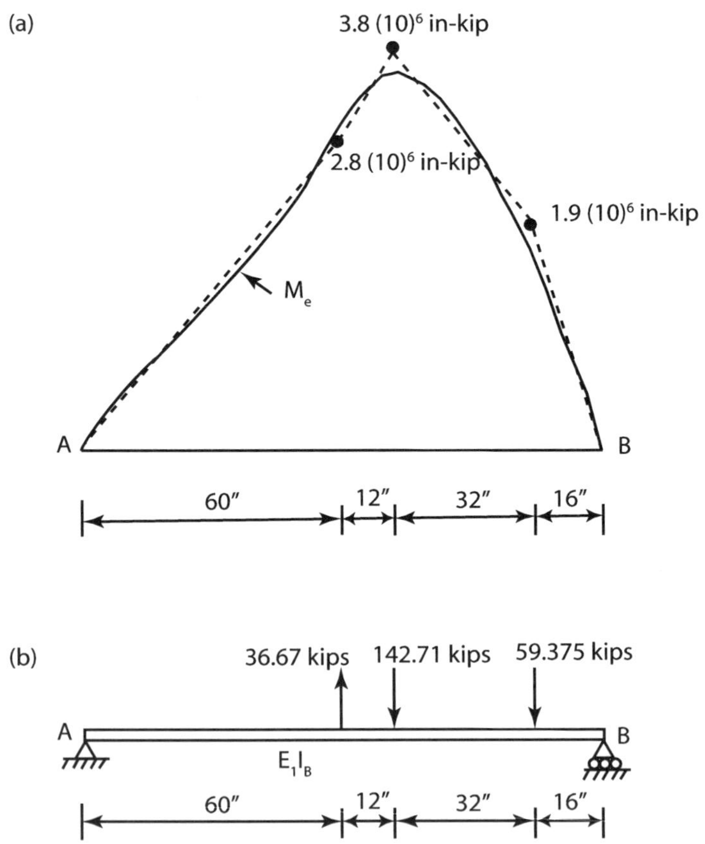

Fig. 4.14. (**a**) M_e diagram with its shape approximated with four straight lines. (**b**) Equivalent system of constant stiffness E_1I_B(1 in. = 0.0254 m, 1 kip = 4.448 kN, 1 in. kip = 112.98 N m)

Table 4.9. Values of rotation, deflection and M_e, at cross-section along the length of the member (1 in. = 0.0254m, 1 in. kip = 113.0 N m)

(1) x (in.)	(2) $M_e(10)^6$ (in. lb)	(3) Rotation (rad)	(4) Deflection (in.)
0	0	0.040458	0
6	−0.2664	0.040127	0.2420
12	−0.5314	0.039136	0.4800
18	−0.7931	0.037491	0.7102
24	−1.0496	0.035201	0.9285
30	−1.2983	0.032284	1.1312
36	−1.5364	0.028762	1.3145
42	−1.7776	0.024658	1.4750
48	−2.0906	0.019856	1.6088
54	−2.4868	0.014174	1.7113
60	−2.9593	0.007421	1.7765
66	−3.4018	−0.000483	1.7977
72	−3.6362	−0.009276	1.7687
78	−3.5337	−0.018244	1.6860
84	−3.1247	−0.026539	1.5513
90	−2.6439	−0.033698	1.3701
96	−2.2099	−0.039712	1.1495
102	−1.8474	−0.044748	0.8958
108	−1.3994	−0.048802	0.6147
114	−0.7940	−0.051553	0.3131
120	0	−0.052571	0

chapters, we obtain the equivalent system of constant stiffness E_1I_B shown in Fig. 4.14b.

The deflections and rotations produced by the equivalent system in Fig. 4.14b, are closely identical to the ones at corresponding points of the original system in Fig. 4.12a. By using the equivalent system and computer assistance, the rotations and vertical deflections at various points along the length of the member were determined and they are shown in Table 4.9. In the fourth column of this table we note that the maximum vertical deflection is 1.7977 in. (0.0457 m) and occurs at x = 66 in. (1.6764 m). From column 3, we note that the maximum rotation occurs at the right support B, where x = 120 in. (3.048 m), and it is equal to −0.052571 radians, or 3.0121°.

The inelastic analysis for the nonprismatic member in Fig. 4.12a is repeated here by varying the magnitude of the uniformly distributed load w. The curves (1), (2), and (3), in Fig. 4.15, illustrate the variation of the vertical deflection at the quarter length points 1, 2, and 3, respectively, with increasing loading w. Sharp increases in deflection are observed for values of $w > 3{,}000\,\text{lb in.}^{-1}$ $(525{,}380.4\,\text{N m}^{-1})$, which indicates that the ultimate capacity of the member to resist stress and deformation is reached.

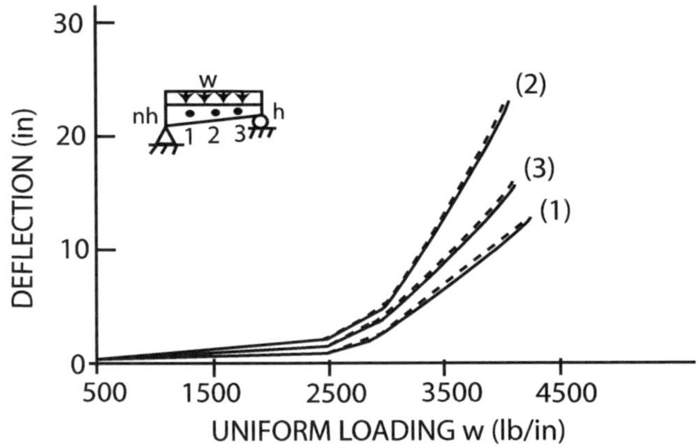

Fig. 4.15. Load–deflection curves for quarter points 1, 2, and 3, in Fig. 4.12a (1 in. = 0.0254 m, 1 lb in.$^{-1}$ = 175.1268 N m^{-1})

The dashed lines in Fig. 4.15, represent the results obtained by using the six-line approximation of the stress–strain curve, and the solid lines illustrate the results obtained by using the four-line approximation of the stress–strain curve. For practical purposes, these results are in close agreement. More about ultimate loads are given later in this chapter and in [2, 3, 84].

4.4 Ultimate Design Loads Using Inelastic Analysis

In this section we will say a few words about ultimate design loads which are very valuable to the practicing design engineer. The design engineer needs to know how much load a structure, or a structural component, can carry before complete failure occurs. This is particularly useful for structures that are subjected to earthquakes, nuclear explosion, heavy loads, and to many other extreme conditions. For such cases, it becomes economically prohibitive to design the structures so that the stresses remain below the elastic limit of its material. If, however, we have to permit the structure to be stressed beyond its elastic limit, then we need to know how far into the inelastic zone should be permitted to go and still be safe. Many of these answers can be answered by following the methodologies discussed in this chapter, in Chapt. 6, and in [2, 3, 5, 15, 84].

The analysis in this section is concentrated in the computation of ultimate loads for structural components. This is a very useful information for the structural design engineer, because once he knows what the ultimate capacity of a structure to resist a given type of load is, then he/she will be able to decide how much load should be permitted to act on the structure and still be safe.

The methodology is illustrated here by determining ultimate loads for single span beams subjected to a uniformly distributed load w over their entire span length. The cross section of the member is assumed to be rectangular, and it can vary along its length. The theory, however, is general, and it can be applied to other types of cross-sectional shapes and load variations. The types of materials considered here are monel, mild steel, and aluminum. The method of the equivalent systems and the reduced modulus E_r concept are used here to carry out the analytical work.

Both methodologies are discussed in detail in the preceding sections of this chapter. The concept of the reduced modulus was used to determine the variation of the modulus E along the length of the member, since the member is permitted to be stressed beyond the elastic limit of its material. Once this is known, then the method of the equivalent systems is used to obtain an equivalent system of constant stiffness that can be used to determine deflections and rotations along the length of the member. In this manner, the equivalent system becomes pseudolinear, and linear methods of analysis can be used to determine deflections and rotations.

Based on this methodology, we can vary the applied load, say the distributed load w, and then determine the variation in the maximum deflection of the member by using a load deflection curve. The ultimate load is reached when we reach a point where even small increases of the load produce large increases in maximum deflection. In other words, when the shape of the load–deflection curve changes abruptly and gets very steep. This procedure provides very reasonable results for practical applications. For convenience, a computer program can be prepared to facilitate the procedure. Such a computer program can be found in Appendix D of [84].

For the computation of the reduced modulus E_r using Eq. (4.16), the shape of the stress–strain curve, for each material, can be approximated with two, three, or more straight-line segments, as discussed in the preceding sections. In the work that follows, the straight-line approximations shown in Fig. 4.16 are used. Figure 4.16a represents the commonly used elastic–perfectly plastic case, and the other two cases in the figure are improvements.

The following examples and beam cases illustrate the application of the methodology.

Example 4.5 By using the tapered cantilever beam in Fig. 4.17b, loaded by a uniformly distributed load w as shown, determine in each case the ultimate load w_u of the member by using various beam length. The material of the member is monel. Use the four-line approximation of the stress–strain curve of monel represented by the solid lines in Fig. 4.16c. The width b = 6 in. (0.1524 m), the depth h_1 = 12 in.(0.3048 m), and the depth h_2 = 8 in.(0.2032 m).

Solution: The variation of the cross-sectional moment of inertia I_x, measured from the fixed support of the cantilever beam is given by the expression

$$I_x = I_B f(x) \tag{4.23}$$

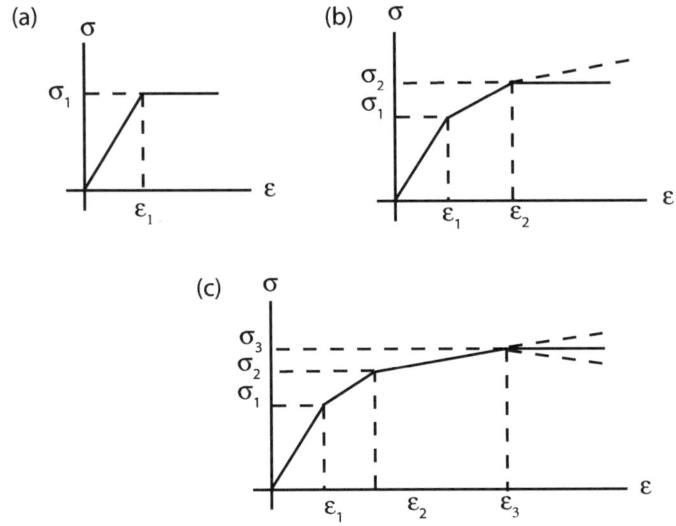

Fig. 4.16. (**a**) Two-line approximation of the stress–strain curve. (**b**) Three-line approximation of the stress–strain curve. (**c**) Four-line approximation of the stress–strain curve

where

$$I_B = \frac{bh_2^3}{12} = \frac{(6)\,(8)^3}{12} = 256\,\text{in.}^4\left(107 \times 10^{-6}\,\text{m}^4\right) \tag{4.24}$$

$$f\,(x) = \left[\frac{(n-1)\,(L-x)+L}{L}\right]^3 \tag{4.25}$$

or, since the taper $n = 1.5$

$$f\,(x) = \left[\frac{1.5\,L - 0.5\,x}{L}\right]^3 \tag{4.26}$$

For monel, the values of the stresses and the strains which correspond to the stress–strain approximation shown in Fig. 4.16c are $\sigma_1 = 48\,\text{ksi}(330,960\,\text{kPa})$, $\sigma_2 = 59\,\text{ksi}(406,805\,\text{kPa})$, $\sigma_3 = 75\,\text{ksi}(517,125\,\text{kPa})$, $\varepsilon_1 = 2.1818 \times 10^{-3}$, $\varepsilon_2 = 24.007 \times 10^{-3}$, and $\varepsilon_3 = 152.007 \times 10^{-3}$.

For the tapered cantilever beam under consideration, the ultimate load w_u for each case is determined by using the vertical deflection y at the free end B and plotting the load w vs. the deflection y curve. As stated earlier, we used the equivalent systems method in combination with the reduced modulus E_r idea to do this. This is amply illustrated in the preceding sections of this chapter.

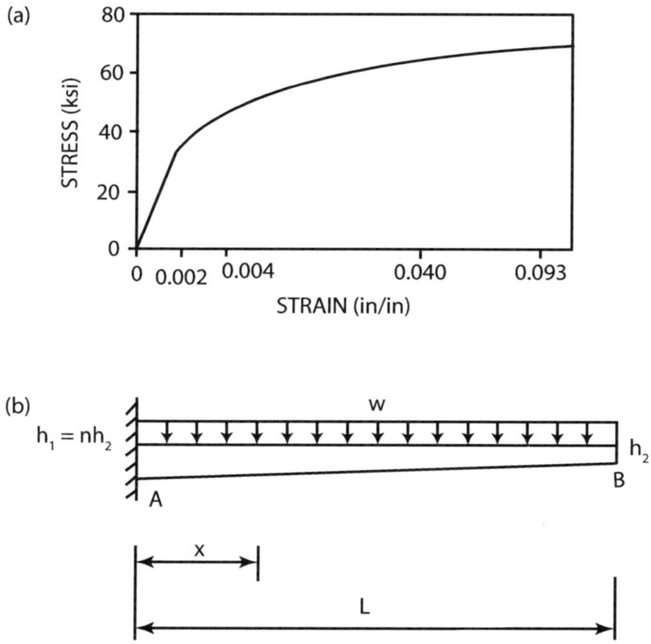

Fig. 4.17. (**a**) Stress–strain curve of monel. (**b**) Tapered cantilever beam loaded with a uniformly distributed load w (1 ksi = 6,895 kPa)

At the ultimate load w_u, the y vs. w curve becomes practically flat, and further increase of the load w cannot be practically materialized. This condition indicates that the ultimate load capacity of the member is reached.

For the tapered cantilever beam in Fig. 4.17b, the lengths $L = 30, 40, 50, 60, 70, 80, 90, 100, 120$, and 140 in. (1 in. = 0.0254 m), were used to perform the inelastic analysis and determine in each case the ultimate load w_u. The y vs. w curves are shown in Fig. 4.18, where y is the deflection at the free end, and w is the uniformly distributed load.

The calculated ultimate load w_u for each length case is shown in the fifth column of Table 4.10. We notice here that for L = 30 in. (0.762 m), the ultimate load $w_u = 34.97$ kips in.$^{-1}$(6,124.18 kN m^{-1}), and it reduces to 1.51 kips in.$^{-1}$(264.44 kN m^{-1}) when L = 140 in. (3.556 m).

In the computation of the ultimate load w_u, the calculated value of w_u for each case of beam length will be more accurate if the stress–strain curve of the material is approximated with more straight-line segments. For example, if we assume that L = 90 in. (2.286 m) and perform the inelastic analysis using the bi-linear stress–strain curve approximation shown in Fig. 4.16a, we find that $w_u = 2.54$ kips in.$^{-1}$(444.82 kN m^{-1}). However, if we use the approximation given in Fig. 4.16b, we find that $w_u = 3.11$ kips in.$^{-1}$(544.64 kN m^{-1}), and we

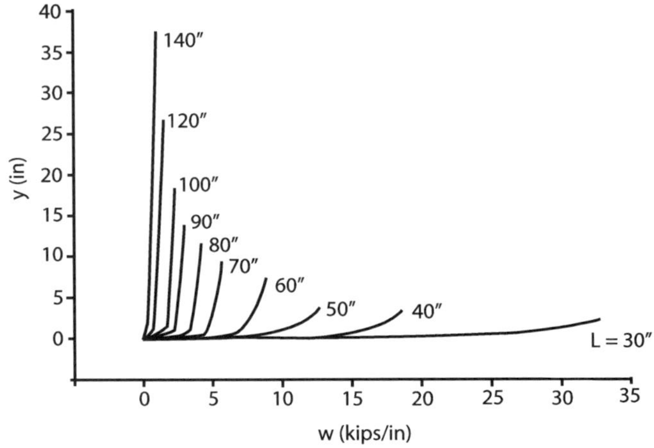

Fig. 4.18. Curves of y vs. w for a tapered cantilever beam of various lengths using the four-line approximation of the stress–strain curve of monel represented by the solid lines in Fig. 4.16c (1 in. = 0.0254 m, 1 kip in.$^{-1}$ = 175.1268 kN m^{-1})

Table 4.10. Ultimate loads w_u for tapered cantilever beam of various lengths L using the straight-line approximation of monel shown by the solid lines in Fig. 4.16c (1 in. = 0.0254 m, 1 kip in.$^{-1}$ = 175.1268 kN m^{-1})

(1) length L (in.)	(2) width b (in.)	(3) h_1 (in.)	(4) h_2 (in.)	(5) ultimate load w_u (kips in.$^{-1}$)
30	6	12	8	34.97
40	6	12	8	19.67
50	6	12	8	12.59
60	6	12	8	8.74
70	6	12	8	6.42
80	6	12	8	4.91
90	6	12	8	3.87
100	6	12	8	3.11
120	6	12	8	2.10
140	6	12	8	1.51

will have w_u = 3.87 kips in.$^{-1}$(677.74 kN m^{-1}), when the approximation in Fig. 4.16c is used, thus verifying the validity of the preceding remark.

These are very important observations for the practicing engineer, because he/she is the one to decide how accurate he/she wants to be. The preceding results tell us that w_u is the lowest when the two-line (elastic–perfectly plastic) approximation is used, and this is very conservative. Many practical design engineers are basing their design on this approximation and they feel comfortable. They do not worry much if their structure is overdesigned. However, there are other types of structures, such as aircraft, where it is extremely

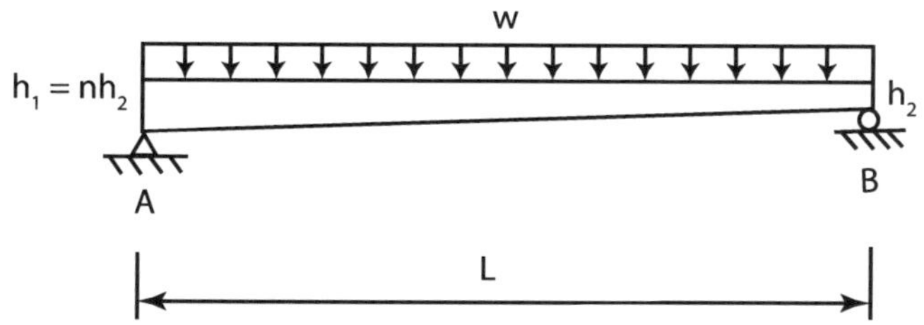

Fig. 4.19. Tapered simply supported beam loaded with a uniformly distributed load w

important to reduce weight and keep them light. In such cases the design engineer needs to have more accurate values in order to make the correct and safe decisions.

Example 4.6 By using the tapered simply supported beam in Fig. 4.19 which is loaded with a uniformly distributed load w as shown, determine in each case the ultimate load w_u of the member by using various beam lengths. The material of the member is mild steel, and its stress–strain curve is shown by the solid line in Fig. 4.12. In the analysis, use the three-line approximation of the stress–strain curve of the mild steel which is shown by the solid lines in Fig. 4.16b. The width $b = 6$ in. (0.1524 m), depth $h_1 = 12$ in. (0.3048 m), and depth $h_2 = 8$ in. (0.2032 m).

Solution: For the three-line approximation of the mild steel shown by the solid lines in Fig. 4.16b, we have $\sigma_1 = 33.5$ ksi (5,866.75 kN m^{-1}) and $\sigma_2 = 54.1$ ksi(9,474.36 kN m^{-1}). The inelastic analysis was performed here using the reduced modulus E_r concept and the method of the equivalent systems. Ultimate load w_u were determined by using the length $L = 80, 100, 120, 160, 200, 240, 280$, and 320 in. (1 in. = 0.0254 m). For each length case, the graphs of the deflection y at the center of the member vs. the distributed load w are shown in Fig. 4.20.

The values of the ultimate load w_u for each length case, are shown in the fifth column of Table 4.11. This table shows that the value of w_u reduces from $w_u = 9.66$ kips in.$^{-1}$(1,691.72 kN m^{-1}) for $L = 80$ in. (2.032 m) to $w_u = 0.50$ kips in.$^{-1}$(87.56 kN m^{-1}) for $L = 320$ in. (8.128 m).

For purposes of comparison, the inelastic analysis was repeated by using the length $L = 100$ in. (2.54 m), width $b = 6$ in. (0.1524 m), and assuming that the depth h is constant along the length of the member and equal to 8 in. (0.2032 m). The three-line stress–strain curve approximation of the mild steel yielded $w_u = 4.033$ kips in.$^{-1}$(706.2864 kN m^{-1}). The analogous value for the tapered beam shown in Table 4.11 is $w_u = 6.18$ kips in.$^{-1}$(1,082.2836 kN m^{-1}). This shows that the taper in the member increased by 53.24 % its ultimate load capacity. This is one of the reasons why structural components often have a variable cross section along their length.

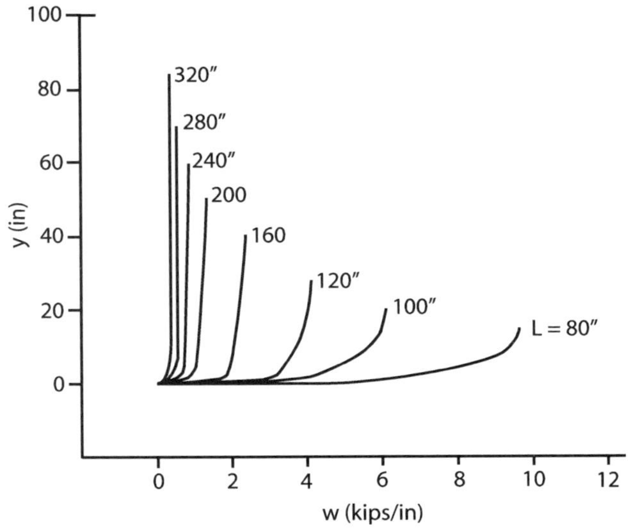

Fig. 4.20. Curves of y vs. w for a tapered simply supported beam of various lengths L, using the three-line approximation of the stress–strain curve of mild steel represented by the solid lines in Fig. 4.16b (1 in. = 0.0254 m, 1 kip in.$^{-1}$ = 175.1268 kN m^{-1})

Table 4.11. Ultimate Loads w_u for mild steel simply supported beam of various lengths L, using the straight-line approximation shown by the solid lines in Fig. 4.16b (1 in. = 0.0254 m, 1 kip in.$^{-1}$ = 175.1268 kN m^{-1})

(1) length L (in.)	(2) width b (in.)	(3) h_1 (in.)	(4) h_2 (in.)	(5) ultimate load w_u (kips in.$^{-1}$)
80	6	12	8	9.66
100	6	12	8	6.18
120	6	12	8	4.24
160	6	12	8	2.30
200	6	12	8	1.41
240	6	12	8	0.93
280	6	12	8	0.66
320	6	12	8	0.50

The preceding results were obtained by considering the deflection y at the center of the simply supported member. For uniform beams the maximum deflection occurs at the center. For tapered beams, the deflection at the center would be slightly less than the maximum one, but reasonable results are obtained for practical design applications. However, if desired, the location of the maximum deflection can easily be taken into account. The method of the equivalent systems can easily provide the location and value of such maximum deflections.

It was also noted that if we run again the inelastic analysis of the uniform simply supported beam of length L = 100 in. (2.54 m), b = 6 in. (0.1524 m), h = 8 in. (0.2032 m), and using the four-line approximation shown by the solid lines in Fig. 4.16c, we find $w_u = 4.088$ kips in.$^{-1}$(715.92 kN m^{-1}). We notice here that the three-line and the four-line approximations of the stress–strain curve of the mild steel give similar results, the difference being only 1.56 %. This suggests that for practical applications, the three-line approximation should be sufficient for the design engineer to make his/her decisions.

Example 4.7 Repeat the problem in Example 4.6 by assuming that the material is 7075-T651 aluminum alloy. The stress–strain curve of this aluminum is shown in Fig. 4.21. Use the three-line approximation of the stress–stain curve to obtain the ultimate loads for the chosen beam length.

Solution: The two-line, three-line, and six-line approximations of the stress–strain curve of the 7075-T651 aluminum alloy yield the values of E shown in Table 4.12 and the corresponding stresses σ are shown in Table 4.13. The three-line approximation of the stress–strain curve of the aluminum alloy is also shown in Fig. 4.16b, where $\sigma_1 = 59.94$ ksi(413, 286 kPa), $\sigma_2 = 79.67$ ksi(549, 325 kPa), $\varepsilon_1 = 6.3084(10)^{-3}$, and $\varepsilon_2 = 8.589(10)^{-3}$.

The analysis is based on the reduced modulus E_r and the method of the equivalent systems. The ultimate loads w_u were calculated for lengths L = 80, 100, 120, 160, 200, 240, 280, and 320 in. (1 in. = 0.0254 m), and the results are shown in the fifth column of Table 4.14. This table shows that w_u is progressively reduced from $w_u = 10.79$ kips in.$^{-1}$(1, 889.62 kN m^{-1}) for L = 80 in. (2.032 m), to $w_u = 0.68$ kips in.$^{-1}$(119.086 kN m^{-1}) for L = 320 in. (8.128 m).

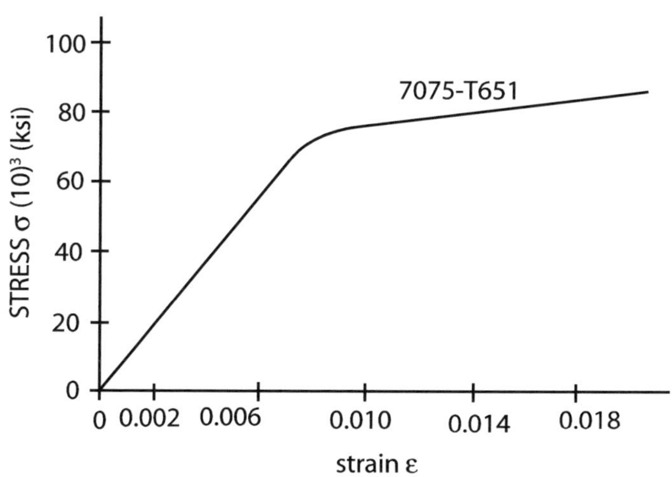

Fig. 4.21. Experimental stress–strain curve of the 7075-T651 aluminum alloy (1 ksi = 6, 895 kPa)

Table 4.12. Values of E for the two-line, three-line, and six-line approximations of the stress–strain curve of the 7075-T651 aluminum alloy (1 psi = 6.895 kPa)

(1) modulus E (psi)	(2) Two-line approximation	(3) Three-line approximation	(4) Six-line approximation
E_1	9.50×10^6	9.51×10^6	9.49×10^6
E_2	2.00×10^5	8.63×10^6	8.65×10^6
E_3	–	1.90×10^5	4.99×10^6
E_4	–	–	1.24×10^6
E_5	–	–	5.00×10^5
E_6	–	–	1.99×10^5

Table 4.13. Values of σ for the two-line, three-line, and six-line approximations of the stress–strain curve of the 7075-T651 aluminum alloy (1 psi = 6.895 kPa)

(1) stress σ (psi)	(2) two-line approximation	(3) three-line approximation	(4) six-line approximation
σ_1	79.79×10^3	59.94×10^3	56.94×10^3
σ_2	–	79.67×10^3	72.93×10^3
σ_3	–	–	77.92×10^3
σ_4	–	–	79.42×10^3
σ_5	–	–	79.92×10^3

Table 4.14. Ultimate loads w_u for tapered aluminum alloy simply supported beam of various lengths L, using the straight-line approximation shown by the solid lines in Fig. 4.16b (1 in. = 0.0254 m, 1 kip in.$^{-1}$ = 175.1268 kN m^{-1})

(1) length L (in.)	(2) width b (in.)	(3) h_1 (in.)	(4) h_2 (in.)	(5) ultimate load w_u (kips in.$^{-1}$)
80	6	12	8	10.79
100	6	12	8	6.91
120	6	12	8	4.80
160	6	12	8	2.70
200	6	12	8	1.73
240	6	12	8	1.20
280	6	12	8	0.88
320	6	12	8	0.68

If we assume that the member is a uniform cantilever beam with width b = 6 in. (0.1524 m) and depth h = 8 in. (0.2032 m), and perform the inelastic analysis for lengths L = 30, 40, 50, 60, 70, 80, 90, 100, 120, and 140 in. (1 in. = 0.0254 m) using the same as above three-line approximation of the aluminum stress–strain curve, the ultimate w_u load for each case would be as shown in the fourth column of Table 4.15.

If we compare some of these results with the analogous ones obtained in Table 4.14 for the tapered simply supported beam, we note that for the length

Table 4.15. Ultimate loads w_u for uniform aluminum cantilever beam of various lengths L, using the straight-line approximation of the stress–strain curve shown by the solid lines in Fig. 4.16b (1 in. = 0.0254 m, 1 kip in.$^{-1}$ = 175.1268 kN m^{-1})

(1) length L (in.)	(2) width b (in.)	(3) depth h (in.)	(4) ultimate load w_u (kips in.$^{-1}$)
30	6	8	16.87
40	6	8	9.48
50	6	8	6.06
60	6	8	4.22
70	6	8	3.09
80	6	8	2.37
90	6	8	1.87
100	6	8	1.52
120	6	8	1.05
140	6	8	0.77

L = 80 in. (2.032 m), Table 4.14 yields w_u = 10.79 kips in.$^{-1}$(1,889.62 kN m^{-1}) and Table 4.15 yields w_u = 2.37 kips in.$^{-1}$(415.05 kN m^{-1}), which is 4.55 times smaller. Similar observations can be made for L = 100, and 120 in. (1 in. = 0.0254 m). These are important practical observations that are useful for the design engineer to realize the magnitude of the losses or the benefits arising from geometric distributions and supporting conditions.

Problems

4.1 Repeat the problem in Example 4.1 for taper n = 1 and 2 and compare the results.

4.2 For Table 4.3 of Example 4.1, verify all the results obtained at the locations x = 6 in. (0.1524 m) and x = 12 in. (0.3048).

4.3 For Table 4.4 of Example 4.1, verify all the results obtained at the locations x = 0, x = 12 in. (0.3048), x = 18 in. (0.4572 m), and x = 60 in. (1.524 m).

4.4 Repeat the problem in Example 4.1 by using the three-line approximation of the stress–strain curve of monel and compare the results.

4.5 For the tapered cantilever beam in Example 4.1, assume that the taper n = 2, and determine an equivalent system of constant stiffness $E_I I_B$, where E_1 is the elastic modulus of its material and I_B is the moment of inertia at the free end B of the member. By using the equivalent system, determine the inelastic deflection and rotation at the free end. Use the three-line approximation of the stress–strain curve of monel to perform the analysis.

4.6 Repeat Problem 4.5 by using the six-line approximation of the stress–strain curve of monel and compare the results.

4.7 Repeat the problem in Example 4.3 by using the two-line approximation of the stress–strain curve of monel and compare the results.

4.8 Repeat the problem in Example 4.3 when $Q = 125$ kips (556 kN).

4.9 For Table 4.5 of Example 4.3, verify all the results at the locations $x = 6$ in. (0.1524), $x = 18$ in. (0.4572 m), $x = 60$ in. (1.524 m), and $x = 96$ in. (2.4384 m).

4.10 For Table 4.6 of Example 4.3, verify all the results at the locations $x = 12$ in. (0.3048 m), $x = 36$ in. (0.9144 m), and $x = 96$ in. (2.4384 m).

4.11 Repeat the problem in Example 4.3 with $Q = 150$ kips (667.2 kN) and taper $n = 2$.

4.12 By assuming that $Q = 140$ kips (622.72 kN), verify the curve in Fig. 4.11 of Example 4.3 by calculating the function g(x) at various locations.

4.13 Repeat Problem 4.12 with $Q = 175$ kips (778.4 kN).

4.14 The uniform simply supported beam in Fig. P4.14, is loaded with a concentrated load $Q = 160$ kips (711.68 kN) at midspan. By assuming that the width $b = 6$ in. (0.1524 m), depth $h = 8$ in. (0.2032 m), and length $L = 120$ in. (3.048 m), calculate the values of Δ, E_r, $g(x)$, $M_{req} = M_x$, and $M_e = M_x/f(x)g(x)$, at various sections along the length of the member. Use the bilinear stress–strain curve approximation of monel to produce these results.

4.15 By using the results obtained in Problem 4.14, determine an equivalent system of constant stiffness E_1I_1 where E_1 is the elastic modulus and I is the constant moment of inertia of the member. Use the equivalent system to determine the vertical deflection at midspan and the rotations at the end supports.

4.16 Repeat Problem 4.14 by assuming that $Q = 180$ kips (800.64 kN).

4.17 Repeat the problem in Example 4.4 by using the six-line approximation of the stress–strain curve of mild steel and compare the results.

4.18 Verify the results in Table 4.8 of Example 4.4, for locations at $x = 54$ in. (1.3716 m), $x = 78$ in. (1.9812 m), and $x = 108$ in. (2.7432 m).

4.19 By using the equivalent system in Fig. 4.14b of Example 4.4 and applying the moment–area method, determine the vertical deflections and the

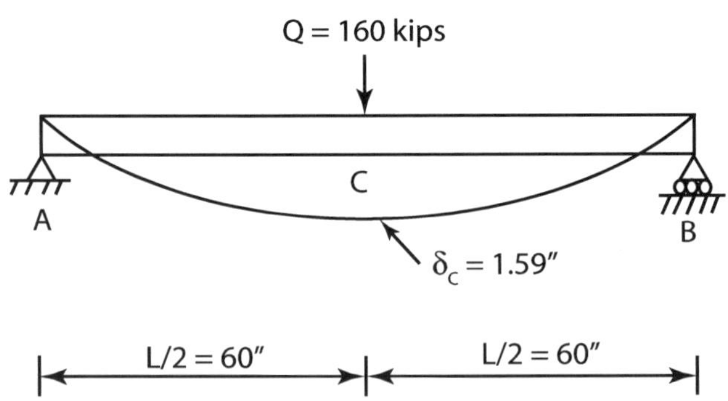

Fig. P4.14.

rotations at the locations x = 30 in. (0.762 m) and x = 90 in. (2.286 m) for support A, and compare them with the ones shown in Table 4.9.

4.20 Repeat the problem in Example 4.4 by assuming that the distributed load $w = 3,000\,\text{lb in.}^{-1}(525,380.4\,\text{N m}^{-1})$ and compare the results.

4.21 Repeat the problem in Example 4.4 by assuming that the depth h at the left support A is 16 in. (0.4064 m) and compare the results.

4.22 Repeat the problem in Example 4.5 by using the two-line approximation of the stress–strain curve of monel (elastic–perfectly plastic) represented schematically by the solid line in Fig. 4.16a. Assume $\sigma_1 = 50(10)^3$ psi(344.75×10^3 kPa), and $E_1 = 26(10)^6$ psi(179.27×10^6 kPa). Perform the inelastic analysis to determine the ultimate load w_u for lengths L = 60 in. (1.524 m) and L = 100 in. (2.54 m) and compare the results.

4.23 Repeat the problem in Example 4.5 by using the six-line approximation of the stress–strain curve of monel and compare the results. Do it only for L = 80 in. (2.032 m) and determine w_u.

4.24 For Table 4.11 of Example 4.6, verify the results for the ultimate load w_u for lengths L = 100 in. (2.54 m), and L = 200 in. (5.08 m).

4.25 Repeat the problem in Example 4.6 by using the four-line approximation of the stress–strain curve of the mild steel and compare the results.

4.26 Repeat the problem in Example 4.7 by using the two-line approximation of the stress–strain curve of the 7075-T651 aluminum alloy and compare the results.

4.27 For Table 4.14 of Example 4.7, verify the results for the ultimate load w_u for L = 100 in. (2.54 m), and L = 200 in. (5.08 m).

4.28 For Table 4.15 of Example 4.7, verify the results for the ultimate load w_u for L = 60 in. (1.524 m), and L = 100 in. (2.54 m).

5

Vibration Analysis of Flexible Structural Components

5.1 Introduction

The main purpose in this chapter is to provide an introduction to the very complex problem regarding the free vibration of flexible members. Extensive discussions on this and other related subjects can be found in the author's work in [2, 3, 5, 84]. The developed methodologies are unique, accurate, and convenient, and they simplify the solution of many complex problems regarding the vibration of flexible members.

Free vibrations, in general, take place from the static equilibrium position of the member. For a flexible member, the static equilibrium position is associated with large static amplitudes, and the differential equation of motion that expresses the free vibration of the member becomes nonlinear. If the vibrational amplitudes, measured from the static equilibrium position, are small, then the free frequencies of vibration are independent of the amplitude of vibration, but they do depend upon the static amplitude that defines the static equilibrium position. If, however, the frequency amplitudes are also large, then the free frequencies of vibration will be dependent on both static and vibrational amplitudes.

In this chapter, and the references provided above, the objective is to derive the general nonlinear differential equations of motion regarding the vibration analysis of flexible members of uniform and/or variable cross section along the length of the member. Unique solution methodologies are also developed which simplify a great deal the solution of the very complex flexible beam problems.

The analysis in this chapter is dealing primarily with small oscillation vibration superimposed on large static displacements which define the static equilibrium position of the flexible member. The methodologies developed in the first chapters of this text, can be used successfully to simplify the solution of the complex vibration problem. The nonlinear differential equation of motion that can be applied for the solution of large amplitude vibrations is also derived in this chapter. The mathematically derived dynamically equivalent system is exact, linearized, and known linear methods of vibration analysis can be rightfully used to solve it. Several methods are used to solve it, by including the Galerkin's finite element method and the energy method of Lord Rayleigh. See also References [2, 3, 5].

5.2 Nonlinear Differential Equations of Motion

Two nonlinear differential equations of motion for flexible members will be derived in this section. The first one is the general nonlinear differential equation of motion which cannot be solved by separation of variables, and where the natural frequencies of vibration are amplitude dependent.

The second one involves the derivation of the nonlinear differential equation of motion of flexible members subjected to small amplitude vibrations. In this case, the complexity of the problem is greatly reduced by taking into consideration that the vibration of the flexible member takes place from its static equilibrium position, and assuming that the vibration amplitudes are small. See also Fertis [2, 3] and Fertis and Afonta [39, 40]. Many practical problems fall into this category.

5.2.1 The general Nonlinear Differential Equation of Motion

We start with the Euler–Bernoulli Law given by Eq. (1.16). This nonlinear differential equation is as follows:

$$\frac{y''}{\left[1+(y')^2\right]^{3/2}} = -\frac{M_x}{E_x I_x} \tag{5.1}$$

By multiplying both sides of Eq. (5.1) by $E_x I_x$, we get

$$E_x I_x \frac{y''}{\left[1+(y')^2\right]^{3/2}} = -M_x \tag{5.2}$$

Differentiating both sides of Eq. (5.2) with respect to x, we get

$$\frac{d}{dx}\left\{E_x I_x \frac{y''}{\left[1+(y')^2\right]^{3/2}}\right\} = -\frac{dM_x}{dx} \tag{5.3}$$

or

$$\frac{d}{dx}\left\{E_x I_x \frac{y''}{\left[1+(y')^2\right]^{3/2}}\right\} = -V(x) \tag{5.4}$$

The shear force V(x) at any cross section must be the same whether it is defined in the reference (undeformed) configuration, or deformed configuration. That is, $V(x_o) = V(x(x_o))$. Thus, Eq. (5.4) may be written as

$$\frac{d}{dx}\left\{E_x I_x \frac{y''}{\left[1+(y')^2\right]^{3/2}}\right\} = -\frac{V(x_0)}{\cos\theta} \tag{5.5}$$

By differentiating Eq. (5.5) with respect to x, we obtain the following expression:

$$\frac{d^2}{dx^2}\left\{E_x I_x \frac{y''}{\left[1+(y')^2\right]^{3/2}}\right\} = -\frac{d}{dx}\left\{\frac{V(x_0)}{\cos\theta}\right\} \tag{5.6}$$

Since the expression for the transverse weight is well defined in the undeformed configuration, we can rewrite Eq. (5.6) as follows:

$$\frac{d^2}{dx^2}\left\{E_x I_x \frac{y''}{\left[1+(y')^2\right]^{3/2}}\right\} = -\frac{1}{\cos\theta}\frac{d}{dx}\left\{\frac{V(x_0)}{\cos\theta}\right\} \tag{5.7}$$

For example, for distributed weight, the relation of the shear force $V(x_o)$ to the load $w(x_o)$, is as follows:

$$\frac{d}{dx_0}V(x_0) = -w(x_0)\cos\theta \tag{5.8}$$

Performing the indicated differentiation on the right-hand side of Eq. (5.7), we obtain

$$\frac{d^2}{dx^2}\left\{E_x I_x \frac{y''}{\left[1+(y')^2\right]^{3/2}}\right\} = -\frac{1}{\cos\theta}w(x_0) \tag{5.9}$$

where, from Eq. (1.75), we have

$$x_o(x) = \int_0^x \sqrt{1+(y')^2}\,dx \tag{5.10}$$

and from Eq. (1.26), we have

$$\cos\theta = \frac{1}{\sqrt{1+(y')^2}} \tag{5.11}$$

For uniformly distributed loading we have $w(x_o) = w_o$, and Eq. (5.9) yields

$$\frac{d^2}{dx^2}\left\{E_x I_x \frac{y''}{\left[1+(y')^2\right]^{3/2}}\right\} = -\frac{1}{\cos\theta}w_0 \tag{5.12}$$

or, by using Eq. (5.11),

$$\frac{d^2}{dx^2}\left\{E_xI_x\frac{y''}{\left[1+(y')^2\right]^{3/2}}\right\} = -\left[1+(y')^2\right]^{1/2} w_0 \tag{5.13}$$

For transverse free vibration, the weight w_o is replaced by the inertia force w_{in}. That is,

$$w_{in} = m\frac{d^2y}{dt^2} \tag{5.14}$$

where m is the uniform mass density, and d^2y/dt^2 is the relative acceleration.

Therefore, the transverse free vibration of a flexible member with a uniformly distributed weight, or a distributed mass m, is given by the equation

$$\frac{d^2}{dx^2}\left\{E_xI_x\frac{y''}{\left[1+(y')^2\right]^{3/2}}\right\} + \sqrt{1+(y')^2}\,m\frac{d^2y}{dt^2} = 0 \tag{5.15}$$

In general, for arbitrarily distributed weight, the mass m is replaced with an appropriate expression $m(x_o)$. Note that m is a function of x_o. Thus, the nonlinear differential equation of motion becomes

$$\frac{d^2}{dx^2}\left\{E_xI_x\frac{y''}{\left[1+(y')^2\right]^{3/2}}\right\} + \sqrt{1+(y')^2}\,m\left(x_0\right)\frac{d^2y}{dt^2} = 0 \tag{5.16}$$

Equation (5.16) is the one, as stated earlier, that cannot be solved by using separation of variables, and where the natural frequencies of vibration are amplitude dependent. A procedure that could be used for the solution of such problems, would be to assume the natural modes of vibration and to determine the natural frequencies of vibration and the corresponding mode shapes by using the perturbation method. For the computation of the fundamental frequency of vibration this approach is realistic, but it becomes extremely difficult for the computation of higher frequencies of vibration.

In order to reduce the difficulties associated with the solution of Eq. (5.16), we take into consideration in the analysis that the vibration of the flexible member is taking place from its static equilibrium position, which we define here as $y_s(x)$, and that the vibrational amplitudes are small. See also Fertis [2,3,5] and Fertis and Afonta [39,40].

The nonlinear differential equation of motion which is derived in the following subsection to represent this problem, takes into consideration that the slope of the dynamic amplitude is negligible small, and that the large static equilibrium configuration of the member is obtained by using static analysis as discussed in the first three chapters of this text.

The weight acting on the deformed segment of the flexible member is approximated by a function of the horizontal displacement, and this is accomplished by replacing the arc length position in the deformed segment by the equation

$$x_0(x) = x + \Delta(x) \tag{5.17}$$

This concept is thoroughly discussed in the first chapter of this text. See for example Eqs. (1.74) and (1.80).

The law of variation of the height h(x) of the cross section of the flexible member in its deformed configuration, is an integral equation which depends on the deformation. This integral will be replaced by a function which contains the horizontal displacement function $\Delta(x)$, as discussed in Chaps. 1–3, of this text.

In the analysis and derivation the time dependent is considered to be harmonic. That is, $\ddot{y}_d = -\omega^2 y_d$, where y_d is the dynamic displacement measured from the static equilibrium position.

In the following subsection we derive the differential equation of motion and the appropriate equivalent system based on the above assumptions. This approach covers a large variety of practical problems and deserves special consideration.

5.2.2 Small Amplitude Vibrations of Flexible Members

We start the derivation of the appropriate nonlinear differential equation of motion of flexible members with small amplitude vibrations by considering the flexible cantilever beam in Fig. 5.1a, where w_o is the uniform weight of the member per unit of length. However, it should be realized that w_o may include other weights which are attached to the beam and participating in its vibratory motion. A practical example of such condition would be the girders of a highway bridge and the bridge deck which is supported by the girders. If, for example, the bridge deck is attached to the girders by shear connectors, then the whole weight of the deck participates in the vibratory motion of the girders and it must be taken into consideration in the vibration analysis of the girders. See also Fertis [5, 16, 84].

The deformation configuration of the member is shown in Fig. 5.1b, where $y_s(x)$ is the large static deformation that defines the static equilibrium position. The dynamic amplitude which represents the small vibration of the member from the static equilibrium position is denoted by $y_d(x, t)$. From the undeformed straight configuration of the member, the amplitude is defined by $y(x, t) = [y_s(x) + y_d(x, t)]$. The rest of the notation is similar to the one used in the first three chapters of this text, and it is self explanatory.

From Fig. 5.1b, the total amplitude $y(x, t)$ of the flexible member may be written as follows:

$$y(x, t) = y_s(x) \pm y_d(x, t) \tag{5.18}$$

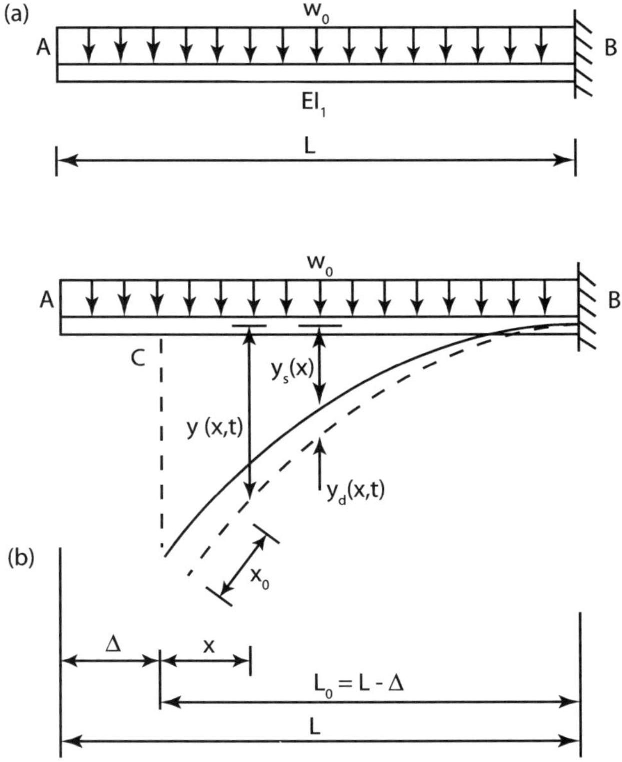

Fig. 5.1. (a) Original undeformed uniform flexible cantilever member. **(b)** Deformed configuration of the uniform flexible cantilever member

In this equation $y_s(x)$, as stated earlier, is the large static deflection which defines the static equilibrium position of the member, and $y_d(x, t)$ is the dynamic amplitude of its free vibration which is considered to be small. The slope of the dynamic amplitude curve is also small when it is compared to the large slope of the static configuration curve. This means that $y_s'(x) \gg y_d(x, t)$. Thus, if we differentiate Eq. (5.18) once with respect to x, we find

$$y'(x, t) = y_s'(x) \tag{5.19}$$

By differentiating Eq. (5.18) one more time with respect to x, we obtain

$$y''(x, t) = y_s''(x) \pm y_d''(x, t) \tag{5.20}$$

Also, if we differentiate Eq. (5.18) twice with respect to time, we obtain

$$\frac{d^2 y(x, t)}{dt^2} = \frac{d^2 y_d(x, t)}{dt^2} \tag{5.21}$$

The substitution of Eqs. (5.20) and (5.21) into Eq. (5.16), yields

$$\frac{d^2}{dx^2}\left\{E_x I_x \frac{y_s''(x) \pm y_d''(x,t)}{\left[1+(y_s')^2\right]^{3/2}}\right\} + \sqrt{1+(y')^2}\, m(x_0)\, \frac{d^2 y_d}{dt^2} = 0 \tag{5.22}$$

or

$$\frac{d^2}{dx^2}\left\{E_x I_x \frac{y_s''(x)}{\left[1+(y_s')^2\right]^{3/2}}\right\} \pm \frac{d^2}{dx^2}\left\{E_x I_x \frac{y_d''(x,t)}{\left[1+(y_s')^2\right]^{3/2}}\right\} + \sqrt{1+(y')^2}\, m(x_0)\, \frac{d^2 y_d(x,t)}{dt^2} = 0 \tag{5.23}$$

From the large static equilibrium configuration, we have the expression

$$\frac{d^2}{dx^2}\left\{E_x I_x \frac{y_s''(x)}{\left[1+(y_s')^2\right]^{3/2}}\right\} = -\frac{w(x_0)}{\cos\theta} \tag{5.24}$$

and from the dynamic equilibrium configuration, we have

$$\frac{d^2}{dx^2}\left\{E_x I_x \frac{y_d''(x,t)}{\left[1+(y_s')^2\right]^{3/2}}\right\} = -\sqrt{1+(y_s')^2}\, m(x_0)\, \frac{d^2 y_d(x,t)}{dt^2} \tag{5.25}$$

Since the time function is harmonic, we can write

$$\frac{d^2 y_d(x,t)}{dt^2} = -\omega^2 y_d(x,t) \tag{5.26}$$

where ω is the free vibration of the flexible member in radians per second. By substituting Eq. (5.26) into Eq. (5.25), we obtain the following differential equation:

$$\frac{d^2}{dx^2}\left\{E_x I_x \frac{y_d''(x,t)}{\left[1+(y_s')^2\right]^{3/2}}\right\} - \left\{\sqrt{1+(y_s')^2}\, m(x_0)\right\} \omega^2 y_d(x,t) = 0 \tag{5.27}$$

From the separation of variables, Eq. (5.27) may also be written as

$$\frac{d^2}{dx^2}\left\{E_x I_x \frac{y_d''(x)}{\left[1+(y_s')^2\right]^{3/2}}\right\} - \left\{\sqrt{1+(y_s')^2}\, m(x_0)\right\} \omega^2 y_d(x) = 0 \tag{5.28}$$

We can also write Eq. (5.28) in terms of an equivalent variable stiffness $I_e(x)$ and an equivalent variable mass density $m_e(x)$ as follows:

$$\frac{d^2}{dx^2}\left\{E_x I_e(x)\, y_d''(x)\right\} - m_e(x)\,\omega^2 y_d(x) = 0 \tag{5.29}$$

where

$$I_e(x) = \frac{I_x}{\left[1 + (y_s')^2\right]^{3/2}} \tag{5.30}$$

$$m_e(x) = \sqrt{1 + (y_s')^2}\, m(x_0) \tag{5.31}$$

and I_x is the moment of inertia of the original member at any $0 \le x \le L_o$.

The equivalent quantities $I_e(x)$ and $m_e(x)$, given by Eqs. (5.30) and (5.31), respectively, take into account the change in moment of inertia and mass caused by the large static deformation. Therefore, the differential equation given by Eq. (5.29) may be thought of as representing a straight beam of length L_o, which vibrates with the same frequencies of vibration as the original member from its static equilibrium position $y_s(x)$.

The variation of its equivalent moment of inertia $I_e(x)$ and equivalent mass $m_e(x)$, along its length L_o, are given by Eqs. (5.30) and (5.31), respectively. Thus, in summary, the transverse vibration of a flexible member subjected to large static deformation $y_s(x)$ with small amplitudes of vibration, is completely defined by the following two differential equations:

$$\frac{d^2}{dx^2}\left\{\frac{E_x I_x y_s''(x)}{\left[1 + (y_s')^2\right]^{3/2}}\right\} = -w_0(x_0)\sqrt{1 + (y_s')^2} \tag{5.32}$$

$$\frac{d^2}{dx^2}\left\{E_x I_e(x)\, y_d''(x)\right\} - m_e(x)\,\omega^2 y_d(x) = 0 \tag{5.33}$$

Equations (5.32) and (5.33) must be solved simultaneously in order to obtain the frequencies of vibration of a flexible member. Equation (5.32) is the static equilibrium differential equation which defines the large static configuration $y_s(x)$ of the member. It is a very difficult equation to solve because it involves a fourth-order nonlinear differential equation. Equation (5.32), however, is equivalent to the Euler–Bernoulli equation

$$\frac{y_s''}{[1 + (y_s'')^2]^{3/2}} = -\frac{M_x}{E_x I_x} \tag{5.34}$$

where M_x is the bending moment of the member, and $E_x I_x$ is its bending stiffness.

Therefore, instead of solving Eq. (5.32), we are now in a position to solve Eq. (5.34), which permits us to utilize the simplifications and equivalent systems analysis used in the first three chapters of this text. Note also the comments and suggestions made at the end of Sect. 5.2.1. We have to realize that both M_x and I_x in Eq. (5.34) are integral equations which depend on the large static deformation of the member. Thus, both M_x and I_x can be simplified by replacing the integral equation with a function of the unknown displacement $\Delta(x)$ as discussed in the first chapter of this text.

With known $y_s(x)$ and $y_s'(x)$, Eq. (5.33) may then be solved to determine the natural frequencies of vibration of the member and its corresponding mode shapes. We can think of Eq. (5.33) as representing the transverse bending vibration of a dynamically equivalent pseudovariable stiffness member of length L_o, and having an equivalent mass density m_e and an equivalent moment of inertia I_e. The quantities $I_e(x)$ and $m_e(x)$ define the geometry and mass of a dynamically equivalent straight member of length L_o, and they can be obtained by using Eqs. (5.30) and (5.31), respectively.

The depth $h_e(x)$ of the dynamically equivalent straight member is given by the following equation:

$$h_e(x) = \frac{h(x)}{\left[1 + (y_s')^2\right]^{1/2}} \tag{5.35}$$

where h(x) represents the variation in depth of the original member. The depth h(x) of the original member may have any arbitrary variation along its length.

The solution of the differential equation of motion given by Eq. (5.33) may be obtained by using known linear methods of analysis for free vibration of beams. The application of the finite difference method, or utilization of the Galerkin's finite element method, should yield reasonable results. The same methods could be used to solve Eq. (5.28), which is the nonlinear differential equation of motion directly representing the original problem. The solution of Eq. (5.29) or (5.33), however, would be the most convenient. It should also be noted that methods such as the ones developed by Rayleigh, Stodola, and others, could be also used to solve the pseudovariable straight member represented by Eq. (5.29). See also [3].

The following section illustrates the application of the theory and method.

5.3 Application of the Theory and Method

The examples that follow illustrate the application of the preceding theory and methodology, and the derivation of equivalent pseudovariable stiffness and mass members.

5.3.1 Free Vibration of Uniform Flexible Cantilever Beams

Example 5.1 For the uniform flexible cantilever beam shown in Fig. 5.1a, derive the dynamically equivalent pseudovariable straight cantilever beam of length L_o by using the methodology discussed in the preceding section.

The uniform stiffness EI_1 of the flexible member is 180×10^3 kip in^2. (516.541×10^3 N m^2), its uniform weight $w_o = 1.5$ lb in^{-1}. (262.6902 N m^{-1}), and its length L = 1,000 in. (25.4 m). The constant width b of the member is 2 in. (0.0598 m), its constant depth h = 3.30 in. (0.0838 m), its modulus of elasticity $E = 30 \times 10^6$ psi (206.84×10^9 Pa), and its constant moment of inertia $I_1 = 6$ in^4. (2×10^{-6} m^4).

Solution: The first thing that needs to be done in this problem is to use Eq. (5.34) and determine the static equilibrium position $y_s(x)$ shown in Fig. 5.1b. This task can be accomplished by using the pseudolinear methodology discussed in Sect. 1.8.1. The static equilibrium position $y_s(x)$ is the one caused by assuming that the weight $w_o = 1.5$ lb in.$^{-1}$ (262.6902 N m^{-1}) is applied to the cantilever beam as a vertical load. Therefore, the problem to solve is a flexible cantilever beam of uniform stiffness $EI_1 = 180 \times 10^3$ kip in.^{2} (516.541×10^3 N m^2), of length L = 1,000 in. (25.4 m), and loaded with a uniformly distributed transverse load $w_o = 1.5$ lb in.$^{-1}$ (262.6902 N m^{-1}). This problem is solved in Sect. 2.2 of the second chapter of this text.

From Sect. 2.2, with $x_o = x + \Delta$, where Δ is the horizontal displacement of the free end A of the member, we find that the slope $y_s'(x)$, at any $0 \leq x \leq L_o$ of the static equilibrium curve $y_s(x)$, can be determined from Eq. (2.14). Note that in its large deformation configuration $y_s(x)$, the initially straight flexible cantilever beam becomes a curved beam with a projected length L_o in the x direction equal to $(L - \Delta)$, as shown in Fig. 5.1b.

Equation (2.14) is rewritten below as

$$y_s{}'(x) = \frac{G(x)}{\left\{1 - [G(x)]^2\right\}^{1/2}} \tag{5.36}$$

where

$$G(x) = \frac{w_0}{12EI_1}\left[2x^3 + 3\Delta x^2 - 2(L - \Delta)^3 - 3\Delta(L - \Delta)^2\right] \tag{5.37}$$

and Δ is the horizontal displacement of the free end A, as shown in Fig. 5.1b.

From Sect. 2.2 with $x_o = x + \Delta$, we have obtained the value $\Delta = 277.25$ in. (7.042 m). Note that anyone of the Eqs. (1.81)–(1.84) could be used to express x_o. This is amply illustrated in the first three chapters of this text. By substituting for Δ, L, w_o, and EI_1 in Eq. (5.37), we obtain

$$G(x) = 0.694444\,(10)^{-9}\left[2x^3 + 831.75x^2 - 1,189,561,532\right] \tag{5.38}$$

With known $y_s'(x)$, the quantities $I_c(x)$, $m_e(x)$, and $h_e(x)$ of the dynamically equivalent pseudovariable straight beam of length L_o that is represented by Eq. (5.29), can be determined by using Eqs. (5.30), (5.31) and (5.35), respectively. Note that $L_o = (1{,}000 - 277.25) = 722.75$ in. (18.358 m).

Table 5.1 shows the values of $y_s'(x)$, $I_e(x)$, $m_e(x)$, and $h_e(x)$ for the indicated values of $0 \leq x \leq L_o$. Note that in this case $m(x_o)$ in Eq. (5.31) is constant and equal to $m_o = w_o/g$, where g is the acceleration of gravity.

Table 5.1. Values of $y_s'(x), I_e(x), m_e(x), h_e(x)$, and $w_e(x)$ of the equivalent pseudovariable beam of length L_o(1 in. = 0.0254 m, 1 lb in.$^{-1}$ = 175.127 N m^{-1})

(1) x (in.)	(2) $y_s'(x)$	(3) $I_e(x)$ (in.^{4})	(4) $m_e(x)$ (lb s^2 in.$^{-2}$)	(5) $h_e(x)$ (in.)	(6) $w_e(x) = gm_e(x)$ (lb in.$^{-1}$)
0	−1.465865	1.073846	0.006889	1.859705	2.6619
100	−1.426914	1.134176	0.006764	1.893900	2.6136
200	−1.296677	1.366519	0.006357	2.015281	2.4563
300	−1.089115	1.856201	0.005740	2.231885	2.2179
400	−0.843542	2.679550	0.005079	2.522421	1.9625
500	−0.589878	3.833873	0.004507	2.842341	1.7415
600	−0.335583	5.112501	0.004095	3.128537	1.5823
700	−0.066818	5.960044	0.003891	3.292657	1.5035
$L_o = 725.75$	0	6.000000	0.003882	3.300000	1.5000

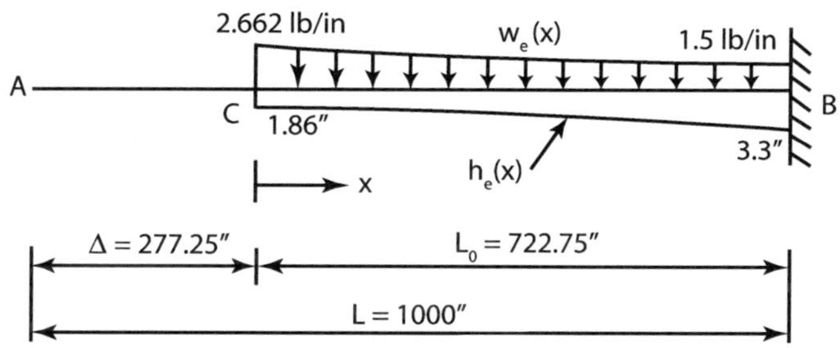

Fig. 5.2. Dynamically equivalent pseudovariable cantilever beam of length L_o undergoing small amplitude transverse vibrations

The acceleration of gravity g = 386.4 in. s^{-2}(9.8146 m s^{-2}). Figure 5.2 illustrates the dynamically equivalent pseudovariable straight cantilever beam of length L_o, which is represented by Eq. (5.29). Known methods of vibration analysis, such as the ones stated earlier, can be used to determine the transverse free frequencies of vibration of the pseudovariable straight beam and the corresponding mode shapes.

For illustration purposes, we will verify some of the values in Table 5.1. For example,
At $x = 0$ we have:

From Eq. (5.38) we find $G(0) = -0.826084$, and from Eq. (5.36), we find

$$y_s{}'(0) = \frac{-0.826084}{\left[1 - (-0.826084)^2\right]^{1/2}} = -1.465865$$

which checks the value obtained in the first row of the second column in Table 5.1.

From Eq. (5.30), we obtain

$$I_e(0) = \frac{6}{\left[1 + (-1.465865)^2\right]^{3/2}} = 1.073846\,\mathrm{in}^4.$$

This value checks the one given in the first row of the third column in Table 5.1.

From Eq. (5.31) we find

$$m_e(0) = \sqrt{1 + (-1.465865)^2}\,(1.5)\left(\frac{1}{386.4}\right) = 0.006889\,\mathrm{lb\,s^2\,in.^{-2}}$$

Again this value checks the one given in the first row of the fourth column in Table 5.1.

From Eq. (5.35) we obtain

$$h_e(0) = \frac{3.30}{\left[1 + (-1.465865)^2\right]^{1/2}} = 1.859705\,\mathrm{in.}$$

which checks the one given in the first row of the fifth column of Table 5.1.

The value of the load $w_e(0)$ can be determined from the equation

$$w_e(0) = gm_e(0) = (386.4)(0.006888) = 2.6615\,\mathrm{lb\,in.^{-1}}$$

which checks the one given in the first row of the last column in Table 5.1.

At x = 300 in we have:

From Eq. (5.38) we have

$$G(300) = 0.694444\,(10)^{-9}\left[(2)(300)^3 + 831.75\,(300)^2 - 1{,}189{,}561{,}532\right] = -0.7366$$

and from Eq. (5.36) we find

$$y_s{}'(300) = \frac{-0.7366}{\left[1 - (-0.7366)^2\right]^{1/2}} = -1.089115$$

This value checks the one given in the fourth row of the second column in Table 5.1.

From Eq. (5.30), we obtain

$$I_e(300) = \frac{6}{\left[1 + (-1.089115)^2\right]^{3/2}}$$
$$= 1.856201\,\text{in.}^4$$

This value checks the one given in the fourth row of the third column in Table (5.1).

From Eq. (5.31) we find

$$m_e(300) = \sqrt{1 + (-1.089115)^2}\,(1.5)\left(\frac{1}{386.4}\right)$$
$$= 0.005740\,\text{lb s}^2\,\text{in.}^{-}2$$

which checks the value given in the fourth row of the fourth column in Table 5.1.

From Eq. (5.35) we obtain

$$h_e(300) = \frac{3.30}{\left[1 + (-1.089115)^2\right]^{1/2}}$$
$$= 2.231885\,\text{in.}$$

This value checks the one given in the fourth row of the fifth column of Table 5.1.

The value of the load w_e (300) can be computed from the equation

$$w_e(300) = gm_e(300) = (386.4)(0.005740)$$
$$= 2.217936\,\text{lb in.}^{-1}$$

which checks the one given in the fourth row of the last column in Table 5.1.

In a similar manner, all the remaining values in Table 5.1 can be checked. The dynamically equivalent pseudovariable cantilever beam of length L_o in Fig. 5.2, depicts the variations of the equivalent load $w_e(x)$ and the equivalent depth $h_e(x)$ which are obtained in Table 5.1.

Example 5.2 The dynamically equivalent pseudolinear straight cantilever beam in Fig. (5.2) of Example 5.1, will be used here to determine the fundamental frequency of vibration of the original flexible member in Fig. 5.1a, by applying the well-known energy method of Lord Rayleigh. All the required data for the solution of this problem is given in Example 5.1. Compare the results with the ones obtained by using Galerkin's finite element method and the finite difference method.

Solution: The dynamically equivalent system in Fig. 5.2 has infinite degrees of freedom. In practice, the vibration analysis of such systems is carried out by reducing the given system into one with finite degrees of freedom. This concept is particularly useful for problem cases where the practicing engineer needs to know only the first few free frequencies of vibration. Many structures that are designed to resist the effects of an earthquake, or the effects of nuclear and conventional explosions, fall into this category.

This is usually accomplished by lumping the weight (or mass) of the member at discrete points along its length. Such an approximation provides reasonable results for the first few free frequencies of vibration, particularly the fundamental. Lord Rayleigh, in his energy method [2,3,5], is using this concept because it makes the application of his method more convenient, and provides accurate values for the computation of the free fundamental frequency of the system. We will apply this concept here to determine the fundamental frequency of the dynamically equivalent system in Fig. 5.2.

According to Lord Rayleigh, the expression to use for this purpose is

$$\omega^2 = g\frac{\sum_{i=1}^{n} W_i y_i}{\sum_{i=1}^{n} W_i {y_i}^2} \tag{5.39}$$

where ω is the fundamental frequency of free vibration, W_i are the lumped weights, and y_i is the total static deflection under the concentrated weights W_i which is produced by the application of all the weights W_i. These weights can include other weights that may be attached to the member for some reason. For example, a beam, in addition to its weight, it may support heavy instrumentation, or machinery, securely attached to it.

For the problem at hand, by using the data in the last column of Table 5.1, we lump the weight $w_e(x)$ shown in Fig. 5.2 at points 1, 2, 3, and 4, as shown in Fig. 5.3. For example, the total weight w_e between points C and D is lumped at point 1 and it is denoted as $W_1 = 511.82$ lb (2,276.5754 N). Mathematically, this is approximately done as follows:

$$\begin{aligned} W_1 &= \frac{2.6619 + 2.4563}{2}(200) \\ &= 511.82\,\text{lb}\,(2,276.5754\,\text{N}) \end{aligned}$$

Note that the value of w_e at C ($x = 0$) is 2.6619 lb in.$^{-1}$, the one at D($x = 200$ in.) is 2.4563 lb in.$^{-1}$, and the distance between C and D is 200 in. (5.08 m).

In a similar manner, the remaining weights W_2, W_3, and W_4, at points 2, 3, and 4, respectively, are determined. A piecewise moment of inertia I_e is also established as shown in Fig. 5.3. For example, the moment of inertia I_1 in the same figure represents the average value of I_e using the values at points $x = 0$ and $x = 200$ in. (5.08 m) from Table 5.1, yielding $I_1 = 1.22$ in.^{4} (1×10^{-6} m^4). In the same manner, the values of I_2, I_3, and I_4, in Fig. 5.3, are obtained.

In order to apply Eq. (5.39), we need to determine the static deflections y_1, y_2, y_3, and y_4, under the weight concentration points 1, 2, 3, and 4,

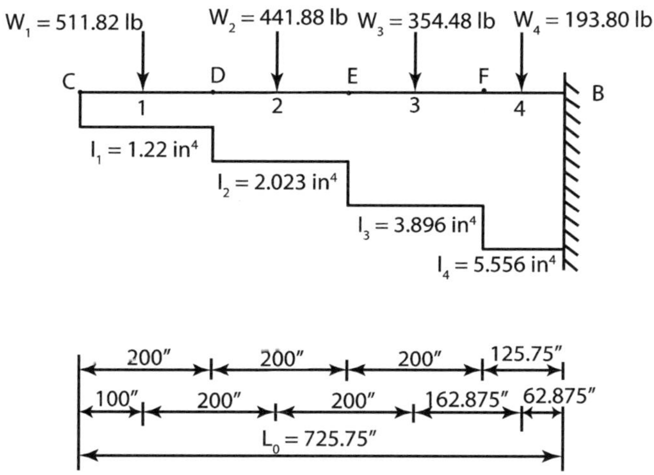

Fig. 5.3. Equivalent pseudovariable straight cantilever beam with lumped weights and piecewise uniform moment of inertia (1 in. = 0.0254 m, 1 lb = 4.448 N)

Table 5.2. Values of M_x, M_x/EI_x, and EI_x for the beam problem in Fig. 5.4 (1 in. = 0.0254 m, 1 in. lb = 0.113 N m, 1 lb in.^{2} = 2.87 × 10^{-3} N m^2)

x (in.)	M_x (in. lb)	$(M_x/EI_x)10^{-6}$		$EI_x \times 10^6$ (lb in.^{2})	
0	0	0		36.6	
100	0	0		36.6	
200	−51,182	−1,398.42	(843.335)	36.6	(60.69)
300	−102,364	−1,686.67		60.69	
400	−197,734	−3,258.10	(1,691.77)	60.69	(116.88)
500	−293,104	−2,507.73		116.88	
600	−423,922	−3,626.98	(2,543.33)	116.88	(166.68)
662.875	−506,174	−3,036.80		166.88	
725.75	−600,610	−3,603.37		166.68	

respectively. This is accomplished by using the moment–area method. In Table 5.2, the values of M_x, M_x/EI_x, and EI_x, at the indicated values of x, are calculated by using the beam problem in Fig. 5.3. By using the values shown in column 3 of Table 5.2, the M_x/EI_x diagram is plotted as shown in Fig. 5.4.

According to the moment–area method, the vertical deflection y_1 at point 1 in Fig. 5.4 is equal to the first moment about point 1 of the M_x/EI_x area

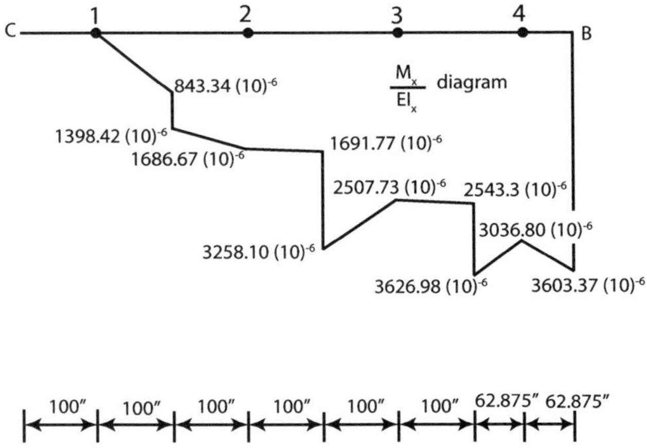

Fig. 5.4. M_x/EI_x diagram for the beam problem shown in Fig. 5.3 (1 in. = 0.0254 m)

between points 1 and B. Thus,

$$
\begin{aligned}
y_1 = 10^{-6} \Big[& \frac{1}{2}(843.34)(100)\left(\frac{2}{3}\right)(100) + (1{,}398.42)(100)(150) \\
& + \frac{1}{2}(288.25)(100)(166.6667) \\
& + (1{,}686.67)(100)(250) + \frac{1}{2}(5.10)(100)(266.6667) + (2{,}507.73)(100)(350) \\
& + \frac{1}{2}(750.37)(100)(333.3333) + (2{,}507.73)(100)(450) \\
& + \frac{1}{2}(35.60)(100)(466.6667) + (3{,}036.80)(62.875)(531.4375) \\
& + \frac{1}{2}(590.18)(62.875)(520.9583) + (3{,}036.80)(62.875)(594.3125) \\
& + \frac{1}{2}(566.57)(62.875)(604.7917) \Big] \\
= {} & 517.766876 \text{ in. } (13.1513 \text{ m})
\end{aligned}
$$

Similarly, the deflection y_2 at point 2 is equal to the first moment of the M_x/EI_x area between points 2 and B, taken about point 2; y_3 at point 3 is the first moment of the M_x/EI_x area between point 3 and B, taken about point 3; and y_4 at point 4 is the first moment of the M_x/EI_x area between points 4 and B, taken about point 4. The final results are

$$
\begin{aligned}
y_1 &= 517.766876 \text{ in. } (13.1513 \text{ m}) \\
y_2 &= 265.975473 \text{ in. } (6.7558 \text{ m}) \\
y_3 &= 82.748407 \text{ in. } (2.1018 \text{ m}) \\
y_4 &= 6.749240 \text{ in. } (0.1714 \text{ m})
\end{aligned}
$$

On this basis, the summations in Eq. (5.39) yield the following

$$\sum_{i=1}^{4} W_i y_i = (511.82)(517.766876) + (441.88)(265.975473)$$
$$+ (354.48)(82.748407) + (193.80)(6.749240)$$
$$= 413,173.34$$

$$\sum_{i=1}^{4} W_i y_i^2 = (511.82)(517.766876)^2 + (441.88)(265.975473)^2$$
$$+ (354.48)(82.748407)^2 + (193.80)(6.749240)^2$$
$$= 170,905,958.79$$

Thus, by using Eq. (5.39) we obtain

$$\omega^2 = g\frac{\sum_{i=1}^{4} W_i y_i}{\sum_{i=1}^{4} W_i {y_i}^2}$$
$$= (386.4)\frac{413,173.34}{170,905,958.79} = 0.934141$$
$$\omega = 0.966509\,\text{rps}$$

The shape of the first mode is characterized by the normalized amplitudes β_i (i = 1, 2, 3, and 4) as follows:

$$\beta_1 = \frac{y_1}{y_4} = \frac{517.766876}{6.749240} = 76.714841$$
$$\beta_2 = \frac{y_2}{y_4} = \frac{265.975473}{6.749240} = 39.408211$$
$$\beta_3 = \frac{y_3}{y_4} = \frac{82.748407}{6.749240} = 12.260404$$
$$\beta_4 = \frac{y_4}{y_4} = \frac{6.749240}{6.749240} = 1.000000$$

The solution of this problem was also performed later in this chapter and in [3], by using the Galerkin's finite element method with 40 elements. The calculated value of the fundamental frequency ω was found to be equal to 0.9468956 rps, which yields a difference of 2.07% compared to the one obtained above using the energy method of Lord Rayleigh. If, on the other hand, we use the finite difference method with 81 elements to solve the same problem, we obtain $\omega = 0.9465$ rps, which yields a difference of 2.10%.

The accuracy in the utilization of Rayleigh's method, if desired, can be improved by lumping the weight $w_e(x)$ in Fig. 5.2 into more concentrated weights W_s along the length of dynamically equivalent system, and also by using a finer stepwise arrangement for $I_e(x)$. In Fig. 5.3 we used only four lumped weights W_i ($i = 1, 2, 3$, and 4) and only four steps for $I_e(x)$ in order to solve the dynamically equivalent pseudovariable system in Fig. 5.2.

The solution of the dynamically equivalent system in Fig. 5.2, as stated earlier, was performed in detail by using Galerkin's finite element method with piecewise equivalent uniform stiffness and piecewise equivalent uniform mass as shown schematically in Fig. 5.5. The first six frequencies of free vibration were calculated by using 10, 20, 40, and 81 elements, and the results are shown in Table 5.3. The finite difference method is used for the 81element case, for comparison purposes. The last column in this table shows the percentage difference between the results obtained by using the 40 and the 81 elements. Notice that the difference is getting larger for the higher frequencies of vibration due to the numerical nature of the methods. However, both the Galerkin's and the finite difference methods yield reasonable results.

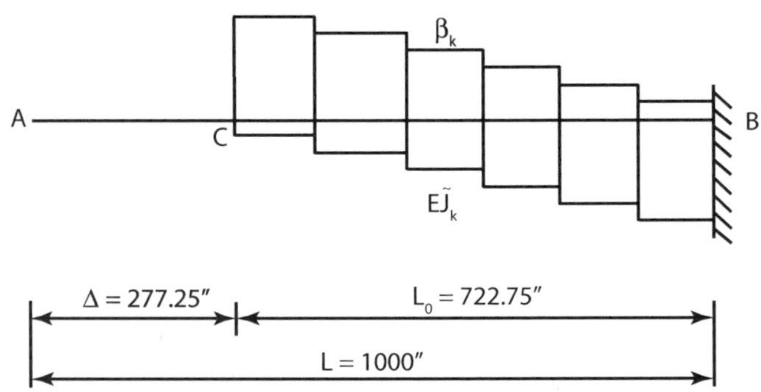

Fig. 5.5. Equivalent piecewise uniform straight cantilever beam with equivalent piecewise uniform mass (1 in. = 0.0254 m)

Table 5.3. Natural frequencies of vibration of a uniform cantilever beam using the approximation $x_o = x + \Delta$(1 in. = 0.0254 m)

	GFEM/equivalent piecewise uniform stiffness and mass				
ω (rps)	10-element	20-element	40-element	FDM 81 elements	rel. diff. (%)
ω_1	0.9428535	0.9453264	0.9468956	0.9464998	0.040
ω_2	4.9057541	4.9192247	4.9261675	4.9216986	0.092
ω_3	13.4414196	13.4236879	13.4253788	13.4040737	0.159
ω_4	26.4260209	26.1823273	26.1596222	26.0850677	0.286
ω_5	43.7238617	43.1523285	43.1515045	42.9671631	0.429
ω_6	63.9175772	64.4496307	64.3845215	64.0141907	0.578

$\Delta = 277.25$ in.

The schematic representation of the first three mode shapes of the free vibration, corresponding to ω_1, ω_2, and ω_3, is shown in Fig. 5.6. Note in this figure that the vibration is taking place with respect to the large static equilibrium position y_s, which in reality tells us that at this position the member behaves like a curved beam subjected to small amplitude vibrations.

Table 5.4, illustrates the results obtained by using the finite difference method with 81 elements to solve the uniform cantilever flexible beam problem for the cases $x_o = x + \Delta$, $x_o = x + \Delta x/L_o$, $x_o = x + \Delta[x/L_o]^{1/2}$, and $x_o = \Delta \sin \pi x/2L_o$. For all these cases of x_o, the first six frequencies of free vibration were calculated and they are shown in the table. We note in this table that the case of $x_o = x + \Delta$ represents an upper bound for the results. The other cases of x_o represent a rather accurate, for practical purposes, representation of the variation of x_o. The results, however, illustrate that the large deformation of the flexible member is not very sensitive to the variation of $\Delta(x)$. See also [3] for more information. We also note from the table that

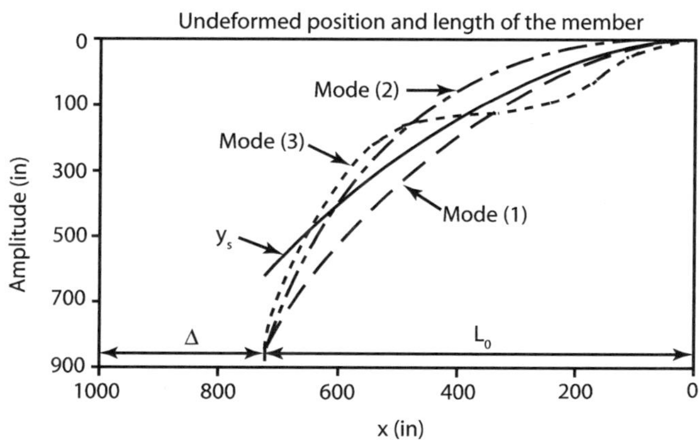

Fig. 5.6. First three mode shapes of the pseudolinear straight cantilever beam using $\Delta(x) = \Delta$ and $w_o = 1.5\,\text{lb in.}^{-1}(1\,\text{lb in.}^{-1} = 175.13\,\text{N m}^{-1})$

Table 5.4. Natural frequencies of vibration of a uniform cantilever beam using the finite difference method with 81 elements and the approximations $x_o = x + \Delta, x_o = x + \Delta x/L_0, x_o = x + \Delta[x/L_o]^{1/2}$, and $x_o = \Delta \sin \pi x/2L_o$

	frequency (rad s^{-1})			
mode	$x_o = x + \Delta$	$x_o = x + \Delta x/L_o$	$x_o = x + \Delta[x/L_o]^{1/2}$	$x_o = \Delta\ \sin \pi x/2L_o$
1	0.9464998	0.9178566	0.9290826	0.9118916
2	4.9216986	4.8487091	4.8703041	4.9044514
3	13.4040737	13.3545284	13.3677883	13.4105587
4	26.0850677	26.0493774	26.0579987	26.1030426
5	42.9671631	42.9428101	42.9477692	42.9932709
6	64.0141907	64.0025330	64.0033722	64.0487671

the effect of $\Delta(x)$ is more indicative in the fundamental frequency of vibration, and it becomes less visible as we move to the higher frequencies of free vibration.

5.3.2 Free Vibration of Flexible Simply supported Beams

In this section we examine the free vibration of flexible simply supported beams. The methodology is general and it applies equally well to members of both uniform and variable cross section and stiffness EI along their length. For illustration purposes, the analysis in this section deals primarily with the free vibration of uniform flexible simply supported beams, such as the one shown in Fig. 2.3 of Section 2.3.

In this figure, the distributed load w acting on the member, represents the distribution of its weight along its length. However, it may include the weight of other possible objects which are securely attached to the member and participate in its vibrational motion. We assume here that the static deflection of the member caused by such weight distribution is large, and from its large static equilibrium position the member behaves like a curved beam during its vibration configuration. This point of view is well discussed in preceding sections of this book. Therefore, we have here a small amplitude free vibration taking place about the large static equilibrium position. The problem is very nonlinear, and its nonlinearity is characterized by the heavily curved large static equilibrium position, as stated in earlier sections of this chapter.

In order to proceed with the free vibration analysis of the uniform simply supported flexible member, we need first to derive the equivalent pseudovariable straight simply supported beam of length L_o, as we have done it previously for the cantilever beam. This task requires the computation of the rotation y_s' of the static equilibrium position curve y_s.

If, for example, we assume that w in Fig. 2.3 is the weight distribution of the simply supported member, then the deflection curve shown in the figure is the static equilibrium position of the member produced by the weight w. As in the preceding sections, the large deformation configuration, and consequently y_s', can be determined by using the Euler–Bernoulli nonlinear differential equation. We rewrite this equation below.

$$\frac{y''}{\left[1+(y_s')^2\right]^{3/2}} = -\frac{M_x}{E_x I_x} \tag{5.40}$$

By following the procedure discussed in Sect.2.3 and assuming in this case that

$$\Delta(x) = \Delta\frac{x}{L_0} \tag{5.41}$$

we find

$$y_s{}' = \frac{G(x)}{\left\{1 - [G(x)]^2\right\}^{1/2}} \tag{5.42}$$

where

$$G(x) = \frac{wL}{24EI(L-\Delta)}\left[-6(L-\Delta)x^2 + 4x^3 + (L-\Delta)^3\right] \tag{5.43}$$

In this equation, the unknown horizontal displacement Δ of the right support A, can be determined by using, as in preceding sections of this text, the equation

$$L = \int_0^{L_o}\left[1 + (y')^2\right]^{1/2} dx \tag{5.44}$$

and applying a trial-and-error procedure.

After having determined Δ, we can find the rotation $y_s{}'$ of the static equilibrium position (or deflection) y_s, at any $0 \le x \le L_0$, by using Eqs. (5.42) and (5.43).

With known $y_s{}'$, we can proceed with the solution of the differential equation of motion which is given by Eq. (5.29). We rewrite this equation for convenience.

$$\frac{d^2}{dx^2}\left\{E_x I_e(x)\, y_d''(x)\right\} - m_e(x)\,\omega^2 y_d(x) = 0 \tag{5.45}$$

In Eq. (5.45), $I_e(x)$ and $m_e(x)$ are given by the following equations:

$$I_e(x) = \frac{I_x}{\left\{1 + (y_s{}')^2\right\}^{3/2}} \tag{5.46}$$

$$m_e(x) = \sqrt{1 + (y_s{}')^2}\, m(x_0) \tag{5.47}$$

where I_x is the moment of inertia of the original member at any $0 \le x \le L_0$.

Note that Eqs. (5.45)–(5.47) are completely defined since $y_s{}'$ can be obtained from Eq. (5.42). The following numerical examples illustrate application of the methodology.

Example 5.3 By assuming that $w = 5\,\text{lb in.}^{-1}$ ($875.634\,\text{N m}^{-1}$) is the weight distribution of the simply supported uniform flexible beam in Fig. 2.3, determine the equivalent pseudovariable straight simply supported beam of length L_0 using the methodology discussed in this section. The length L of the member is 1,000 in. (25.4 m) and its uniform stiffness $EI = 75 \times 10^3\,\text{kip in.}^2$ ($215{,}224\,\text{N m}^2$). Assume also that the uniform depth $h = 2.466$ in. (0.0626 m), the uniform width $b = 2$ in. (0.0508 m), and the modulus of elasticity $E = 30 \times 10^3$ ksi (206.84×10^9 Pa).

Solution: By substituting in Eq. (5.43) the appropriate values of EI, L, and w, we obtain

$$G(x) = \frac{2.777778}{(1{,}000 - \Delta)^6}\left[-6x^2(1{,}000 - \Delta) + (1{,}000 - \Delta)^3 + 4x^3\right] \tag{5.48}$$

The unknown horizontal displacement Δ of the right support A in Eq. (5.48), can be determined by using Eq. (5.44) and applying a trial-and-error procedure, as discussed in Sect. 2.3 and in other places of this text. This is done by assuming values of Δ until Eq. (5.44) is satisfied. This procedure yields 411.335 in. (10.4479 m). Therefore,

$$\begin{aligned} L_0 &= L - \Delta = 1{,}000 - 411.335 \\ &= 588.665\,\text{in.}(14.9521\,\text{m}) \end{aligned}$$

By substituting for Δ in Eq. (5.48), we find

$$G(x) = 4.718776\,(10)^{-9}\left[-3,531.99x^2 + 4x^3 + 203,988,012\right] \tag{5.49}$$

The values of the rotation $y_s{}'$ at any $0 \leq x \leq L_0$ can be determined from Eq. (5.42), since G(x) is given by Eq. (5.49). For example, at $x = 0$, Eq. (5.49) yields

$$G(0) = 0.962574$$

and from Eq. (5.42) we find

$$\begin{aligned} y_s{}'(0) &= \frac{0.962574}{\left[1 - (0.962574)^2\right]^{1/2}} \\ &= 3.551671 \end{aligned}$$

This value corresponds to the rotation

$$\theta_B = \tan^{-1}\left[y_s{}'(0)\right] = 74.275°$$

which is the rotation θ_B at the left support B of the flexible simply supported beam in Fig. 2.3.

In Table 5.5 we include the values of $y_s{}'(x)$, $I_e(x)$, $m_e(x)$ and $h_e(x)$ at the indicated x positions, using, respectively, Eqs. (5.42), (5.46), (5.47), and Eq. (5.35). The last column of this table provides the values of the equivalent pseudovariable weight distribution

$$w_e(x) = gm_e(x)$$

where $g = 386.4\,\text{in.}\,\text{s}^{-2}$ ($9.81\,\text{m}\,\text{s}^{-2}$) is the acceleration of gravity. The uniform moment of inertia I of the original member in Fig. 2.3 is $2.5\,\text{in.}^4$ ($1.04 \times 10^{-6}\,\text{m}^4$).

The dynamically equivalent pseudovariable straight simply supported beam of length $L_0 = 588.665$ in. (14.9521 m) is illustrated in Fig. 5.7. We verify some of the numbers shown in this figure and in Table 5.5. For example, at $x = 0$, we use Eq. (5.42) and we find

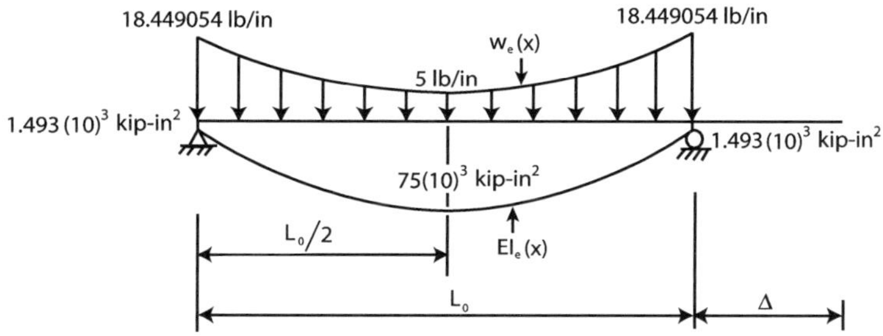

Fig. 5.7. Dynamically equivalent pseudovariable straight simply supported beam of length L_0(1 in. = 0.0254 m, 1 lb in.$^{-1}$ = 175.127 N m^{-1}, 1 kip in.^{2} = 2.87 N m^2)

Table 5.5. Values of $y_s'(x)$, $I_e(x)$, $m_e(x)$, and $h_e(x)$ of the equivalent pseudovariable straight beam of length L_0(1 in. = 0.0254 m, 1 lb in.$^{-1}$ = 175.127 N m^{-1})

(1) x (in.)	(2) y_s'	(3) $I_e(x)$ (in.^{4})	(4) $m_e(x)$ (lb s^2 in.$^{-1}$)	(5) $h_e(x)$ (in.)	(6) $w_e(x)$ (lb in.$^{-1}$)
0	3.551671	0.049767	0.047746	0.668335	18.449054
100	1.405361	0.487193	0.022319	1.429706	8.624062
$L_0/4 = 147.166$	0.882705	1.053451	0.017260	1.848777	6.669264
200	0.499573	1.789771	0.014465	2.206034	5.589276
$L_0/2 = 294.33$	0	2.500000	0.012940	2.466000	5.000000
300	−0.027821	2.497101	0.012945	2.464093	5.001948
400	−0.571349	1.636473	0.014903	2.141160	5.758519
$3L_0/4 = 441.499$	−0.882705	1.053451	0.017260	1.848777	6.669264
500	−1.578208	0.383322	0.024176	1.319880	9.341606
$L_0 = 588.665$	−3.551671	0.049767	0.047746	0.668325	18.449054

$$y_s'(0) = \frac{0.962574}{\left[1 - (0.962574)^2\right]^{1/2}} = 3.551671$$

From Eq. (5.46), we find

$$I_e(0) = \frac{2.5}{\left[1 + (3.551671)^2\right]^{3/2}} = 0.049767\,\text{in.}^4$$

From Eq. (5.35), we find

$$h_e(0) = \frac{2.466}{\left[1 + (3.551671)^2\right]^{1/2}} = 0.668335\,\text{in.}$$

From Eq. (5.47) we obtain

$$m_e(0) = \left[1 + (3.551671)^2\right]^{1/2}\left(\frac{5}{386.4}\right) = 0.047745\,\text{lb}\,\text{s}^2\,\text{in.}^{-1}$$

Therefore,

$$w_e(0) = m_e(0)\,g = (0.047745)(386.4) = 18.448668\,\text{lb}$$

Also,

$$\begin{aligned} EI_e(0) = EI_e(L_O) &= (30)(10)^3(0.049767) \\ &= 1.493\,(10)^3\ \text{kip in.}^2 \end{aligned}$$

$$EI_e(L_O/2) = (30)(10)^3(2.5) = 75\,(10)^3\ \text{kip in.}^2$$

In a similar manner, all other values in Table 5.5 can be verified.

Known methods of vibration analysis, such as the Galerkin's and Rayleigh's methods, can be used to solve the dynamically equivalent pseudovariable system in Fig. 5.7.

Example 5.4 By applying Raleigh's energy method to the dynamically equivalent pseudovariable system in Fig. 5.7, determine the fundamental frequency of its free vibration. The dynamically equivalent system was derived as explained in Example 5.3.

Solution: We start by using the data in column 6 of Table 5.5 and lumping the equivalent weight distribution $w_e(x)$ at points 1,2,3, and 4, as shown in Fig. 5.8a. For example, the lumped weight $W_1 = 1,848.28$ lb (8,221 N) at point 1 is obtained by using the average value of w_e of the two points located at $x = 0$ and $x = 147.166$ in. (3.738 m) and multiplying it by the distance BD = 147.166 in. (3.738 m). In a similar manner, the lumped weights $W_2, W_3,$ and W_4 are obtained. The constant moment of inertia I_1 in the same figure is the average value of $I_e(x)$ using points at $x = 0$ and $x = 147.166$ in., I_2 is the average value of $I_e(x)$ using the points at $x = 147.166$ in. and $x = 294.33$ in., and so on.

The static deflections $y_1, y_2, y_3,$ and y_4, of the equivalent straight member in Fig. 5.8a, are obtained by using the moment–area method. The values of $M_x, EI_e,$ and M_x/EI_e at the indicated values of x, are shown in Table 5.6. The M_x/EI_e diagram of the straight member is plotted as shown in Fig. 5.8b. By applying the moment–area method, the vertical deflection y_1 at point 1 is equal to the first moment of the M_x/EI_e area between points B and C in Fig. 5.8b, taken about point B, minus the first moment of the M_x/EI_e area between points 1 and C taken about point 1. This yields $y_1 = 160.164263$ in. (4.068 m). The vertical deflection y_2 at point 2 is equal to the first moment of the M_x/EI_e area between points B and C, taken about point B, minus the first moment of the M_x/EI_e area between points 2 and C, taken about point 2. This yields $y_2 = 330.843327$ in. (8.403 m). Because of symmetry, we have $y_3 = y_2$, and $y_4 = y_1$.

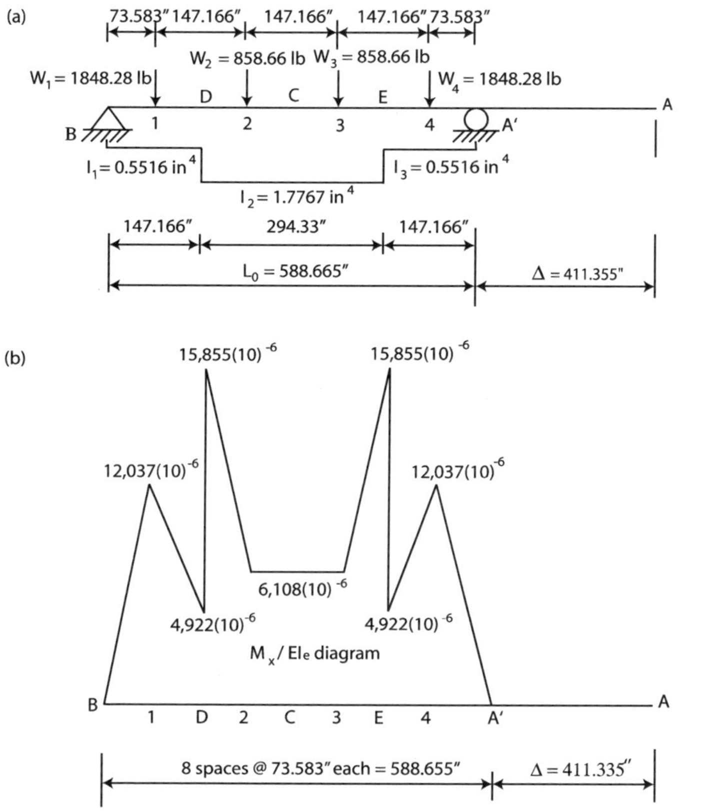

Fig. 5.8. (a) Equivalent pseudovariable straight simply supported beam with lumped weights and piecewise uniform moment of inertia. **(b)** M_x/EI_e diagram of the piecewise pseudovariable beam (1 in. = 0.0254 m, 1 lb = 4.448 N)

Table 5.6. Values of M_x, EI_e, and M_x/EI_e for the beam problem shown in Fig. 5.8a (1 in. = 0.0254 m, 1 in. lb = 0.113 N m, 1 lb in.^{2} = 2.87 × 10^{-3} N m^2)

x (in.)	M_x (in. lb)	$EI_e \times 10^6$ (lb in.^{2})	$M_x/EI_e \times 10^6$
0	0	16.548	0
73.583	199,185	16.548	12,037
147.166	262,368	16.548 (53.301)	15,855 (4,922)
220.749	325,550	53.301	6,108
294.330	325,545	53.301	6,108
367.913	325,550	53.301	6,108
441.496	262,368	53.301 (16.548)	4,922 (15,855)
515.079	199,185	16.548	12,037
588.665	0	16.548	0

On this basis, we have

$$\sum_{i=1}^{4} W_i y_i = (2)(1848.28)(160.164263) + (2)(858.66)(330.843327)$$
$$= 1,160,220.67$$
$$\sum_{i=1}^{4} W_i y_i^2 = (2)(1848.28)(160.164263)^2 + (2)(858.66)(330.843327)^2$$
$$= 282,799,564.81$$

Thus, by using Eq. (5.39), we obtain

$$\omega^2 = g\frac{\sum_{n=1}^{4} W_i y_i}{\sum_{n=1}^{4} W_i y_i^2}$$
$$= (384.6)\,\frac{1,160,220.67}{282,799,564.81}$$
$$= 1.585254$$
$$\omega = 1.2591\,\text{rps}$$

The β_i amplitudes which characterize the shape of the fundamental mode of the free vibration of the flexible member are

$$\beta_1 = \frac{y_1}{y_1} = 1.00$$
$$\beta_2 = \frac{y_2}{y_1} = \frac{330.843327}{160.164263} = 2.065650$$
$$\beta_3 = \beta_2 = 2.065650$$
$$\beta_4 = \beta_1 = 1.00$$

The dynamically equivalent system in Fig. 5.7 is also solved using the Galerkin's finite element method with 10 elements. The value obtained for the fundamental frequency of vibration is $\omega = 1.311$ rps, yielding a difference of −3.96%. The same problem was also solved using the finite difference method with 81 elements, and the value obtained is 1.3058 rps, a difference of −3.58 %. The indicated difference, when comparing to the value obtained using Rayleigh's method, can become much smaller if a finer arrangement for $w_e(x)$ and $I_e(x)$ is used in Fig. 5.7. Most often, however, values within 5 or 10% percent are sufficiently accurate for the practicing design engineer to satisfy his/her design requirements.

The main objective of this discussion, however, is to illustrate the main advantages and convenience in using the dynamically equivalent pseudovariable and pseudolinear system, which permits the use of linear methods of analysis to solve the complicated nonlinear initial problem. This is particularly useful to the finite element method, because the method can deviate from using the nonlinear element which is the problem source for the method. It can use a pseudolinear element where all nonlinear conditions are appropriately satisfied.

5.4 The Effect of Mass Position Change During the Vibration of Flexible Members

It is already established that the free vibration of a flexible member takes place with respect to the static equilibrium position y_s as shown in Fig. 5.6. At this position, the geometric distribution of the mass $m(x_o)$ of the flexible member is changed substantially when it is compared to its initial straight configuration.

In order to clarify this point further, we recall Eq. (5.27) and we rewrite it as follows:

$$\frac{d^2}{dx^2}\left\{E_x I_x \frac{y''_d(x,t)}{\left[1+(y'_s)^2\right]^{3/2}}\right\} - \left(\sqrt{1+(y'_s)^2}\right) m(x_0)\,\omega^2 y_d(x,t) = 0 \quad (5.50)$$

In the second term of Eq. (5.50), the quantity $[1+({y_s}')^2]^{1/2}$ is the one that takes into consideration the geometric configuration of the mass $m(x_o)$ at the large static equilibrium position y_s. If the static deformation y_s is small, or moderately large, the quantity $[1+({y_s}')^2]^{1/2}$ could be assumed to be equal to unity, and Eq. (5.50) becomes

$$\frac{d^2}{dx^2}\left\{E_x I_x \frac{y''_d(x,t)}{\left[1+(y'_s)^2\right]^{3/2}}\right\} - m(x_0)\,\omega^2 y_d(x,t) = 0 \quad (5.51)$$

Equations (5.50) and (5.51) will be used here to demonstrate the error introduced in the vibration analysis of flexible members if the changes of the mass geometry during large deformation are not taken into consideration. The following numerical example illustrates this particular point of view.

Example 5.4 Determine the free frequencies of vibration and the corresponding mode shapes for the tapered cantilever beam in Fig. 5.9a by using Eqs. (5.50) and (5.51) and compare the results. The weight

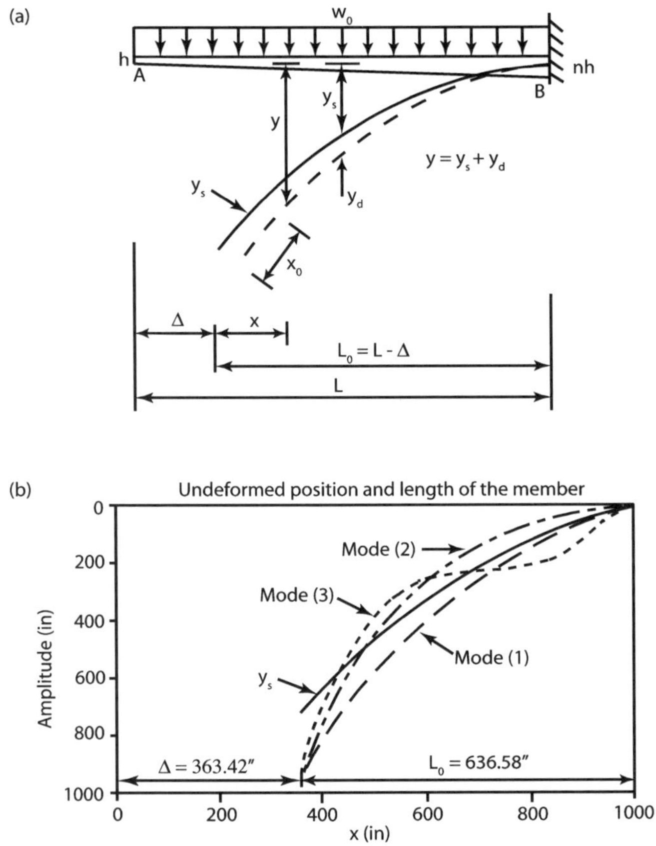

Fig. 5.9. **(a)** Tapered cantilever beam. **(b)** First three mode shapes of the member (1 in. = 0.0254m)

$w_o = 5\,\text{lb in.}^{-1}(875.59\,\text{N m}^{-1})$ represents the weight of the beam and may be other additional weights that are securely attached to the member and participate in its vibrational motion. The length L = 1,000 in. (25.4 m), the stiffness $EI_A = 180,000\,\text{kip in.}^2\,(516,541\,\text{N m}^2)$, and the taper n = 1.5.

Solution: The computation of the slope $y_s'(x)$ of the static equilibrium position curve $y_s(x)$ was carried out in Example 3.2 of Sect. 3.3, and the value of $y_s'(x)$ can be computed by using Eq. (3.29), where G(x) in this equation is given by Eq. (3.31). The fourth column of Table 3.4 contains the calculated values of $y_s'(x)$ for the indicated values of x. Note in this example that the calculated value of the horizontal displacement Δ of the free end of the member is 363.0 in. (9.22 m), and $L_o = 1{,}000 - 363 = 637.0$ in (16.18 m).

With known $y_s{}'$, Eqs. (5.50) and (5.51) may be used to determine the frequencies of vibration of the flexible member and its corresponding mode

Table 5.7. Values of the first six free frequencies of vibration of a tapered flexible cantilever beam by using Eqs. (5.50) and (5.51)

(1) ω (rps)	(2) FDM Eq. (5.50) 81 elements	(3) FEM Eq. (5.50) 10 elements	(4) % difference	(5) FDM Eq. (5.51) 81 elements	(6) % error
ω_1	0.8954	0.8957	0.0335	1.3707	53.08
ω_2	3.8808	3.8896	0.2268	5.3082	36.78
ω_3	9.8536	9.8990	0.4607	13.0459	32.40
ω_4	18.7886	18.9436	0.8250	24.6507	31.20
ω_5	30.6729	31.1738	1.6330	40.1203	30.80
ω_6	45.4691	46.7509	2.8191	59.4082	30.66

shapes. Any known method of analysis that is appropriate for such types of problems may be used for this purpose. For the results obtained here, the Galerkin's finite element method and the finite difference method are used, and the results are compared. The finite element method was carried out by using 10 elements, and 81 elements were used to carry out the finite difference method. The results for the first six free frequencies of vibration are shown in Table 5.7. This table shows that both the finite element and the finite difference methods yield similar results for the solution of Eq. (5.50). See fourth column of the table. The relative difference for practical purposes is rather small. The fifth column of Table 5.7 shows the results obtained by solving Eq. (5.51), where the change in mass geometry during deformation is not taken into consideration. The finite difference method with 81 elements was used for this purpose. The last column in this table shows the large errors involved when Eq. (5.51) is used. Therefore, it may be concluded that the change in mass geometry during the large static deformation of a member, *must* be taken into consideration when vibration analysis is considered. The fundamental free frequency has the largest error, and this is important to know for practical applications.

The first three mode shapes of the member which are obtained from the solution of Eq. (5.50) are shown in Fig. 5.9b. In this figure, the curve y_s represents the static equilibrium position of the member.

5.5 Galerkin's Finite Element Method (GFEM)

The Galerkin's finite element method (GFEM) may be used rather conveniently to solve Eq. (5.29), which, as stated earlier, may be thought of to represent the transverse bending vibration of a pseudovariable stiffness member of equivalent mass density $m_e(x)$ and equivalent moment of inertia $I_e(x)$. The quantities $m_e(x)$ and $I_e(x)$ define, respectively, the mass and geometry of an equivalent pseudolinear straight member of length L_0.

In order to apply this method, we need to calculate the interpolation functions. If we want to use the consistent Galerkin's finite element method, the interpolation functions can be determined by using the following differential equation:

$$\frac{d^2}{dx^2}\left\{E_x I_x \frac{y_d''(x)}{\left[1+(y_s')^2\right]^{3/2}}\right\} = 0 \tag{5.52}$$

We note, however, that y_d in Eq. (5.52) cannot be solved explicitly, thus creating a situation which complicates the procedure for the solution by the Galerkin's method. In this section, we avoid this problem by developing an equivalent uniform stiffness and mass approach which permits the utilization of uniform shape functions. In this approach, an equivalent uniform stiffness and an equivalent uniform mass are defined for each element. Then, based on the differential equation of the kth element, the equivalent uniform stiffness and mass are defined using Galerkin's finite element method with uniform shape functions.

We should have in mind here that the solution of Eq. (5.27) involves the solution of the original member behaving as a curved beam because of its large static deformation. On the other hand, the solution of Eq. (5.29) represents the solution of an equivalent straight beam of length L_0, which has an equivalent mass $m_e(x)$ given by Eq. (5.31), and an equivalent moment of inertia $I_e(x)$ given by Eq. (5.30). This is a much more convenient problem to solve.

The approach regarding Galerkin's method with uniform shape functions is as follows:

We divide the beam into M elements, thus having a total of M+1 node points. We assume here that each element has two degrees of freedom per node, which are rotation and vertical translation. For the kth element, the differential equation of motion is written as

$$E\tilde{J}y_d'''' - \beta_k \omega^2 y_d = 0, \qquad k = 1, 2, 3, \ldots, M \tag{5.53}$$

where $\tilde{J}_k$ is the equivalent uniform moment of inertia of the kth element, and β_k is the equivalent uniform mass of the same element.

With Eqs. (5.30) and (5.31) as a guidance, the expressions for $\tilde{J}_k$ and β_k in Eq. (5.53) may be written as follows:

$$\tilde{J}_k = \frac{1}{2} I_1 \left\{ \frac{f(x_i)}{\left\{1+[y_s'(x_i)]^2\right\}^{3/2}} + \frac{f(x_j)}{\left\{1+[y_s'(x_j)]^2\right\}^{3/2}} \right\} \tag{5.54}$$

$$\beta_k = \frac{1}{2} m(x_0) \left\{ \sqrt{1+[y_s'(x_i)]^2} + \sqrt{1+[y_s'(x_j)]^2} \right\} \tag{5.55}$$

Note that I_1 in Eq. (5.54) is a reference value for the moment of inertia I_x, see Eq. (5.30), and f(x) represents its variation. By examining Eq. (5.55), we also not that $\beta_k = m(x_0)$ if the change due to the large static curvature is ignored.

In this section, Galerkin's finite element method is used to obtain an approximate solution to the given differential equation of motion. We accomplish this task by requiring that the error between the approximate solution and the true one be orthogonal to the function used in the approximation.

For example, if we can start with the differential equation given by Eq. (5.53) and approximate the solution $y_d(x)$ by the expression

$$y_d(x) = \sum \mathcal{L}_i U_i \tag{5.56}$$

and then substitute into Eq. (5.53), we obtain

$$E\tilde{J}_k [\mathcal{L}_i U_i]'''' - \beta_k \omega^2 [\mathcal{L}_i U_i] = \varepsilon \tag{5.57}$$

where ε in this equation is the residual, or error, because the solution is only approximate,

$\mathcal{L}_i$ are the basis functions that need to be evaluated, and U_i are the nodal coordinates.

The purpose here is to make the error as small as possible. We can achieve this task by requiring that

$$\int_R \mathcal{L}_i \varepsilon \, dR = 0 \tag{5.58}$$

for each of the basis function $\mathcal{L}_i$. This integral states that the basis function must be orthogonal to the error.

By substituting into Eq. (5.58) the expression for ε given by Eq. (5.57), and making use of Eq. (5.56), we obtain

$$\int_0^{L_0} \mathcal{L}_i \left[E\tilde{J}_k y_d'''' - \beta_k \omega^2 y_d\right] dx = 0 \tag{5.59}$$

The region R is defined by using the projected length L_0 of the flexible member between the limits 0 and L_0.

Since the interpolation function for y_d is defined over a single element, Eq. (5.59) must be written in terms of a summation. That is,

$$\sum_{k=1}^{M} \int_0^{L_0} \mathcal{L}_i \left[E\tilde{J}_k y_d'''' - \beta_k \omega^2 y_d\right] dx = 0 \tag{5.60}$$

The integral must be reduced into one containing first and second derivatives before we are in a position to define the stiffness and mass matrices.

If we integrate by parts and element-by-element, we find

$$\begin{aligned} &E\tilde{J}_k \left\{ \left[\mathcal{L}_i^k \frac{d}{dx} [y_d''] \right]_0^{L_k} - \left[\frac{d}{dx} \mathcal{L}_i^k [y_d''] \right]_0^{L_k} + \int_0^{L_k} \frac{d^2}{dx^2} \mathcal{L}_i^k [y_d''] \, dx \right\} \\ &- \omega^2 \beta_k \int_0^{L_k} \frac{d^2}{dx^2} \mathcal{L}_i^k [y_d] \, dx = 0 \end{aligned} \tag{5.61}$$

When the summation over the elements is completed, the first two terms in Eq. (5.61) will drop out, and the equation reduces to the following equation:

$$E\tilde{J}_k \int_0^{L_k} \frac{d^2}{dx^2} [\mathcal{L}_k] \{y_d''\} \, dx - \omega^2 \beta_k \int_0^{L_k} [\mathcal{L}_k] \{y_d\} \, dx = 0 \tag{5.62}$$

If we assume that $y_d(x) = [\mathcal{L}]\{U\}$, then Eq. (5.62), by substitution, may be written as

$$E\tilde{J}_k \int_0^{L_k} [\mathcal{L}_k'']^t [\mathcal{L}_k''] \{U\} \, dx - \omega^2 \beta_k \int_0^{L_k} [\mathcal{L}_k]^t [\mathcal{L}_k] \{U\} \, dx = 0 \tag{5.63}$$

Now we can write Eq. (5.63) as

$$[K]\{U\} - \omega^2 [M]\{U\} = 0 \tag{5.64}$$

where the stiffness matrix [K] and the mass matrix [M] are defined as

$$[K^k] = E\tilde{J}_k \int_0^{L_k} [\mathcal{L}_k'']^t [\mathcal{L}_k''] \, dx \tag{5.65}$$

$$[M^k] = \beta_k \int_0^{L_k} [\mathcal{L}_k]^t [\mathcal{L}_k] \, dx \tag{5.66}$$

We can derive the element stiffness and mass matrices as follows:

Consider an element in bending vibration where $E\tilde{J}_k$ and β_k are the equivalent uniform bending stiffness and the equivalent uniformly distributed mass density, respectively, for the kth element. The interpolation or shape function may be obtained by solving the following differential equation:

$$E\tilde{J}_k \frac{d^4 y_d(x)}{dx^4} = 0 \tag{5.67}$$

If we integrate Eq. (5.67) four times, we find

$$y_d(x) = C_1 x^3 + C_2 x^2 + C_3 x + C_4 \tag{5.68}$$

In Equation (5.68), C_1, C_2, C_3, and C_4 are the constants of integration. These constants are determined by using the geometric boundary conditions at each node. We have two degrees of freedom per node, that is, vertical translation and rotation. These boundary conditions are as follows:

$$y_d(0) = y_1 \tag{5.69}$$

$$\frac{dyd(0)}{dx} = \Theta_1 \tag{5.70}$$

$$y_d(L_k) = y_2 \tag{5.71}$$

$$\frac{dy_d(L_k)}{dx} = \Theta_2 \tag{5.72}$$

By applying the above four boundary conditions to Eq. (5.68), and solving, simultaneously, the resulting four equations for C_1, C_2, C_3, and C_4, we find

$$C_1 = \frac{\Theta_1}{L_k^2} + \frac{\Theta_2}{L_k^2} + \frac{2y_1}{L_k^3} - \frac{2y_2}{L_k^3} \tag{5.73}$$

$$C_2 = -\frac{2\Theta_1}{L_k} - \frac{\Theta_2}{L_k} - \frac{3y_1}{L_k^2} + \frac{3y_2}{L_k^2} \tag{5.74}$$

$$C_3 = \Theta_1 \tag{5.75}$$

$$C_4 = y_1 \tag{5.76}$$

By substituting for C_1, C_2, C_3, and C_4, in Eq. (5.68) and simplifying, we find

$$\begin{aligned} y_d(x) = &\left[1 + \frac{2x^3}{L_k^3} - \frac{3x^2}{L_k^2}\right] y_1 + \left[x - \frac{2x^2}{L_k} + \frac{x^3}{L_k^2}\right] \Theta_1 \\ &+ \left[\frac{3x^2}{L_k^2} - \frac{2x^3}{L_k^3}\right] y_2 + \left[-\frac{x^2}{L_k} + \frac{x^3}{L_k^2}\right] \Theta_2 \end{aligned} \tag{5.77}$$

If we use the notation

$$\mathcal{L}_1(x) = 1 - 3\left(\frac{x}{L_k}\right)^2 + 2\left(\frac{x}{L_k}\right)^3 \tag{5.78}$$

$$\mathcal{L}_2(x) = \frac{x}{L_k} - 2\left(\frac{x}{L_k}\right)^2 + \left(\frac{x}{L_k}\right)^3 \tag{5.79}$$

$$\mathcal{L}_3(x) = 3\left(\frac{x}{L_k}\right)^2 - 2\left(\frac{x}{L_k}\right)^3 \tag{5.80}$$

$$\mathcal{L}_4(x) = -\left(\frac{x}{L_k}\right)^2 + \left(\frac{x}{L_k}\right)^3 \tag{5.81}$$

and substitute into Eq. (5.77), we find

$$y_d(x) = \mathcal{L}_1(x)\, y_1 + \mathcal{L}_2(x)\, L_k\Theta_1 + \mathcal{L}_3(x)\, y_2 + \mathcal{L}_4(x)\, L_k\Theta_2 \tag{5.82}$$

The quantities $\mathcal{L}_1(x), \mathcal{L}_2(x), \mathcal{L}_3(x)$ and $\mathcal{L}_4(x)$ in Eq. (5.82) are given by Eqs. (5.78)–(5.81), respectively, and they are known as the Hermite cubics. For more information on this subject consult [3, 5] listed at the end of this text.

The stiffness matrix $[K^k]$ and mass matrix $[M^k]$ for the kth element are derived here by using Eqs. (5.65) and (5.66), respectively. A great deal of additional information regarding the derivation of stiffness and mass matrices for truss, beam and plate elements is given in the sixth chapter of [5].

In our problem here, an important factor in the procedure is to note that both the stiffness and mass matrices for each element have the same form, and this simplifies the solution. The stiffness and mass matrices for the kth element are as follows:

$$[K^k] = \frac{E\tilde{J}_k}{L_k^3}\begin{bmatrix} 12 & 6 & -12 & 6 \\ & 4 & -6 & 2 \\ & & 12 & -6 \\ \text{symm} & & & 4 \end{bmatrix} \tag{5.83}$$

$$[M^k] = \frac{\beta_k}{420L_k}\begin{bmatrix} 156 & 22 & 54 & -13 \\ & 4 & 13 & -3 \\ & & 156 & -22 \\ \text{symm} & & & 4 \end{bmatrix} \tag{5.84}$$

where $E\tilde{J}_k$ and β_k are the equivalent uniform bending stiffness and equivalent uniformly distributed mass density, respectively, for the kth element.

After the individual stiffness and mass matrices are formed, then they are assembled. This is accomplished by adding the contributions from all the elements. That is,

$$[K_s] = \sum_{k=1}^{M} [K^k] \tag{5.85}$$

$$[M_s] = \sum_{k=1}^{M} [M^k] \tag{5.86}$$

On this basis, the equation of motion has the following form:

$$[K_s]\{U\} - \omega^2 [M_s]\{U\} = 0 \tag{5.87}$$

In order to compute the natural free frequencies of vibration ω and the corresponding mode shapes, we must solve the eigenvalue problem,

$$[K]\{U\} = \omega^2 [M]\{U\} \tag{5.88}$$

where $[K]$ is the stiffness matrix, $[M]$ is the mass matrix, and $\{U\}$ represents the nodal coordinates. By multiplying both sides of Eq. (5.88) by $[K]^{-1}$, we obtain

$$\{U\} = \omega^2 [K]^{-1} [M] \{U\} \tag{5.89}$$

Equation (5.89) may be also written in a more standard form as follows:

$$[A - \Lambda I] \{U\} = 0 \tag{5.90}$$

where

$$A = [K]^{-1} [M]$$

$$I = \text{unit diagonal matrix} = [K]^{-1} [K] = \begin{bmatrix} 1 & 0 & 0 \\ 0 & 1 & 0 \\ 0 & 0 & 1 \end{bmatrix}$$

$$\Lambda = \frac{1}{\omega^2} \tag{5.91}$$

A canned eigensolver may be used to determine the eigenvalues and the corresponding eigenvectors. Equation (5.90) is used in this section to obtain the natural frequencies of vibration and the corresponding mode shapes for the case under consideration.

The following example illustrates the application of the preceding methodology.

Example 5.5 The uniform flexible cantilever beam in Fig. 5.1a has a uniform stiffness $EI_1 = 180 \times 10^3$ kip in.^{2}(516.54×10^3 N m^2), a uniform weight $w_0 = 1.5$ lb in^{-1}.(262.6902 N m^{-1}) along its length, and a length $L = 1{,}000$ in.(25.4 m). By using Galerkin's finite element method (GFEM), determine the free frequencies of vibration and the corresponding mode shapes of the flexible beam.

Solution: Before we are in a position to solve Eq. (5.29) using GFEM, we have to find first the rotation expression $y_s'(x)$ of the static equilibrium deflection curve y(s) of the flexible cantilever beam. We need to know $y_s'(x)$ because $I_e(x)$ and $m_e(x)$, which are given by Eqs. (5.30) and (5.31), respectively, are functions of $y_s'(x)$. For the problem at hand, this part of the work is already executed in Example 5.1, Sect. 5.3.1, of this text. By referring to this example, we find out that

$$y_s'(x) = \frac{G(x)}{\left\{1 - [G(x)]^2\right\}^{1/2}} \tag{5.92}$$

where G(x) is given by the expression

$$G(x) = 0.694444\,(10)^{-9}\left[2x^3 + 831.75\,x^2 - 1,189,561,532\right] \tag{5.93}$$

By substituting Eq. (5.92) into Eqs. (5.30), (5.31), and (5.34), we find

$$I_e(x) = I_1 \left\{1 - [G(x)]^2\right\}^{3/2} \tag{5.94}$$

$$m_e(x) = \frac{m_0}{\left\{1 - [G(x)]^2\right\}^{1/2}} \tag{5.95}$$

$$h_e(x) = h_1 \left\{1 - [G(x)]^2\right\}^{1/2} \tag{5.96}$$

where I_1, m_0, and h_1 are, respectively, the uniform moment of inertia, uniform mass per unit length, and uniform depth of the initial flexible member in Fig. 5.1a. The quantity $G(x)$ at any $0 \leq x \leq L_0$ is given by Eq. (5.93). Note that m_0 in Eq. (5.95) is equal to w_0/g, where g is the acceleration of gravity. The dynamically equivalent pseudovariable straight cantilever beam undergoing small amplitude vibrations is shown in Fig. 5.2. In Example 5.1 we have also found that $\Delta = 277.25$ in.(7.0422 m) and $L_0 = 722.75$ in.(18.3579 m).

The pseudovariable stiffness member in Fig. 5.2, may be solved by using the GFEM with piecewise equivalent uniform stiffness and piecewise equivalent uniform mass, as shown schematically in Fig. 5.5.

In order to apply GFEM with equivalent uniform stiffness $E\tilde{J}_k$ and equivalent uniform mass β_k for each element, the beam is subdivided into M elements. For the kth element, the expressions for $\tilde{J}_k$ and β_k are defined by Eqs. (5.54) and (5.55), respectively. By substituting for y_s' the expression given by Eq. (5.92), we find that $\tilde{J}_k$ and β_k are as follows:

$$\tilde{J}_k = \frac{1}{2} I_1 \left\{ f(x_i) \left[1 - G^2(x_i)\right]^{3/2} + f(x_j) \left[1 - G^2(x_j)\right]^{3/2} \right\} \tag{5.97}$$

$$\beta_k = \frac{1}{2} \left\{ \frac{m(x_i)}{\left[1 - (G(x_i))^2\right]^{1/2}} + \frac{m(x_j)}{\left[1 - (G(x_j))^2\right]^{1/2}} \right\} \tag{5.98}$$

The stiffness and mass matrices for the kth element are obtained using Eqs. (5.83) and (5.84), respectively. The element stiffness and mass matrices are then assembled by using Eqs. (5.85) and (5.86), respectively, and applying the boundary conditions of zero vertical displacement and zero rotation at the fixed end. The eigenvalue problem is then solved to determine the natural frequencies of vibration of the member and the corresponding mode shapes. The results are shown in Table 5.3 by using $M = 10, 20$, and 40 elements. The schematic representation of the first three modes of vibration is depicted in Fig. 5.6. Note in this figure that the vibration is taking place with respect to the static equilibrium position y_s.

The pseudovariable system in Fig. 5.2 is also solved by using the finite difference method (FDM) with 81 elements. The results are shown in Tables 5.3 and 5.4, which they show that both methods are in good agreement for the results obtained. In Table 5.4, we note that the case of $\Delta(x) = \Delta$ represents an upper bound for the results, while all other cases of $\Delta(x)$ represent a more correct variation of $\Delta(x)$. The results, however, indicate that the large deformation of the flexible bar is not very sensitive to the variation of $\Delta(x)$. Note in

Table 5.3 that the column designated as FDM, provides the results obtained using the finite difference method with 81 elements.

5.6 Vibration of Tapered Flexible Simply Supported Beams Using Galerkin's FEM

In this section, Galerkin's FEM with equivalent uniform stiffness and mass will be used to calculate the free frequencies of vibration and the corresponding mode shapes for tapered flexible simply supported beams. The following example illustrates the procedure and methodology.

Example 5.6 For the tapered flexible simply supported beam in Fig. 3.5, determine its free frequencies of vibration and the corresponding mode shapes by using Galerkin's FEM with equivalent uniform stiffness and mass. Assume that the uniform weight $w = 10\,\mathrm{lb\ in.}^{-1}(1,751.27\,\mathrm{N\ m}^{-1})$ represents the distributed weight of the member, as well as other weights securely attached to the member and participating in its vibrational motion. The length $L = 1000\,\mathrm{in.}$ (25.4 m), the taper $n = 1.5$, the stiffness EI_A at the left support A is $75 \times 10^3\,\mathrm{kip\ in.}^2(215,224\,\mathrm{N\ m}^2)$, and the constant modulus of elasticity $E = 30 \times 10^6\,\mathrm{psi}(206.85 \times 10^6\,\mathrm{kPa})$.

Solution: As discussed in the preceding section, we need to determine first the expression $y_s'(x)$ of the static equilibrium deflection curve $y_s(x)$ of the flexible tapered simply supported member. We need to do this because $I_e(x)$ and $m_e(x)$, which are given by Eqs. (5.30) and (5.31), respectively, are functions of $y_s'(x)$. This part of the work, for the problem at hand, is already carried out in Sect. 3.5 of this text by using the Euler–Bernoulli nonlinear differential equation.

On this basis, by referring to Sect. 3.5 and using Eqs. (3.53), (3.54), (3.50), and (3.52), we may write the expression

$$y_s'(x) = \frac{G(x)}{\sqrt{1 - [G(x)]^2}} \tag{5.99}$$

where

$$G(x) = \Phi(x) + C \tag{5.100}$$

$$\begin{aligned}\Phi(x) = \frac{w(L-\Delta)^3}{2EI_A}\Bigg\{&\frac{1}{(n-1)^3}\Bigg[\ell n\left((L-\Delta)+(n-1)x\right)\\&+\frac{2(L-\Delta)}{(L-\Delta)+(n-1)x}-\frac{(L-\Delta)^2}{2\left[(L-\Delta)+(n-1)x\right]^2}\Bigg]\\&+\frac{\Delta}{(n-1)^2}\Bigg[-\frac{1}{(L-\Delta)+(n-1)x}+\frac{(L-\Delta)}{2\left[(L-\Delta)+(n-1)x\right]^2}\Bigg]\Bigg\}\end{aligned} \tag{5.101}$$

$$C = \frac{w(L-\Delta)^3}{2EI_A}\left\{\frac{1}{(n-1)^3}\left[\ell n\left((L-\Delta)+(n-1)x_D\right)\right.\right.$$
$$+\frac{2(L-\Delta)}{(L-\Delta)+(n-1)x_D}-\left.\frac{(L-\Delta)^2}{2\left[(L-\Delta)+(n-1)x_D\right]^2}\right] \tag{5.102}$$
$$\left.+\frac{\Delta}{(n-1)^2}\left[-\frac{1}{(L-\Delta)+(n-1)x_D}+\frac{(L-\Delta)}{2\left[(L-\Delta)+(n-1)x_D\right]^2}\right]\right\}$$

In the above equations the quantity x_D defines the position $x = x_D$ where the maximum vertical displacement occurs. At this position, the rotation $y'_s(x) = 0$. Also, in the above equations, Δ represents the horizontal displacement Δ_B of the left support A and EI_A is the stiffness at support A. For the problem at hand, by referring to Sect. 3.5 and Table 3.7, we find, for example, that $\Delta = \Delta_B = 326.59$ in. (8.295 m), and $L_0 = L - \Delta = 1{,}000 - 326.59 = 673.41$ in. (17.105 m). Also at the position $x = x_D = 297.14$ in. (7.55 m), the maximum static vertical displacement δ_D of the number is 326.98 in. (8.3053 m), and at the left support ($x = 0$) the rotation $\theta_A = 73.49°$.

With known G(x) and $y'_s(x)$ we substitute Eq. (5.99) into Eqs. (5.30), (5.31), and (5.34) to obtain

$$I_e(x) = I_x\left\{1-[G(x)]^2\right\}^{3/2} \tag{5.103}$$

$$m_e(x) = \frac{m_0}{\left\{1-[G(x)]^2\right\}^{1/2}} \tag{5.104}$$

$$h_e(x) = h(x)\left\{1-[G(x)]^2\right\}^{1/2} \tag{5.105}$$

In the above three equations the moment of inertia I_x at any $0 \le x \le L_0$ is given by the expression, see Sect. 3.5,

$$I_x = I_A f(x) \tag{5.106}$$

where

$$I_A = \frac{bh^3}{12} \tag{5.107}$$

$$f(x) = \left[1+(n-1)\frac{x}{L-\Delta}\right]^3 \tag{5.108}$$

and b and h are, respectively, the width b and depth h at the left support A of the member. The width b, although not required, is assumed to be constant along the length of the member. In Eq. (5.104) the quantity $m_0 = w/g$ is the mass of the weight w carried by the original member, and in Eq. (5.105) the quantity h(x) represent the variation of the depth of the original tapered member.

The dynamically equivalent pseudovariable straight simply supported beam of length $L_0 = 673.41$ in. (17.1046 m) is shown schematically in Fig. 5.10a. This equivalent system may be solved by using Galerkin's FEM with piecewise equivalent uniform stiffness and piecewise equivalent uniform mass, as shown schematically in Fig. 5.10b.

For the application of the Galerkin's FEM with equivalent uniform stiffness $E\tilde{J}_k$ and equivalent uniform mass β_k for each k element, we subdivide the member into M elements as shown in Fig. 5.10b. For the kth element the expressions for $\tilde{J}_k$ and β_k are given by Eqs. (5.54) and (5.55), respectively. By using these two expressions and substituting for y_s' the expression given by Eq. (5.99), we find

$$\tilde{J}_k = \frac{1}{2} I_A \left\{ f(x_i) \left[1 - G^2(x_i)\right]^{3/2} + f(x_j) \left[1 - G^2(x_j)\right]^{3/2} \right\} \tag{5.109}$$

$$\beta_k = \frac{1}{2} \left\{ \frac{m(x_i)}{\left[1 - G^2(x_i)\right]^{1/2}} + \frac{m(x_j)}{\left[1 - G^2(x_j)\right]^{1/2}} \right\} \tag{5.110}$$

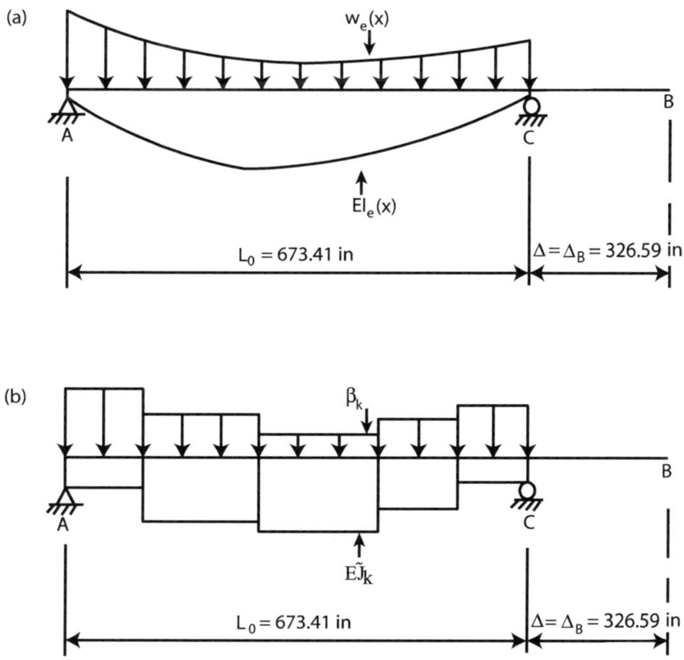

Fig. 5.10. **(a)** Dynamically equivalent straight simply supported beam of length L_0. **(b)** Dynamically equivalent piecewise uniform straight simply supported beam with equivalent piecewise uniform mass

Table 5.8. Free frequencies of vibration for a tapered flexible simply supported beam by assuming that Δ(x) is constant and equal to Δ

	GFEM				
ω (rps)	12 elements	22 elements	40 elements	FDM	% difference
ω_1	1.0955	1.0940	1.0927	1.0893	0.315
ω_2	3.0713	2.9992	2.9816	2.9632	0.618
ω_3	6.8513	6.6673	6.6194	6.5661	0.810
ω_4	12.3459	11.7284	11.6385	11.5034	1.174
ω_5	19.8877	18.2372	18.1244	17.8279	1.663
ω_6	29.9358	26.1894	26.1040	25.5183	2.295

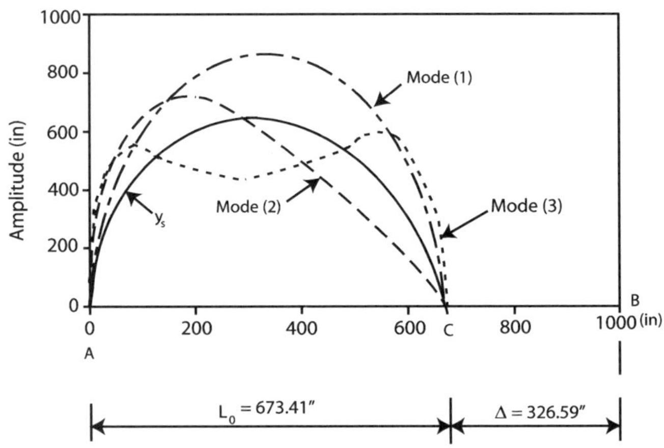

Fig. 5.11. The first three mode shapes of the tapered flexible simply supported beam

By following the procedure discussed in Sect. 5.5 and in Example 5.5, we find that the first six free frequencies of vibration, in radians per second (rps), are as shown in Table 5.8. The most accurate results were obtained by using 40 elements. In the same table the finite difference method (FDM) with 81 elements was also used to calculate these six frequencies and the results are compared. Note the percentage difference in the last column of the table. In Fig. 5.11 the corresponding mode shapes of the first three free frequencies of vibration are shown. The curve y_s represents the static equilibrium position of the member.

5.7 Concluding Remarks

The results and discussions in this chapter show that the undamped free frequencies of vibration of uniform and variable stiffness flexible members and their corresponding mode shapes may be determined rather conveniently in three main steps. Since free vibrations take place from the static equilibrium position of the flexible member, the first step would be the utilization of the Euler–Bernoulli nonlinear differential equation to determine the rotations y_s' and establish the static equilibrium position y_s. For this purpose, the utilization of appropriate equivalent systems as discussed in the first three chapters of this text will greatly simplify the solution of this part of the problem.

From the static equilibrium position, since the static deflection y_s is large, the member vibrates as a curved beam. Therefore, the purpose of this second step is to derive a dynamically equivalent straight beam of length $L_o = (L-\Delta)$, which vibrates transversely in the same way as the original member does about its static equilibrium position y_s. The moment of inertia $I_e(x)$, mass $m_e(x)$ and depth $h_e(x)$ of the equivalent straight beam of length L_o are appropriately defined so that the equivalent straight member vibrates with the same frequencies of vibration as the initial flexible member. An equivalent differential equation of motion given by Eq. (5.33), or Eq. (5.29) is derived for this purpose. Although the static deflection y_s is large, the amplitudes of vibration are assumed to be small and damping is neglected. This is not a severe limitation, because many engineering structures and their components are associated with small amplitude vibrations. However, if desired, these effects could be incorporated in the analysis.

The third and final step is to solve the equivalent differential equation of motion of the dynamically equivalent straight beam. The methods used here for this purpose are the Galerkin's finite element method with equivalent uniform stiffness and uniform mass for each element, and Rayleigh's energy method. The finite difference method is also used in order to compare results. The results illustrate that all three methods are reliable and they are in excellent agreement for practical applications. The results also show that the effect of mass position change during the large static deformation y_s *must* be taken into consideration. The error is very large if this is not included in the analysis (see Sect. 5.4. Additional work regarding this topic may be also found in the work of the author in [2, 3, 5]. In [2, 3] the Galerkin's consistent finite element method is also discussed in detail. All these methods are reliable and accurate because they solve the dynamically equivalent straight beam which is pseudolinear. In other words the initial nonlinear problem is reduced mathematically to an equivalent linear one, which permits the application of linear methods of vibration analysis.

Problems

5.1 For the uniform flexible cantilever beam shown in Fig. 5.1a, determine the dynamically equivalent pseudovariable straight cantilever beam of length L_o by using the methodology discussed in Sect. 5.2.2 and 5.3. The uniform stiffness EI of the flexible member is 180×10^3 kip in.^{2}(516, 541 N m^2), its uniform weight $w_o = 1$ lb in.$^{-1}$(175.1268 N m^{-1}), and its length $L = 1{,}000$ in. (25.4 m). Use $\Delta(x) = \text{constant} = \Delta$.

5.2 By using the dynamically pseudovariable straight cantilever beam derived in Problem 5.1, determine its fundamental frequency of vibration ω by applying Rayleigh's energy method. The frequency obtained by using the Galerkin's finite element method with piecewise uniform stiffness and mass is $\omega = 1.0564$ rps.

5.3 Solve Problem 5.1 by using $\Delta(x) = \Delta(x/L_o)^{1/2}$ and compare the results.

5.4 Repeat Problem 5.1 by using the dynamically equivalent pseudovariable straight cantilever beam derived in Problem 5.3.

5.5 Solve Problem 5.1 with a weight distribution $\omega_o = 2.5\,\text{lb in.}^{-1}$ ($437.817\,\text{N m}^{-1}$) and $\Delta(x) = \Delta \sin(\pi x/2L_o)$.

5.6 By using the dynamically equivalent pseudovariable straight cantilever beam derived in Problem 5.3 and applying Rayleigh's energy method, determine its fundamental frequency of vibration ω. The value obtained by using Galerkin's finite element method with piecewise uniform stiffness and mass is $\omega = 0.7808$ rps.

5.7 The tapered flexible cantilever beam shown in Fig. 5.7a has a uniform weight distribution $w_o = 2.0\,\text{lb in.}^{-1}(350.2536\,\text{N m}^{-1})$ as shown. The stiffness EI_A at the free end A of the member is 180×10^3 kip in.^{2}($516,541\,\text{N m}^2$), the length L = 1,000 in. (25.4 m), and the taper n = 1.5. By using a dynamically equivalent straight beam of length L_o and the energy method of Lord Rayleigh, determine the fundamental free frequency of vibration of the flexible member. Assume that $\Delta(x) = \Delta$. The value obtained by Galerkin's finite element method with piecewise uniform stiffness and mass is $\omega = 1.1649$ rps.

5.8 Solve Problem 5.7 by assuming that $\Delta(x) = \Delta(x/L_o)$ and compare the results.

5.9 Solve Problem 5.7 by assuming a weight distribution $w_o = 10.0\,\text{lb in.}^{-1}$ ($1,751.268\,\text{N m}^{-1}$). The value obtained by Galerkin's finite element method with uniform stiffness and mass is $\omega = 0.7964$ rps.

5.10 Solve Problem 5.7 by assuming a weight distribution $w_o = 5.0\,\text{lb in.}^{-1}$ ($875,634\,\text{N m}^{-1}$), $\Delta(x) = \Delta(x/L_o)$, and taper n = 2. The value obtained by Galerkin's finite element method with piecewise uniform stiffness and mass is $\omega = 1.0657$ rps.

5.11 The uniform flexible simply supported beam in Fig. 2.3 has a uniform weight distribution $w = 2.0\,\text{lb in.}^{-1}(350.2536\,\text{N m}^{-1})$, stiffness EI = 75,000 kip in.^{2}($215,224\,\text{N m}^2$), and its length L = 1,000 in. (25.4 m). By using a dynamically equivalent straight beam of length L_o and the energy method of Lord Rayleigh, determine the fundamental frequency of vibration of the flexible member. The value obtained by using Galerkin's finite element method is $\omega = 1.3935$ rps. Assume $\Delta(x) = \text{constant} = \Delta$.

5.12 Solve Problem 5.11 by using the horizontal displacement functions $\Delta(x) = \Delta(x/L_o)$, and $\Delta(x) = \Delta(x/L_o)^{1/2}$ and compare the results.

5.13 Solve Problem 5.11 by using the horizontal displacement functions $\Delta(x) = \Delta(x/L_o)$, and $\Delta(x) = \Delta \sin(\pi x/2L_o)$ and compare the results.

5.14 Solve Problem 5.11 by using a uniform weight distribution $w = 5.0\,\text{lb in.}^{-1}(875.634\,\text{N m}^{-1})$. The value obtained by using Galerkin's finite element method is $\omega = 1.1064$ rps.

5.15 Solve Problem 5.14 by using $\Delta(x) = \Delta(x/L_o)$, and $\Delta(x) = \Delta(x/L_o)^{1/2}$ and compare the results.

5.16 Solve Problem 5.14 by using $\Delta(x) = \Delta(x/L_o)^{1/2}$ and $\Delta(x) = \Delta\sin(\pi x/2L_o)$ and compare the results.

5.17 Solve Problem 5.11 by using a uniform weight distribution $w = 10.0\,\text{lb in.}^{-1}(1{,}751.268\,\text{N m}^{-1})$. The value obtained by using Galerkin's finite element method is $\omega = 0.9767$ rps.

5.18 Solve Problem 5.17 by using $\Delta(x) = \Delta(x/L_o)$, and $\Delta(x) = \Delta(x/L_o)^{1/2}$ and compare the results.

5.19 The tapered flexible simply supported beam shown in Fig. P5.19, is assumed to have the attached uniform weight distribution $w = 2.0\,\text{lb in.}^{-1}(350.2536\,\text{Nm}^{-1})$, stiffness $EI_A = 75{,}000\,\text{kip in.}^2$ $(215{,}224\,\text{N m}^2)$, length $L = 1{,}000\,\text{in.}$ (25.4 m), and taper $n = 1.5$. By using an equivalent straight beam of length L_o and the method of Lord Rayleigh, determine the fundamental frequency of the flexible member. Assume that $\Delta(x) = \Delta$. The value obtained by Galerkin's finite element method is $\omega = 1.7410$ rps.

5.20 Solve Problem 5.19 for $\Delta(x) = \Delta$ and $\Delta(x) = \Delta x/L_o$ and compare the results.

5.21 Solve Problem 5.19 by using a weight distribution $w = 5.0\,\text{lb in.}^{-1}$ $(875.634\,\text{N m}^{-1})$. The value obtained by using Galerkin's finite element method is $\omega = 1.2824$ rps.

5.22 Solve Problem 5.21 by using $\Delta(x) = \Delta$ and $\Delta(x) = \Delta x/L_o$ and compare the results.

5.23 Repeat the solution of Problem 5.7 by using Galerkin's finite element method with piecewise uniform stiffness and mass.

5.24 Repeat the solution of Problem 5.7 by using Galerkin's finite element method and a weight distribution $w_0 = 10.0\,\text{lb in.}^{-1}(1{,}751.27\,\text{N m}^{-1})$.

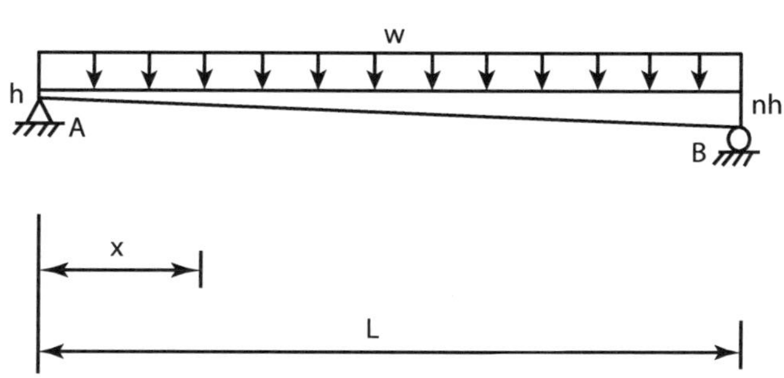

Fig. P5.19.

5.25 Repeat the solution of Problem 5.11 and determine the first six frequencies of its free vibration by using Galerkin's finite element method with piecewise uniform stiffness and mass.

5.26 Repeat Problem 5.25 by using a uniform weight distribution $w = 5.0\,\mathrm{lb\ in.}^{-1}(875.63\,\mathrm{N\ m}^{-1})$ and compare the results.

5.27 Repeat the solution of Problem 5.19 by using Galerkin's finite element method with piecewise uniform stiffness and mass.

5.28 Repeat the solution of Problem 5.20 by using Galerkin's finite element method with piecewise uniform stiffness and mass.

6

Suspension Bridges, Failures, Plates, and Other Types of Nonlinear Structural Problems

6.1 Introduction

In the past five chapters of this text, the nonlinear analysis was limited to one dimensional structures and structural components such as beams with various loading conditions and cross-sectional geometric variations throughout their length. Convenient and unique methods of analysis have been developed to simplify the solution of very complex static and dynamic nonlinear problems. This process helped to explain and understand the fundamentals and the principles associated with the one-dimensional general beam problem and prepare ourselves for in-depth comprehension of the underlined principles associated with the more complicated two-dimensional and three-dimensional nonlinear problems.

In this chapter, we expand this nonlinear process to include selected multi-dimensional practical problems, such as suspension bridges and their failures, rectangular plates of uniform and variable thickness, and multistory building subjected to earthquake excitations, as well as columns loaded eccentrically, and the inelastic analysis of members with axial restraints. The nonlinear problem is complicated, and it needs the development of new improved and unique methods of analysis for a satisfactory and safe design for our nonlinear structures.

6.2 Brief Discussion on Fundamental Aspects of Suspension Bridges

The following items are intended to pinpoint some of the problems, concerns, suggestions, and experimental investigations that need to be taken into consideration regarding the design of suspension bridges:

1. *Drag forces.* Suspension bridges must be designed to withstand drag forces that are produced by severe winds.

2. *Aeroelastic effects.* Suspension bridges must be capable of resisting aeroelastic effects, which involve vortex-induced oscillation, flutter, torsional divergence or lateral buckling, galloping, and buffeting caused by the presence of self-excited forces. Tunnel tests can provide the required information to perform such studies.
3. *Wind action.* The action of the wind must be taken into consideration both for the completed bridge and for its state during the construction stage.
4. *Precautions during construction.* Temporary ties and damping mechanisms may be needed for partially completed suspension bridges, and when possible, construction of suspension bridges should take place at times and seasons where the probability of severe storms is low.
5. *Other parts of the structure.* Aeroelastic studies must be also performed for suspension bridge towers, hangers, and cables, in addition to the ones performed for the bridge deck, because these bridge elements can be effected by the various aeroelastic phenomena mentioned in item 2 above.
6. *Wind tunnel tests.* These tests are used extensively by experimental researchers to obtain the necessary data for calculating the aerodynamic response of suspended-span bridges. The following test models are currently used for such studies
   (a) *Full-bridge models.* Scales in the order 1/300 are used to prepare models of the full bridge. Such scales are geometrically similar to the full bridge and satisfy similarity requirements with regard to mass distribution, reduced frequency, mechanical damping, and mode shapes. Such full-scale models are very costly and very elaborate to construct.
   (b) *Taut strip models.* These models are prepared in such a way that they will respond in a manner similar to the center span of the suspension bridge when they are placed in the laboratory wind flow. Two wires are stretched across the wind tunnel and are internally clad to represent the geometry of a given bridge, and also to permit the duplication to model frequency scale of the fundamental bending and torsional frequencies of the bridge.
   (c) *Section models.* Section models are usually inexpensive, and they can be constructed to scales of the order of 1/50 to 1/25 in order to reduce the discrepancies in Reynolds number between the full-scale bridge and the scale model. They consist of representative spanwise sections of the deck that are supported at the ends by springs that allow both torsional and vertical motion. They are usually enclosed between end plates in order to reduce the end aerodynamic effects. Based on simple tests, they are used to make initial assessments regarding the aeroelastic stability of the shape of the bridge deck, and also to make measurement of fundamental aerodynamic characteristics of the bridge deck that are required for comprehensive analytical studies. These aerodynamic characteristics involve measurements of the steady-state drag, lift and moment coefficients, as well as motional

aerodynamic coefficients that characterize the self-excited forces acting on the oscillating bridge, and the Strouhal number that describes the vortex-shedding phenomenon.

7. *Torsional divergence or lateral buckling.* When the deck of a suspension bridge is subjected to a slight twist, the drag load and the self-excited aerodynamic moment will cause a torsional divergence instability. When the wind acts on a bridge, a drag force, a lift force, and a twisting moment will be developed, and these must be resisted by the bridge. The twisting moment increases with increasing wind velocity and increasing angle of attack of the wind relative to the structure, thus demanding additional reactive moment from the structure. A velocity, however, may be reached where the magnitude of the wind-induced moment, together with the tendency of the twist to require additional structural reaction, develops an unstable condition that causes the structure to twist to destruction. This can be thought of as a problem of instability analogous to the one associated with column buckling. Torsional divergence will occur at some critical divergence velocity of the wind, while column buckling will occur when a critical value of the compressive column load is reached. Such conditions do not depend upon ultimate structural strength, but they do depend upon the structural flexibility and the way the aerodynamic moments develop with twist.
   In the case of a full bridge, the analysis can be carried out by using the experimentally measured moment coefficients and the torsional flexibility matrix of the bridge deck. An iteration procedure may be used such that, for any chosen velocity less than the critical divergence velocity, the procedure will converge. The iteration procedure will approach the critical divergence velocity in an asymptotic manner. This instability, for wind speeds attainable in practice, is found to occur only for torsional weak bridges.
8. *Aeroelastic stability.* The aeroelastic stability of a suspension bridge is controlled by three important factors, namely, geometry of the bridge deck, frequencies of vibration of the bridge, and mechanical damping of the bridge. Regarding the first factor, unstable shapes are considered to be solid girders of deck form, very bluff cross section, and open-truss deck sections with closed or unvented roadways. Streamlined forms and open-truss sections with vents or grills through the roadway surface enhance the stability of the bridge deck. Regarding the second factor, stability is enhanced with high torsional frequencies and high ratio of torsional-to-bending frequency. The aerodynamic stability of a suspension bridge is also enhanced if its mechanical damping ratios are high.
9. *Galloping.* The influence of galloping on suspension bridge decks may be evaluated by examining the graphs of the lift and drag coefficients that are plotted as functions of the angle α between the horizontal plane and the plane of the bridge deck. Bridge deck shapes with regions of strongly negative slopes of the lift curve should be avoided.

10. *Buffeting.* Buffeting is defined as the unsteady loading of a structure caused by velocity fluctuations in the oncoming flow. Wide bridge decks generally exhibit higher buffering response. Also, bridges with deck sections that have high drag, or that have steep slopes in their lift and moment curves, tend to be more susceptible to buffering. Reduction of buffering response may be obtained by increased mechanical damping.

Additional work on this subject may be found in the literature; see for example [98].

6.3 The Collapse of the Tacoma Narrows Suspension Bridge

The collapse of the first Tacoma Narrows suspension bridge triggered the great concern regarding the dynamic performance and response of suspension bridges. The first Tacoma Narrows suspension bridge was built near Seattle, Washington, between 23 November 1938 and 1 July 1940, at a cost of approximately $6.4 million at that time. The total length of the bridge was 5,000 ft (1,524 m) accommodating two lanes of traffic, the height of the side girders was 8 ft (2.44 m), and the two end spans were each 1,100 ft (335.4 m) long.

Problems with this bridge were apparent even during construction, as large vertical oscillations were exhibited at that time. After completion, the deck of both the side spans and the main span of the bridge oscillated in a twisting fashion under the action of relatively light winds. Finally, on 7 November 1940, the suspension bridge collapsed. It is now generally assumed that the collapse of the bridge was due to a combination of torsional and vertical oscillations of the deck, and that this mode of failure was not taken into consideration by the designer.

It is interesting to note the various different explanations that were given by investigators regarding the failure of the Tacoma Narrows bridge. Six such explanations are given below:

1. Board of Investigation, Tacoma Narrows Bridge, L.J. Svendrup, Chairman, 26 June 1941. Primary cause lies in the general proportions of the bridge and the type of stiffening girders and floors. Ratio of width to length of main span much smaller, and vertical stiffness mush less compared to previously constructed bridges.
2. A Report to the Honorable John M. Carmody, Administrator, Federal Works Agency, Washington, DC, 28 March 1941. The vortex formation and frequency is determined by the oscillation of the structure rather than the oscillatory motion is induced by the vortex formation.
3. *Bridges and Their Builders,* D. Steinman and S. Watson, Putnam's Sons, New York, 1941. Small undulation, once started, the resultant effect of a wind tends to cause a building up of vertical undulations with a tendency

to change to a twisting motion, until the torsional oscillations reach destructive proportions.

4. University of Washington Engineering Experiment Station Bulletin No. 116, 1952. The motions were a result of vortex shedding.
5. *Bridges and Men*, J. Gies, Doubleday and Co., New York, 1963. Long narrow, shallow, and therefore very flexible structure standing in a wind ridden valley. Its stiffening support was a solid girder, which, combined with a solid floor, produced a cross section peculiarly vulnerable to aerodynamic effects.
6. *Wind Forces on Buildings and Structures*, E. Houghton and N. Carruthers, John Wiley & Sons, New York, 1976. Aerodynamic instability was responsible for the failure of the Tacoma Narrows Bridge in 1940. The oscillations are caused by the periodic shedding of vortices on the leeward side of the structure, a vortex being shed first from the upper section and then the lower section.

As stated above, the construction of the first Tacoma Narrows bridge was finished on 1 July 1940. The construction of a second bridge at the same location was finished on 14 October 1950. On 7 November 1940, at 10 a.m., winds had reached 42 mph, at 10:30 collections of tolls were discontinued, and at 11:08 a.m. the first Tacoma Narrows suspension bridge collapsed. Figure 6.1a and b shows photographs of the failing suspension bridge. These two figures show the predominant influence of twisting during failure.

To this date, we do not really have a complete explanation regarding the exact factors that caused the collapse of the Tacoma Narrows suspension bridge. It is, however, proven that this bridge was significantly different in many ways from other suspension bridges built at that time. The cross section of the deck consisted of two very narrow I beams, 39 ft (11.89 m) apart and 8 ft (2.44 m) deep. The deck was braced laterally to resist static wind loads. It was unusually narrow and less deep compared to previous suspension bridges. The deck was stiffened by girders (flanges) rather than trusses, and such girders do not provide good aeroelastic stability. After work on the bridge had started, the deck itself was redesigned in order to save on the structural steel reinforcement of the concrete roadway. These factors must have contributed significantly to the torsional stiffness reduction of the bridge deck.

Another factor that is believed to have contributed significantly to the failure of this suspension bridge is the relative closeness of the normal vertical and torsional frequencies. The normal vertical frequency represents the motion in which the amplitudes of oscillation of the two sides are equal in magnitude and direction. The torsional frequency represents the motion in which the amplitudes of the two sides are equal in magnitude but opposite in direction. It was reported [99] that the calculated value of the torsional frequency was 10 cycles per minute (cpm), while that of the normal vertical frequency was 8 cpm, giving a torsional-to-vertical frequency ratio of 1.25. This is very low compared to general practices for other existing suspension bridges, Table 6.1,

(a)

(b)

Fig. 6.1. (**a**) Fatal torsional oscillation mode of Tacoma Narrows Bridge. (**b**) Collapsing of the Tacoma Narrows Bridge. Photos by Farguharson, negative nos. 4 & 12 Special Collections Division University of Washington Libraries

Table 6.1. Vertical f_v and torsional f_t oscillations of suspension bridges (1 ft = 0.3048 m)

Bridge	Span length (ft)	f_v (cpm)	f_t (cpm)	f_v/f_t
Verrazano	4,260	6.2	11.9	1.92
Golden Gate	4,200	5.6	11.0	1.96
Sevem	3,240	7.7	30.6	3.97
First Tacoma Narrows	2,800	8.0	10.0	1.25

indicates that the first Tacoma Narrows Bridge was torsionally flexible. Since the calculated values of these two frequencies are only approximate, the probability of resonant vibration and the creation of divergence instability is rather high. See also the author's work in [[2], Chap. 12].

The collapse of the Tacoma Narrows bridge, however, created a great deal of interest in research and discussion of the aerodynamic and general dynamic behavior of suspension bridges. A great deal of this research concentrated on suspension bridge aerodynamics, identifying the aerodynamic behavior of bridge sections by experimentally determining motional aerodynamic coefficients of freely oscillating models. Studies in turbulent flow and wake effects were performed by using bridge section models, and full-scale investigations were used to improve the design of suspension bridges. Extensive work was also performed in obtaining motional aerodynamic coefficients to be used in analytical studies and nonlinear aerodynamic investigations.

The controversy regarding the causes that contributed to the collapse of the first Tacoma Narrows Bridge still continues today, because no satisfactory answer to these questions has yet been obtained.

6.4 Other Failures and What We Learn from Them

The Tacoma Narrows bridge is not the only suspension bridge that has failed. Table 6.2 lists 10 additional suspension bridges that failed before the first Tacoma Narrows bridge. Note, however, that the main span of the Tacoma Narrows bridge is much longer than the span lengths of all other bridges that have failed.

The failure of the first Tacoma Narrows suspension bridge, as well as many other structural failures, has taught us that a structure may be considered adequately designed and may provide problem-free service for decades, while still having a serious design flaw present, a flaw that will eventually result in a sudden collapse. For example, the elevated walkways of the Kansas City Hyatt Regency Hotel were considered safe to carry pedestrian crowds until they collapsed on the afternoon of 17 July 1981. The Mianus River bridge near Greenwich, Connecticut, was considered to be adequately designed to

Table 6.2. Collapsed suspension bridges (1 ft = 0.3048 m)

Bridge	Designer	Span length (ft)	Failure date
Dryburg (Scotland)	J. and H. Smith	260	1,818
Union (England)	Sir Samuel Brown	449	1,821
Nassau (Germany)	L. and W (Lossen and Wolf)	245	1,834
Brighton (England)	Sir Samuel Brown	255	1,836
Montrose (Scotland)	Sir Samuel Brown	432	1,838
Menai (Wales)	T. Telford	580	1,839
Roche (France)	Le Blanc	641	1,852
Wheeling (USA)	C. Ellet	1,010	1,854
Niagara (USA)	E. Serrell	1,041	1,864
Niagara (USA)	S. Keefer	1,260	1,889
Tacoma Narrows	L. Moisseff	2,800	1,940

carry the heavy traffic of Interstate 95, but it suddenly collapsed early in the morning of 28 June 1983.

Failure analysis, coupled with comprehensive knowledge of the causes of failure, can help to eliminate design errors and prevent the repetition of such catastrophic failure. We, the engineers, as Lev Zetlin [100] has said, "have to be somewhat paranoiac during the design stage of a structure." We should consider and imagine that the impossible could happen. We should not feel secure that the structure will be safe and sound merely because all the requirements of the design handbooks and manuals have been satisfied. If we do not remember the past and we do not learn from past mistakes, then we will be condemned to repeat them.

John Augustus Roebling, a well-known suspension bridge design engineer, realized the economical suspension bridge principle and he did not abandon it because of past failures. Instead, he used the concept of failure analysis to learn what mistakes should be avoided and eliminate in his new designs the weaknesses that had caused the collapse of past suspension bridges. He used this knowledge in his successful design of the Niagara Gorge Bridge, between western New York State and Canada, and the Brooklyn Bridge. In fact the span of the Brooklyn Bridge was twice as long as that of the Niagara Gorge Bridge.

Civil engineering structures are not the only ones that have failed. Other engineering fields, such as electrical, mechanical, computer, manufacturing, aeronautical, aerospace, and so on, have also recorded catastrophic failures. Such examples, to name just a few, are the 1979 loss of coolant at the Three Mile Island Nuclear Plant near Harrisburg, Pennsylvania, and the 1986 explosion and fire at the Chernobyl Nuclear Plant near Kiev, Ukraine, where radioactive material was spread well beyond its point of release. Another devastating failure was the 1984 release of toxic gas from a Union Carbide insecticide plant at Bhopal, India, which killed more than 2,000 people and harmed about 150,000 more.

The decisions regarding the design of engineering structures are made by people, and people are responsible when errors are made and the structure fails. Therefore, in order to develop successful designs that protect the safety and welfare of people, we must learn from past mistakes and also develop sound new technologies that help to improve the functioning and service of our infrastructure systems. This is the lesson we learn from past failures.

6.5 Eccentrically Loaded Columns

In practical applications the compressive loads on columns are not usually perfectly axial. Such situations can develop for many reasons. One reason is that the construction of structural and mechanical components is not perfect. Construction imperfections can produce column eccentricities that were never intended to exist; these can result in axial loads that are no longer axial. Construction imperfections can also result in a column that is no longer straight, and the assumption that requires the column to be straight is violated. In other situations, the design of the total structure is performed in a way that permits compressive loads not to be axial. With such eccentricities, the column will bend even with very small compressive loads, and it will not be initially straight as it is assumed for ideal columns. As the compressive load increases, bending of the column will increase, and the critical load for such columns will be different than the one obtained for ideal columns.

The preceding discussion can be illustrated graphically by observing the load–deflection curves shown in Fig. 6.2. For an ideal column at $P = P_{cr}$, the column may be subjected to any lateral small deflection y that is represented by curve 1 in the figure. The deflection must be small, because small deflection

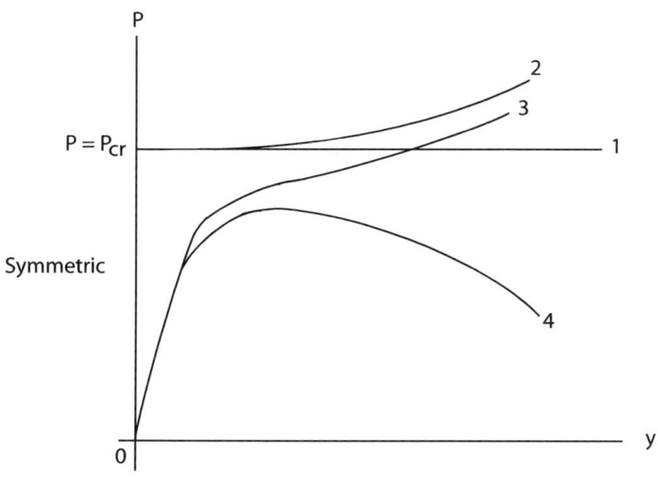

Fig. 6.2. Load–deflection curves for buckled columns

theory is used to determine P_{cr}. If deflections are permitted to be large and large deflection theory is used to determine P_{cr}, then the ideal column will follow curve 2, as shown in Fig. 6.2, indicating that when an elastic column starts to buckle it requires increasingly larger compressive loads to cause an increase in deflection.

If the column is not constructed perfectly, as discussed above, the load–deflection curve follows the shape shown by curve 3 in Fig. 6.2. For small deflections, it approaches curve 1, but when the deflections become large, it approaches curve 2. However, the degree of closeness to either curve 1 or curve 2 depends on the degree of the imperfections. When the imperfections are very small, curve 3 approaches curve 1, and curve 3 will move further to the right if the imperfections are larger.

If the compressive stresses are permitted to exceed the proportional limit of the material of the column and Hooke's law does not apply, then the load–deflection curve would be as shown by curve 4 in Fig. 6.2. Initially curves 3 and 4 coincide, since the response is elastic, but curve 4 departs from curve 3 as the column response becomes inelastic. It continues upward, reaches a maximum value for the compressive load P, and then turns downward. Stockier columns fall into the inelastic response category and follow curve 4, while only very slender columns may remain elastic up to $P = P_{cr}$. The disadvantage with curve 4 is that after the maximum compressive load is reached, larger deflections can be produced with smaller P. On the other hand, the critical load increases as the deflections become large, and the maximum compressive load is smaller than the one for the other three curves. Thus, if a structure is permitted to respond inelastically, different design criteria may have to be followed by the practicing design engineer regarding allowable limits for compressive loads.

For eccentrically loaded columns, we consider the column shown in Fig. 6.3a in order to determine its response in terms of critical load and lateral deflection. By using the free-body diagram in Fig. 6.3b and applying statics, the bending moment M, at any distance x from end A, is

$$M = P(e + y) \tag{6.1}$$

For small deflection theory, the Euler–Bernoulli equation is given by the following second order differential equation:

$$EIy'' = -M \tag{6.2}$$

By substituting Eq. (6.1) into Eq. (6.2), we obtain

$$EIy'' = -P(e + y) \tag{6.3}$$

or

$$y'' + k^2y = -k^2e \tag{6.4}$$

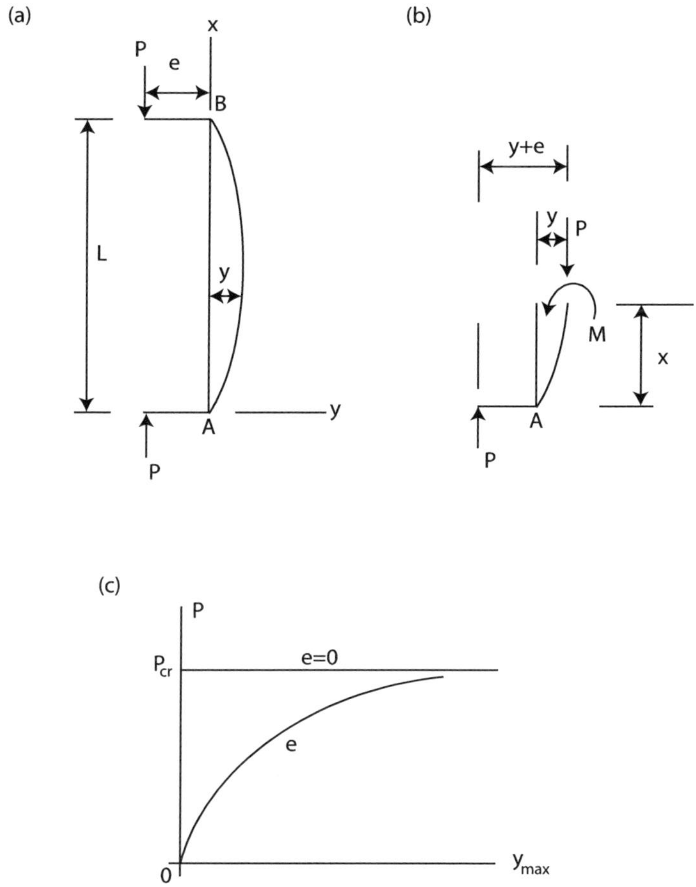

Fig. 6.3. (**a**) Eccentrically loaded column. (**b**) Free-body diagram. (**c**) Plot of P vs. y_{max} for values of $e = e$ and $e = 0$

where

$$k^2 = \frac{P}{EI} \tag{6.5}$$

Equation (6.4) is a nonhomogeneous differential equation in y, and its solution y is

$$y = y_c(x) + y_p(x) \tag{6.6}$$

where the complementary solution $y_c(x)$ is the solution of the homogeneous equation $y'' + k^2 y = 0$, and it is given by the expression

$$y = A \sin kx + B \cos kx \tag{6.7}$$

where A and B are constants that can be determined by using the boundary conditions of the column. The particular solution $y_p(x)$ may be assumed to be equal to a constant C, since k^2e in Eq. (6.4) is constant. By substituting into Eq. (6.4), we obtain $k^2C = k^2e$, or $C = -e$. Thus, $y_p = -e$. On this basis, Eq. (6.6) yields

$$y = A \sin kx + B \cos kx - e \tag{6.8}$$

The constants A and B can be determined by using the boundary conditions that the deflection y is zero at $x = 0$ and $x = L$, yielding

$$B = 0 \tag{6.9}$$

$$A = \frac{e(1 - \cos kL)}{\sin kL} = \frac{e \tan kL}{2} \tag{6.10}$$

By substituting Eqs. (6.9) and (6.10) into Eq. (6.8), the general solution for y is as follows:

$$y = e\left(\tan \frac{kL}{2} \sin kx + \cos kx - 1\right) \tag{6.11}$$

Equation (6.11) may be used to determine the deflection y at any distance x from support A for a given value of the eccentricity e. The maximum deflection y_{max} occurs at $x = L/2$, and it is as follows:

$$y_{max} = e\left(\sec \frac{kL}{2} - 1\right) \tag{6.12}$$

Equation (6.12) shows that $y_{max} = 0$ when $e = 0$, or when $P = 0$.

We note here that the behavior of eccentrically loaded columns is not similar to the response of ideal columns. Ideal columns are initially straight, and they stay straight until the critical value of P is reached. At $P = P_{cr}$, the static equilibrium becomes neutral, and the lateral deflection changes from zero to undefined. On the other hand, for a given value of P, eccentrically loaded columns are subjected immediately to definite values of the deflection y. Also, if we select a value for the eccentricity e and plot P vs. y_{max}, we note that this curve is nonlinear and, consequently, the superposition principle cannot be used to calculate deflection when the loads are more than one.

The plot of P vs. y_{max} for a given value of e is shown in Fig. 6.3c. This graph shows that as P approaches the value $P = P_{cr} = \pi^2 EI/L^2$, the secant term in Eq. (6.12) approaches infinity, which indicates that y_{max} increases without limit as $P \to P_{cr}$, and the horizontal line representing $e = 0$ becomes an asymptote for any curve e. Thus, the ideal column is the limiting case of an eccentrically loaded column. Note, however, that the differential equation used in the derivation applies only to small deflections, and large nonlinear deflection theory must be used for large deflections.

6.6 Inelastic Analysis of Members with Axial Restraints Using Equivalent Systems

In Chap. 4, the inelastic analysis was performed for members of uniform and variable thickness, and with loading conditions which permit consideration of coaxiality between neutral and centroidal axes. Such types of problems often confront the practicing engineer, and they deserve this special treatment. There are, however, situations where the assumption of coaxiality between neutral and centroidal axes is no longer valid, and the shifting of the neutral axis at cross sections along the length of the member must be taken into consideration. Members subjected to axial compressive or tensile loads in addition to transverse loadings, or statically indeterminate problems involving axial restraints, are examples of such types of problems.

The inelastic analysis developed in this section takes into consideration such types of loading conditions, and the computation of the reduced modulus E_r, and, consequently, the derivation of the equivalent system of constant stiffness E_1I_1, takes into consideration the shifting of the neutral axis at cross sections along the length of the member. Figure 6.4 illustrates a loading condition for a member of rectangular cross section, where coaxiality between

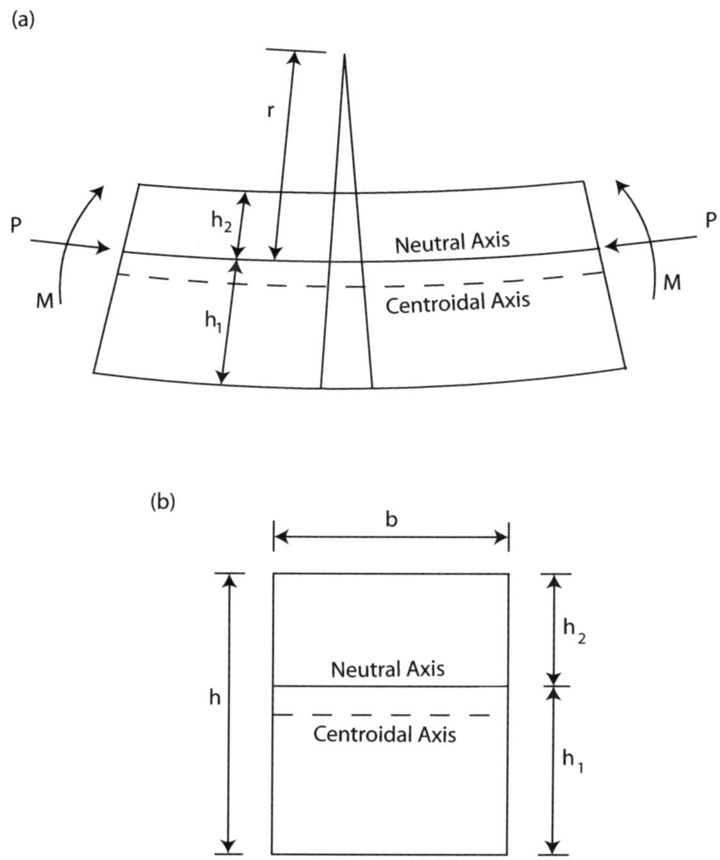

Fig. 6.4. (**a**) Free-body diagram of a portion of a member where neutral and centroidal axes do not coincide. (**b**) A rectangular cross-section of the member

neutral and centroidal axes does not exist. Prismatic and nonprismatic members with axial restraints and transverse loadings are discussed in this section. Both elastic and inelastic ranges are considered for comparison purposes, but it should be emphasized, however, that axial restraints become more important in the inelastic range where deformations are larger. Also, in this range, the variations in the axial restraints and in the position of the neutral axis at cross sections along the length of the member are more distinctive. The derivation of equivalent systems for the inelastic analysis of such members is as follows.

When a member is subjected to external loads that produce axial restraints, or to external loads acting in both transverse and axial directions, the strains ε_1 and ε_2 in Fig. 4.1a or 6.4a will not be equal, and the neutral axis will not coincide with the centroidal axis of the rectangular cross section. Under these loading conditions, the computation of the reduced modulus E_r and, consequently, the derivation of the constant stiffness equivalent system, should take into consideration the shifting of the neutral axis at sections along the length of the member.

Let it be assumed that in addition to the transverse loading, the member is also subjected to an axial tensile or comprehensive force P. In Fig. 6.4, h_1 and h_2 are the distances from the neutral axis to the lower and upper surfaces of the member, respectively. Thus, $\varepsilon_1 = h_1/r$, $\varepsilon_2 = h_2/r$, and $r = h/\Delta$, where r is the radius of curvature. At a distance y from the centroidal axis of a cross section of the member, the total longitudinal strain ε may be expressed as

$$\varepsilon = \varepsilon_0 + \frac{y}{r} \tag{6.13}$$

where

$$\varepsilon_0 = \frac{P}{bhE} \tag{6.14}$$

b is the width of the cross section, y/r is the strain produced by bending only, and E is the modulus of elasticity that can be either elastic or inelastic. Thus, by using Eq. (6.13), we can write,

$$y = r(\varepsilon - \varepsilon_0) \tag{6.15}$$

$$dy = r\ d\varepsilon \tag{6.16}$$

On this basis, we have

$$P = b\int_{-h_2}^{h_1} \sigma\ dy = br\int_{\varepsilon_2}^{\varepsilon_1} \sigma\ d\varepsilon = \frac{bh}{\Delta}\int_{\varepsilon_2}^{\varepsilon_1} \sigma\ d\varepsilon \tag{6.17}$$

The integral on the right-hand side of Eq. (6.17) represents the area under the stress–strain curve between the limits ε_1 and ε_2.

The expression for the internal bending moment at a cross section of the member may be written as

$$M = b \int_{-h_2}^{h_1} \sigma y \, dy = br^2 \int_{\varepsilon_2}^{\varepsilon_1} \sigma (\varepsilon - \varepsilon_0) \, d\varepsilon \tag{6.18}$$

or, by rearranging, we may rewrite Eq. (6.18) as follows:

$$M + br^2 \varepsilon_0 \int_{\varepsilon_2}^{\varepsilon_1} \sigma \, d\varepsilon = br^2 \int_{\varepsilon_2}^{\varepsilon_1} \sigma \varepsilon \, d\varepsilon \tag{6.19}$$

In Eq. (6.19), M is the bending moment with respect to the centroidal axis, and

$$br^2 \int_{\varepsilon_2}^{\varepsilon_1} \sigma \varepsilon_0 \, d\varepsilon \tag{6.20}$$

represents the additional moment produced by the axial load that causes the shifting of the neutral axis. Therefore, Eq. (6.19) gives the bending moment at any cross section of the member, with respect to the neutral axis. Due to the presence of the axial force P, the neutral axis will not coincide with the centroidal axis.

By using Eq. (6.17), we can rewrite Eq. (6.19) as follows:

$$M + \varepsilon_0 r P = br^2 \int_{\varepsilon_2}^{\varepsilon_1} \sigma \varepsilon \, d\varepsilon \tag{6.21}$$

$$M + \varepsilon_0 \frac{h}{\Delta} P = br^2 \int_{\varepsilon_2}^{\varepsilon_1} \sigma \varepsilon \, d\varepsilon \tag{6.22}$$

$$M + \varepsilon_0 \frac{h}{\Delta} P = \frac{I}{r} \frac{12}{\Delta^3} \int_{\varepsilon_2}^{\varepsilon_1} \sigma \varepsilon \, d\varepsilon \tag{6.23}$$

or

$$M + \varepsilon_0 \frac{h}{\Delta} P = \frac{I}{r} E_e \tag{6.24}$$

where

$$E_e = \frac{12}{\Delta^3} \int_{\varepsilon_2}^{\varepsilon_1} \sigma \varepsilon \, d\varepsilon \tag{6.25}$$

$$I = \frac{bh^3}{12} \tag{6.26}$$

In Eq. (6.24), the quantity E_e may be thought of as an equivalent modulus that corrects for the shifting of the neutral axis and allows us in the analysis to maintain coaxiality between neutral and centroidal axes when the equivalent system of constant stiffness $E_1 I_1$ is used. In order to determine E_e, however, the position of the neutral axis at sections along the length of the member must be known.

The position of the neutral axis may be determined by using a trial-and-error procedure. For example, we can start the procedure by assuming a shape for the vertical deflection curve v of the member. As a first step, the assumed curve may be the deflection curve that is obtained from the inelastic analysis with the axial force being assumed as equal to zero. The procedure is discussed in Chap. 4. This is usually an excellent first trial deflection and makes the procedure to converge rather rapidly. The next reasonable step would be to assume values of $\varepsilon_1, \varepsilon_2$, and adjust Δ in Fig. 6.5a so that Eq. (6.17) is satisfied. However, the values of $\varepsilon_1, \varepsilon_2$, and Δ should be proportioned so that the sum of the two terms on the left-hand side of Eq. (6.23) is equal to the value obtained from the right-hand side of the equation. This procedure should be repeated until both Eqs. (6.17) and (6.23) are satisfied.

Note that M in Eq. (6.23), or Eq. (6.24), is the bending moment produced by the external load by including the moment Pv of the axial load. The term $\varepsilon_0 hP/\Delta$ in Eq. (6.23) or (6.24) is the additional bending moment produced by the axial load due to the shifting of the neutral axis. In reality, since $\varepsilon_0 h/\Delta$ is the distance between neutral and centroidal axes, it represents the moment of the axial load P about the neutral axis of the cross section considered. The axial load at sections along the length of the member will be somewhat

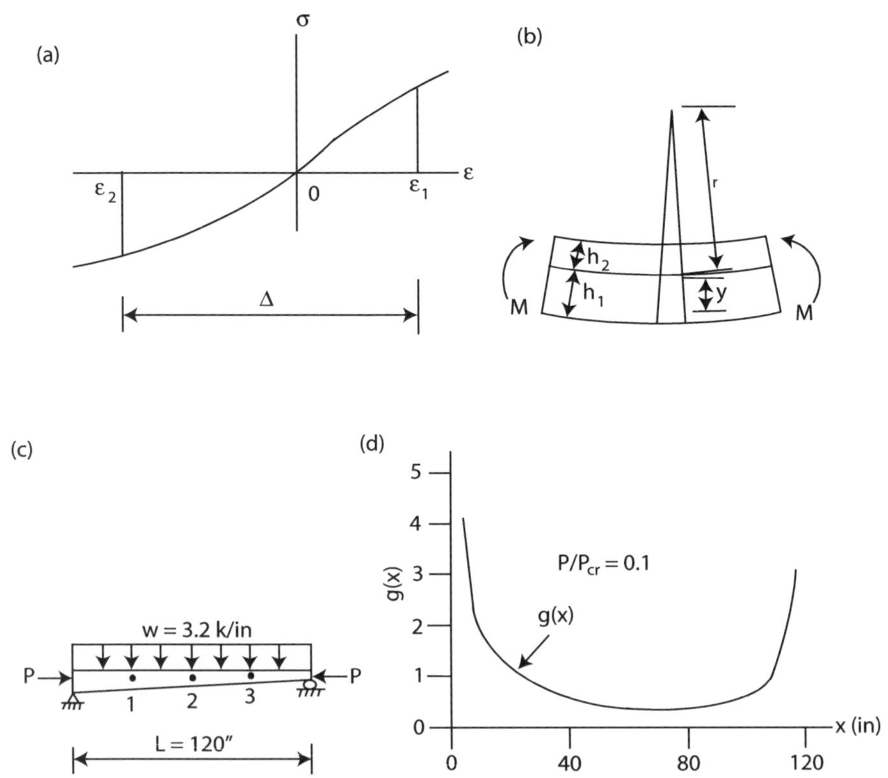

Fig. 6.5. (**a**) Stress-strain curve. (**b**) Free-body diagram showing r, h_1, h_2 and y. (**c**) Variable thickness member loaded with a uniformly distributed load $w = 3.2\,\text{kips in.}^{-1}$ and a compressive axial load P. (**d**) Variation of modulus function g(x) along the length of the member ($1\,\text{in.} = 0.0254\,\text{m}$, $1\,\text{kip in.}^{-1} = 175.1268\,\text{kN m}^{-1}$)

variable, since ε, Δ, and h may be variable. The equivalent modulus E_e may be obtained from Eq. (6.25).

With known equivalent modulus E_e, the modulus function g(x) may be determined from the equation

$$g(x) = \frac{E_e}{E_1} \tag{6.27}$$

where E_1 in this equation is the reference elastic modulus value. With known g(x), the equivalent system of constant stiffness E_1I_1 may be determined as in Chaps. 1 and 4. For small deflection theory, for example, the moment diagram M_e of the equivalent system of constant stiffness E_1I_1 may be determined from Eq. (1.115) or Eq. (1.142), where f(x) and g(x) are, respectively, the moment of inertia and modulus of elasticity functions. The shear force V_e and loading w_e of the equivalent system can be obtained from Eqs. (1.144) and (1.145), respectively.

The solution of the equivalent system of constant stiffness E_1I_1 yields a new deflection curve v that can be used to repeat the procedure. The procedure may be repeated until the final deflection curve is as close as desired with the deflection curve obtained from the preceding trial. Although the procedure converges very rapidly, the use of a digital computer will facilitate the procedure. The solution becomes more convenient if we approximate the shape of M_e with straight-line segments as it was done in preceding chapters of this text. See, for example, Chaps. 1 and 4.

The following example illustrates the application of the theory.

Example 6.1 The tapered simply supported member in Fig. 6.5c is loaded with a uniformly distributed load w = 3.2 kips in.$^{-1}$ (560.63 kN m^{-1}), as shown. At the left support of the member the depth h = 12 in. (0.3048 m), and it is 8 in. (0.2032) at the other support. The constant width b = 6 in. (0.1524 m). The member is also subjected to an axial force P = 386 kips (1,716.93 kN). The material of the member is monel, having a stress–strain curve as shown in Fig. 4.3a. By using the three-line approximation of the stress–strain curve of monel and applying the preceding methodology, determine the maximum rotation and the maximum vertical deflection of the member.

Solution: We start the solution by using the trial-and-error procedure, discussed earlier in this section, to determine the position of the neutral axis at cross-sections along the length of the member. We can do this by assuming a shape for the vertical deflection v of the member. Since the correct shape of v that takes into consideration the effect of the axial force P is not known, it would be reasonable for a first trial to assume P = 0 and determine the deflection shape v that is produced by using only the distributed load w = 3.2 kips in.$^{-1}$ (560.63 kN m^{-1}). Since P = 0 in this first trial, the deflection curve v can be obtained by using the methodology discussed in the fourth chapter.

The next step is to use the trilinear approximation of the stress–strain curve of monel, assume values for ε_1 and ε_2, and adjust Δ so that Eq. (6.17) is satisfied. The assumed values for ε_1, ε_2, and Δ should also satisfy Eq. (6.23) by making the sum of the two terms on the left-hand side of the equation to be equal to the term on the right-hand side of the equation. In other words, Eqs. (6.17) and (6.23) should be simultaneously satisfied. When this procedure is completed, we determine the equivalent modulus M_e by using Eq. (6.25), and the modulus function g(x) by using Eq. (6.27). In Eq. (6.27) the reference value $E_1 = 22\times10^6$ psi (151.685×10^9 Pa), and it is shown in Table 4.1. The reference moment of inertia $I_1 = bh^3/12 = (6)(8)^3/12 = 256\,\text{in.}^4(107 \times 10^{-6}\,\text{m}^4)$, and it is located at the right support of the member. This procedure is performed for all selected cross sections along the length of the member.

When f(x) and g(x) are determined for the selected cross-sectional locations of the member, the corresponding values of the moment M_e of the equivalent system of constant stiffness E_1I_1 can be determined by using Eq. (1.115) or Eq. (1.142). The approximation of the shape of M_e with straight-line segments leads to the derivation of the equivalent system of uniform stiffness E_1I_1, loaded with equivalent concentrated loads. By solving the equivalent system we obtain new deflection curve v, and we compare it with the one determined in the preceding trial. If they are not as closely identical as we would desire, then we repeat the procedure by using the new shape of the deflection curve.

For the fourth trial, the values of the rotations and deflections at length intervals of 6 in. (0.1524 m) are shown in the second and third columns, respectively, of Table 6.3. In this table, x is measured from the left support of the member. The values of the deflection v that are obtained from the fourth trial will be used here to repeat the procedure by assuming values for the strains ε_1 and ε_2 and adjusting Δ until Eqs. (6.17) and (6.23) are satisfied. For example, by using the beam cross section at x = 60 in. (1.524 m), we assume $\varepsilon_1 = 2.840 \times 10^{-3}$, $\varepsilon_2 = 3.7077 \times 10^{-3}$, and $\Delta = \varepsilon_1 + \varepsilon_2 = 6.5477 \times 10^{-3}$.

Schematically, the trilinear stress–strain curve approximation of monel would be as shown in Fig. 6.6a. For the assumed values of $\varepsilon_1, \varepsilon_2$, and Δ, we have the shape shown in Fig. 6.6b. In Fig. 6.6a or b, the elastic strain ε_e is

$$\varepsilon_e = \frac{\sigma_1}{E_1} = \frac{(48)(10)^3}{(22)(10)^6} = 2.1818 \times 10^{-3}$$

At the position where the stress $\sigma = 59{,}000$ psi (406.79×10^6 Pa), the strain ε is

$$\begin{aligned}\varepsilon = \varepsilon_e + \frac{\sigma_2 - \sigma_1}{E_2} &= 2.1818 \times 10^{-3} + \frac{(11)(10)^3}{(504)(10)^3} \\ &= 24.0028 \times 10^{-3}\end{aligned}$$

From Fig. 6.6b, the stress σ_A at position A, is

$$\sigma_A = -48 - (59 - 48)\left(\frac{3.7077 - 2.1818}{24.0068 - 2.1818}\right) = -48.7691\,\text{ksi}\,(336.25 \times 10^6\,\text{Pa})$$

Table 6.3. Rotations and vertical deflections of a variable stiffness member subjected to a distributed load w and an axial compressive force P (1 in. = 0.0254 m)

x (in.)	Rotation $(10)^{-3}$ (rad) fourth trial	Deflection (in.) fourth trial	Final deflection (in.)
0	22.4258	0	0
6	22.2370	0.1341	0.1341
12	21.6697	0.2659	0.2658
18	20.7258	0.3930	0.3930
24	19.4101	0.5134	0.5134
30	17.7306	0.6249	0.6248
36	15.6953	0.7251	0.7251
42	13.2988	0.8121	0.8121
48	10.5172	0.8836	0.8835
54	7.3084	0.9371	0.9370
60	3.6266	0.9699	0.9698
66	−0.5336	0.9791	0.9791
72	−5.0538	0.9624	0.9623
78	−9.6737	0.9182	0.9181
84	−14.0864	0.8469	0.8469
90	−18.0787	0.7504	0.7504
96	−21.5720	0.6315	0.6314
102	−24.5293	0.4932	0.4931
108	−26.8768	0.3389	0.3389
114	−28.4567	0.1729	0.1729
120	−29.0459	0	0

and at position B, the stress σ_B is

$$\sigma_B = 48 + (59 - 48)\left(\frac{2.84 - 2.1818}{24.0068 - 2.1818}\right) = 48.3317\,\text{ksi}\,(333.24 \times 10^6\,\text{Pa})$$

By using Fig. 6.6b, the area of the stress–strain curve under portion A0B of the curve, is

$$\begin{aligned}\int_{\varepsilon_2}^{\varepsilon_1} \sigma\, d\varepsilon = &\left(\frac{1}{2}\right)(48)(10)^3(2.1818)(10)^{-3} + (48)(10)^3(2.24 - 2.1818)(10)^{-3}\\
&+ \left(\frac{1}{2}\right)(2.84 - 2.1818)(10)^{-3}(48.3317 - 48.0)(10)^3\\
&- \left(\frac{1}{2}\right)(2.1818)(10)^{-3}(48)(10)^3 - (3.7077 - 2.1818)(10)^{-3}(48)(10)^3\\
&- \left(\frac{1}{2}\right)(3.7077 - 2.1818)(10)^{-3}(48.7691 - 48.0)(10)^3\\
= &-42.1272\end{aligned}$$

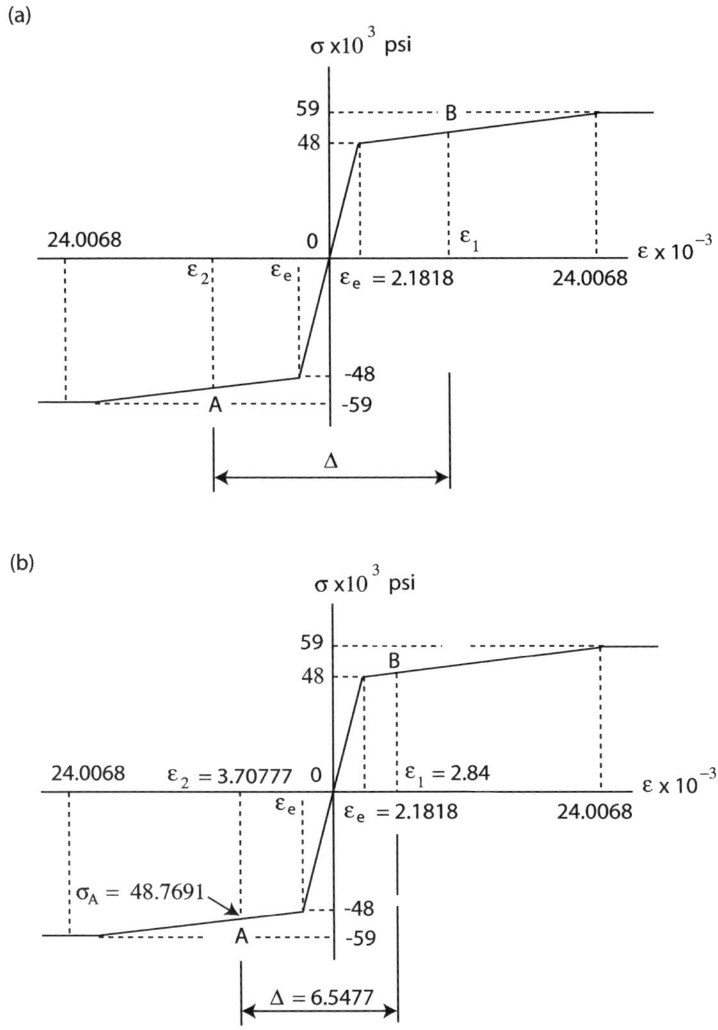

Fig. 6.6. (**a**) Trilinear approximation of the stress–strain curve of monel. (**b**) Assumed values of ε_1, ε_2, and Δ (1 psi = 6,894.757 Pa)

Thus, by using Eq. (6.17), we find

$$P = \frac{bh}{\Delta}\int_{\varepsilon_2}^{\varepsilon_1} \sigma \, d\varepsilon = \frac{(6)\,(10)}{(6.5477)\,(10)^{-3}}\,(-42.1272)$$
$$= -386.0336 \times 10^3\,\text{lb}\,\left(1,717.08 \times 10^3\,\text{N}\right)$$

The actual value of the axial compressive force $P = 386 \times 10^3$ lb (1,716.93 N) and, consequently, Eq. (6.17) is closely satisfied for practical purposes and we accept it.

By using again Fig. 6.6b, we calculate the first moment about the origin 0 of the area of the stress–strain curve under portion A0B of the curve. This area is

$$\int_{\varepsilon_2}^{\varepsilon_1} \sigma\varepsilon\, d\varepsilon = \left(\frac{1}{2}\right)(48)(10)^3(2.1818)(10)^{-3}\left(\frac{2}{3}\right)(2.1818)(10)^{-3}(2)$$
$$+(48)(10)^3(2.84-2.1818)(10)^{-3}(2.5109)(10)^{-3}$$
$$+\left(\frac{1}{2}\right)(2.84-2.1818)(10)^{-3}(48.3317-48.0)(10)^3(2.6206)(10)^{-3}$$
$$+(3.7077-2.1818)(10)^{-3}(48)(10)^3(2.9448)(10)^{-3}$$
$$+\left(\frac{1}{2}\right)(3.7077-2.1818)(10)^{-3}(48.3317-48.0)(10)^3(3.1991)(10)^{-3}$$
$$= 449.5064\times 10^{-3}$$

By using Eq. (6.25), we find that the equivalent modulus E_e, is

$$E_e = \frac{12}{\Delta^3}\int_{\varepsilon_2}^{\varepsilon_1}\sigma\varepsilon\, d\varepsilon = \frac{12}{(6.5477\times 10^{-3})^3}(449.5064)(10)^{-3}$$
$$= 19.2155\times 10^6 \text{ psi } \left(132.4862\times 10^9 \text{ Pa}\right)$$

We check now to find out if Eq. (6.24) is satisfied. For the term on the right-hand side of the equation, we have

$$\frac{E_e I}{r} = \frac{E_e I\Delta}{h} = \frac{(19.2155)(10)^6(6)(10)^3(6.5477)(10)^{-3}}{(12)(10)}$$
$$= 6.2909\times 10^6 \text{ in. lb } \left(0.7109\times 10^9 \text{ N m}\right)$$

From the two terms on the left-hand side of Eq. (6.24), we find

$$M + \varepsilon_0\frac{h}{\Delta}P = 5760.0 + (386)(0.9699)$$
$$+\frac{386}{(22)(10)^3(6)(10)}\frac{10}{(6.5499)(10)^{-3}}(386)$$
$$= 5760.0 + 374.38 + 172.39$$
$$= 6306.77 \text{ in. kips } (712,665 \text{ N m})$$

Note that the moment M in the above equation is composed of the bending moment of 5,760 in. kips (650,880 N m) produced by the applied load $w = 3.2$ kips in.$^{-1}$(560.63 kN m^{-1}), and the moment $Pv = (386)(0.9699) = 374.38$ in. kips (42,305 N m), where $v = 0.9699$ in. (0.0246 m) is the vertical deflection at $x = 60$ in. (1.524 m) of the fourth trial, as shown in Table 6.3.

The above shows that the two sides of Eq. (6.24) are closely identical and, consequently, the equation is satisfied for practical purposes. For other cross sections along the length of the member, we can follow the same procedure to determine E_e and satisfy Eqs. (6.17) and (6.23) or (6.24).

The procedure is repeated here for various sections along the length of the member at intervals of 6.0 in. (0.1524 m). The values of f(x), g(x), $\varepsilon_1, \varepsilon_2, \Delta, E_e, M_{req}$, and M_e are shown in Table 6.4. The values of the moment M_e of the equivalent system of constant stiffness E_1I_1, where $E_1 = 22 \times 10^6$ psi (151.58×10^6 kPa) and $I_1 = 256$ in.^{4} (106.555×10^{-6} m^4), are shown in the last column of the table. The moment diagram M_e approximated with four straight-line segments is plotted as shown in Fig. 6.7a. The equivalent system of constant stiffness E_1I_1 loaded with concentrated loads is shown in Fig. 6.7b.

Since M_e incorporates the effects of the axial load P, the equivalent loads on the equivalent system include the effect of P also. Therefore, the constant stiffness equivalent system in Fig. 6.7b may be used to determine the deflections and rotations of the initial variable stiffness member in Fig. 6.5c by applying linear analysis. For example, if the moment–area method is used, the deflection Δ_C at a point C of distance x = 72.0 in. (1.8288 m), would be equal to 0.9623 in. (0.02444 m), which is almost identical to the one obtained in the third column of Table 6.3. The computer results of this final trial are shown in the last column of Table 6.3. The maximum deflection occurs at x = 66 in. (1.6764 m) and it is equal to 0.9791 in. (0.0249 m).

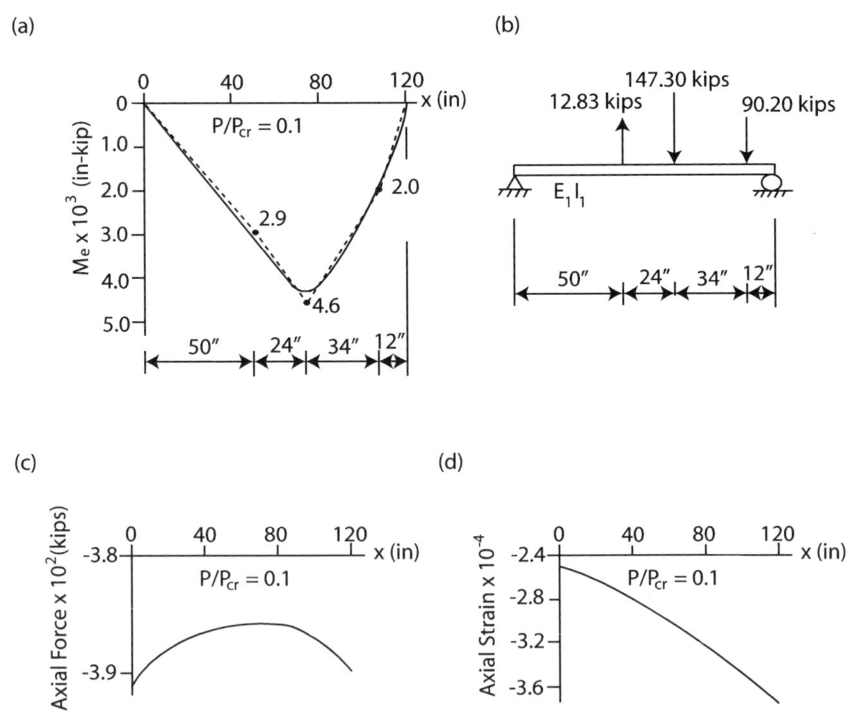

Fig. 6.7. Moment diagram M_e of the equivalent system of constant stiffness E_1I_1. (**b**) Equivalent system of constant stiffness E_1I_1. (**c**) Axial force variation for $P/P_{cr} = 0.1$. (**d**) Axial strain variation for $P/P_{cr} = 0.1$ (1 in. = 0.0254 m, 1 kip = 4,448 N, 1 in. kip = 113.0 N m)

Table 6.4. Values of f(x), ε_1, ε_2, Δ, E_e, g(x), M_{req}, and M_e for a simply supported variable stiffness member subjected to a uniform vertical load w and an axial compressive load P (1 in. = 0.0254 m, 1 in. b = 0.113 N m)

x (in.)	h (in.)	f(x)	$\varepsilon_1 = (h_2/r)(10)^{-3}$ (in.in.$^{-1}$)	$\varepsilon_2 = (h_2/r)(10)^{-3}$ (in.in.$^{-1}$)	$\Delta(10)^{-3}$ (in.in.$^{-1}$)	$E_e(10)^6$ (psi)	$g(x) = E_e/E_1$	$M_{req} = M(10)^6$ (in.lb)	$M_e(10)^6$ (in.lb)
0	12.0	3.3750	–	–	–	0	–	0	–
6	11.8	3.2090	0.1244	−0.6201	0.7445	51.3655	2.3348	2.6622	0.3550
12	11.6	3.0486	0.4793	−0.9835	1.4628	29.9509	1.3614	2.9477	0.7100
18	11.4	2.8936	0.8186	−1.3316	2.1502	25.8673	1.1758	3.6142	1.0620
24	11.2	2.7440	1.1395	−1.6617	2.8012	24.4036	1.1093	4.2875	1.4080
30	11.0	2.5996	1.4391	−1.9708	3.4098	23.7147	1.0779	4.8922	1.7450
36	10.8	2.4604	1.7191	−2.2623	3.9814	23.2642	1.0575	5.4018	2.0760
42	10.6	2.3262	1.9883	−2.5728	4.5610	22.6453	1.0293	5.8026	2.4230
48	10.4	2.1970	2.2533	−2.9169	5.1702	21.7772	0.9899	6.0890	2.7990
54	10.2	2.0727	2.5402	−3.3020	5.8422	20.5892	0.9359	6.2573	3.2250
60	10.0	1.9531	2.8400	−3.7077	6.5477	19.2637	0.8756	6.3067	3.6870
66	9.8	1.8383	3.1036	−4.0705	7.1742	18.1063	0.8230	6.2377	4.1220
72	9.6	1.7280	3.2079	−4.2299	7.4378	17.6604	0.8027	6.0528	4.3630
78	9.4	1.6222	3.0913	−4.1020	7.1933	18.1024	0.8228	5.7529	4.3090
84	9.2	1.5209	2.7789	−3.7145	6.4934	19.4291	0.8831	5.3391	3.9750
90	9.0	1.4238	2.3959	−3.2279	5.6238	21.1195	0.9600	4.8102	3.5190
96	8.8	1.3310	2.0106	−2.7391	4.7497	22.6623	1.0301	4.1678	3.0390
102	8.6	1.2423	1.5779	−2.2595	3.8374	24.1155	1.0962	3.4221	2.5130
108	8.4	1.1576	1.0649	−1.7611	2.8260	26.1163	1.1871	2.6039	1.8940
114	8.2	1.0769	0.4243	−1.1376	1.5619	35.8741	1.6306	1.8837	1.0720
120	8.0	1.0000	–	–	–	–	–	0	0

The variation of the axial force along the length of the member and the strain produced by this force are shown in Fig. 6.7c and d, respectively. The presence of the axial force causes the neutral axis to move away from the centroidal axis. With reference to the bottom of the beam, the variation $h_1(x)$ of the position of the neutral axis at cross-sections along the length of the member is shown in Fig. 6.8a. The variation of the function g(x) is shown in Fig. 6.5d. Note that the neutral axis position approaches infinity at cross

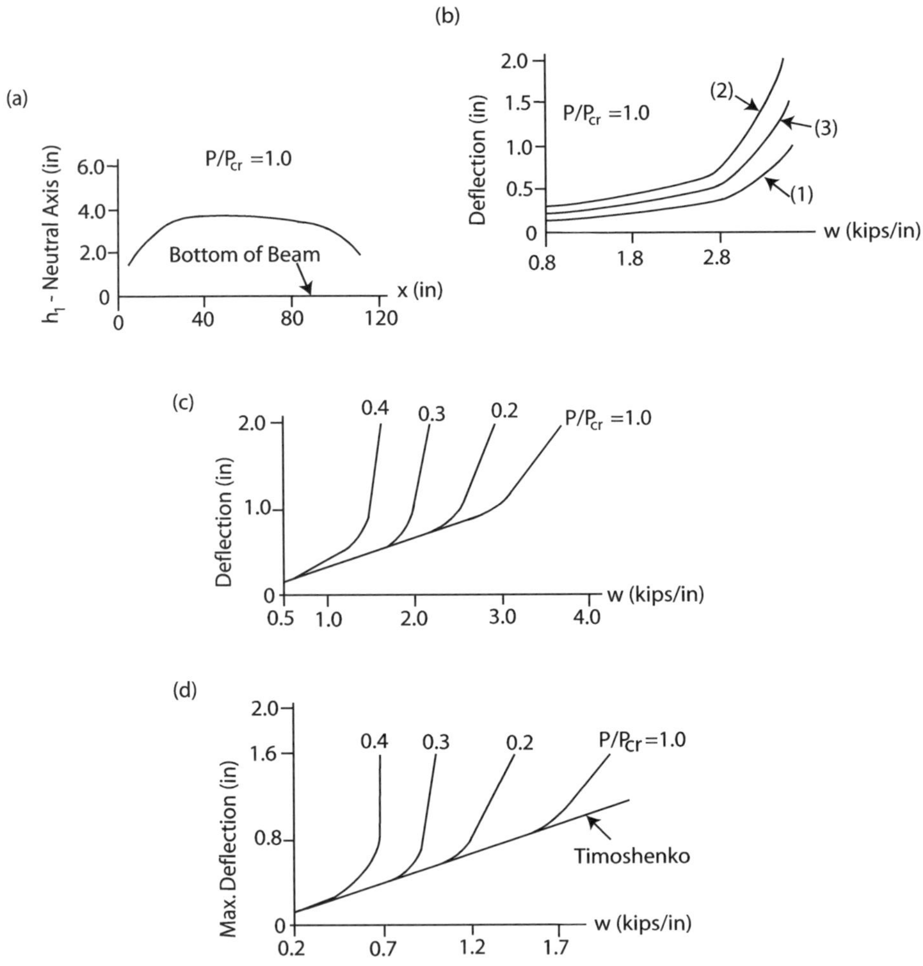

Fig. 6.8. (**a**) Variation of neutral axis location for $P/P_{cr} = 0.1$. (**b**) Variation of vertical deflection at points 1, 2, 3, Fig. 6.5c, with increasing load w for $P/P_{cr} = 0.1$. (**c**) Variation of vertical deflection at midspan with increasing w for $P/P_{cr} = 0.1$, 0.2, 0.3, and 0.4. (**d**) Variation of maximum vertical deflection with increasing w for $P/P_{cr} = 0.1$, 0.2, 0.3, and 0.4 (1 in. = 0.0254 m, 1 kip = 4,448 N, 1 in. kip = 113.0 N m).

sections close to the end supports. The variation of the vertical deflection at points 1, 2, and 3, with increasing load w, Fig. 6.5c, is shown in Fig. 6.8b. Note that the curves become more nonlinear with increasing w, thus establishing a critical value of w where small increases beyond this value of w result in a very large deformation.

Figure 6.8c shows the vertical deflection at midspan with increasing values of the distributed load w, for $P/P_{cr} = 0.1, 0.2, 0.3$, and 0.4. Note that these curves become steeper with increasing values of the ratios P/P_{cr}, and for each case there is a value of the load w that becomes critical. Any small deviation from this critical value of w results in large increases in the vertical deflection, which indicates that the ultimate capacity of the member to resist load and deformation is reached. Figure 6.8d shows the analogous curves for a prismatic beam of depth h = 8 in. (0.2032 m), width b = 6 in. (0.1524 m), and length L = 120 in. (3.048 m). The curve for $P/P_{cr} = 0.1$ is also compared with the results obtained by Timoshenko [101], where linear analysis was used. The deviation from the Timoshenko's straight line becomes more remarkable as the member gets stressed further into the inelastic rang.

6.7 The Longest Cable-Stayed Suspension Bridge in the World

Today, the longest cable-stayed suspension bridge in the world is in Greece, and spans the stretch of water between Antirion in the main land of Greece and Rion, which is on the Peloponisos side. The bridge is commonly known as the Rion–Antirion bridge. The construction started on 19 July 1998, where Costas Simitis, Prime Minister of Greece, in the presence of Constantinos Stephanopoulos, President of Greece, cast the first stone for the Rion–Antirion bridge. The project was completed and put to traffic on August 2004, just before the start of the Athens 2004 Olympic Games.

Figure 6.9 shows a photograph of the Rion–Antirion cable-stayed suspension bridge which connects the Rion and the Antirion sides of the Gulf. The picture is taken by looking at the bridge from the Antirion seaside of the Gulf. In the background you see some of the rugged mountains which predominate in the Rion–Antirion environment. Figure 6.10 shows an other closer view of the Rion–Antirion cable-stayed suspension bridge and the entrance to the bridge from the Antirion side of the Gulf, with the tolls.

By comparing the Rion–Antirion bridge with other world famous bridges, we find out that the Tatara bridge in Japan and the Normandy bridge in France are the cable-stayed bridges with the longest spans in the world, respectively, 890 m and 856 m. The Rion–Antirion bridge, with reference to its largest span of 560 m, shall rank in the top 10 list of the world's longest span for cable-stayed bridges. However, by considering its four pylons, it becomes the cable-stayed bridge with the longest suspended

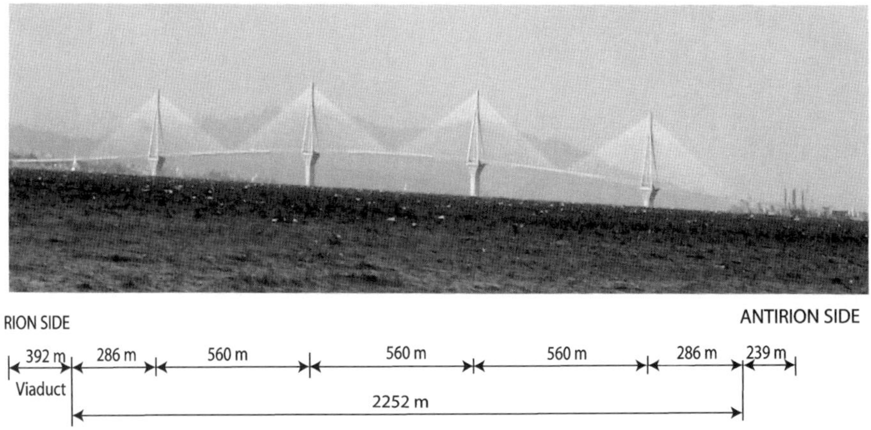

Fig. 6.9. Photograph of the Rion–Antirion cable stayed suspension bridge connecting the Rion side with Antirion side of the Gulf

Fig. 6.10. An other view of the Rion–Antirion cable-stayed suspension bridge showing the entrance to the bridge from its Antirion side of the tolls

deck of 2,252 m in the world. This deck length is longer than the one of 1966 m of the well-known Golden Gate suspension bridge.

Some of the physical data that confront this project are water depth up to 65 m, absence of hard seabed subsoil, strong seismic activity, and possible tectonic movements.

The seabed profile is characterized by steep slopes on each side and by a long horizontal plateau of about 60 m below sea level. The studies have concluded that there is no bedrock up to a depth of 100 m below seabed.

Based on further geological studies, it is also believed that the thickness of the thick layers of clay that are mixed in some areas with fine sand and silt, is greater than 500 m thick and consequently, there is no bedrock even at depths of 500 m beneath the seabed. This meant that the bridge had to be founded on soil rather than rock.

The soil consists of a cohesionless layer of sand and gravel extending from the mud line to a depth of 4–7 m in most locations. Beneath this layer, the soil profile, which is rather erratic and heterogeneous, includes strata of sand, silty sand, and silty clay. At depths beyond 30 m the soils are more homogeneous and consist mainly of silty clays or clays.

Having in mind the nature of the soil, it appeared that liquefaction will not present a problem to the design team. On the other hand, on the mainland, the first 20 m are seen as susceptible. This led the engineers to design very deep pile foundations for the approach viaduct on the Antirion side.

On top of this, the Greek state has imposed stringent design seismic loading additions which involve a peak ground acceleration of 0.48g and a maximum spectral acceleration of 1.20g between 0.2 and 1.0 s. These specifications are more severe than the acceleration recorded on 17 August 1999, during the Izmit Turkey, 7.4 Richter scale earthquake.

The cable-stayed technique that was developed in Europe during the sixties is used for the Rion-Antirion bridge. With this technique, the deck of the bridge is suspended through stay cables to pylon in a balanced and aesthetic way. The equilibrium of the structure lies on each and any pylon, thus making it possible for cable-stayed bridges to have one, two, or more pylons. The Rion–Antirion suspension bridge has four pylons. Starting from either end of the bridge, we find that the individual lengths of its five spans are 286, 560, 560, 560, and 286 m. It also has two approach viaducts at the two ends, measuring 392 m on the Rion side, and 239 m on the Antirion side.

The upper soil layers are reinforced with 2-m diameter hollow steel pipes 25–30 m long driven at a regular spacing of 7 m. Approximately two hundred pipes are used at each pier location. A gravel surface about 3 m thick was also appropriately placed at the top. The foundations consist of 90-m diameter reinforced concrete caissons which rest on the gravel layer. The lower part of the pier is conical in shape and its diameter ranges from 38 m at the bottom to 26 m at the top, thus forming the upper part of the underwater foundation.

Because of the large size of the foundation base, the caissons are strengthened by radial beams, each one meter thick. The radial beams decrease in height from 13.5 m at the center of the caisson to 9 m at its edge and are 26 m long. See Fig. 6.11.

The upper shaft of the pier carries a 15-m high reverse pyramid that supports a square base with a side that is 38 m long. Each pylon is composed of four high-strength reinforced concrete legs having a 4×4 m cross section, that is embedded in the pylon head to form a monolithic structure, in order to support the asymmetrical service loads and seismic forces.

Fig. 6.11. The caissons here are strengthened by 32 radial beams, each 1 m thick, which can be seen in the cutaway view in the figure

In some more detail, the four legs angle toward each other and are joined together at their top to impart the rigidity necessary to support the asymmetrical service loads and seismic forces. The tops of the legs are rigidly embedded in a pier head that is 35 m high and comprises a steel core connected to two vertical concrete walls that are 2.5 m thick. Each pier forms a monolithic structure extending up to 230 m from the bottom of the sea to the top of the pier head.

It should be pointed out that such foundations are very large and very costly, but they can support long spans and, consequently, only a few of them are required to do it. Other types of suspension bridges were investigated, but the design team concluded that the best solution was to adapt a continuous deck of 2,252 m long that is fully suspended from pylon heads. This design provides an isolation system that reduces the effect of the seismic forces on the deck, and also imparts sufficient flexibility to absorb relative movements between piers.

The stay cables incline to the sides of the deck in a fan shape arrangement and they are made of parallel galvanized strands where each strand in the system is protected by its own polyethylene sheath. They have their lower anchorage on the deck sides, see Fig. 6.12, while their upper anchorage is at the 35-m high pylon head. The thickest cable consists of 70 strands each having a diameter of 15 mm.

Fig. 6.12. A closer view of the Rion–Antirion bridge showing the arrangement of the cables

The bridge deck is continuous and it is fully suspended throughout its length. Four damping devices are used to connect the deck to the top of each pier in order to limit the pendulum movement of the bridge deck during a possible earthquake. For a possible seismic action, the design is based on a relative dynamic movement of $\pm 1.30\,\mathrm{m}$ and for velocities that may exceed $1\,\mathrm{m\,s^{-1}}$.

The deck is 27.2 m wide with two traffic lanes in each direction. In addition, it includes a safety lane and a pedestrian walkway in each direction. It is a composite structure with a steel frame which is made up of two longitudinal steel plate girders (one on each side). These girders are 2.2 m high. The design also includes transverse plate girders which are also 2.2 m high and spaced at 4 m from each other. The top slab is made out of precast concrete panels. On each side, an impressive transition pier links together the deck of the cable-stayed bridge with the deck of the approaching viaducts.

In the design concept, the major concern was the design of the bridge to withstand severe seismic activity. In addition, the bridge should be capable of sustaining the impact of an 180,000-ton tanker sailing at 18 knots, winds at high speeds, and also the impact of moving trucks and passenger vehicles.

Regarding the possible tilting of the bridge, caused by an earthquake action, the studies have revealed that large shallow foundations is the most satisfactory solution and they should also include in the design a reinforcement for the top 20 m of the seabed subsoil. We can achieve this by using 90-m diameter pier bases and also inserting metallic inclusions in the seabed subsoil.

In addition, structural isolation systems that are designed to react favorably against seismic forces were systematically investigated. Based on these studies, the design of a 2,252-m fully suspended continuous deck was selected, which permits the deck to move as a pendulum during an earthquake action and it is also designed to permit significant movement between adjacent piers.

The deck, when suspended, becomes free to accommodate all thermal and tectonic movements in the longitudinal direction, and in the transverse direction the deck will behave like a pendulum. Its movements during a large seismic event will be buffered by four hydraulic dampers connected to each pylon base. The capacity of each damper in either tension or compression, is approximately 3,500 kN.

The allowed relative dynamic movement between the deck and the pylons during an extreme seismic event is about 3.5 m, with velocities up to $1.6\,\mathrm{m\,s^{-1}}$. An additional system was also designed in order to keep the deck in place when it is under the action of strong winds. For this purpose, an horizontal strut of 10,000 kN capacity connects the deck to each pylon base. During a strong seismic event the strut will break, and in this manner the dampers will be activated. This concept was validated by tests performed by the California Department of Transportation at the University of California testing facility in San Diego.

The connection of the main bridge with the approach viaducts at each end, is designed to accommodate the large movements of the cable-stayed deck which can be produced by seismic, thermal, and tectonic forces. The longitudinal movement could reach 2.5 m, but the transverse movements are negligible. However, under extreme conditions, movements of up to 5 m in either direction may occur. In order to resist such large movements, the ends of the cable-stayed deck are supported by a vertical steel frame, 14 m high, which is designed to accommodate uplift loads and longitudinal deck movements. Between the deck and the steel frame, their connection design is made similar to the one used between the deck and the pylon bases which involves two dampers and a strut service.

Construction wise, the methods used for the foundations are similar to those used for offshore concrete platforms. Each foundation base was built to a height of 15 m in a dry dock and after this, it was towed and moved to a wet dock. The foundation was then completed with the construction of the conical shaft. After this, it was towed away to its final position in the gulf, and then it was lowered into the water.

The exceptional features regarding the construction process for the foundations are (a) The dry dock near the site is 200 m long, 100 m wide, and 14 m deep, and provides the simultaneous construction of two bases. (b) During the construction process for the conical shaft at the wet dock, the foundations remained afloat and moored, and their balance was very sensitive to wind and the currents. (c) In their final position, the foundations were filled with water in order to accelerate settlement. This preloading condition was retained during the in situ construction of the pier shafts and pyramids, in order to be able to make corrections for differential settlements before we start the construction of the pylon leg. (d) The steel core of the pylon heads is lifted into position by using a large floating crane of 170 m maximum height.

The Rion–Antirion cable-stayed suspension bridge encompasses many exceptional features regarding its design and construction. Many of these features are derived from the exceptional efforts of the design team in order to meet the imposed seismic resistance requirements. See also [102] at the end of this text. The bridge is the winner of the 2005 Outstanding Civil Engineering Achievement (OCEA) award.

6.8 Inelastic Analysis of Thin Rectangular Plates

Figure 6.13a shows a thin simply supported rectangular plate of uniform thickness and Fig. 6.13b illustrates a differential element of the plate which includes all the shear forces and moments acting on it. The rigidity D of the plate is given by the equation

$$D = \frac{Eh^3}{12\,(1 - \nu^2)} \tag{6.28}$$

where ν in this equation is the Poisson ratio, E is the modulus of elasticity, and h is the thickness of the plate.

The thickness h of the thin plate can vary in both x and y directions of the plate and, therefore, the rigidity D in Eq. (6.28) can be also variable. For example, if we assume that the thickness h varies only in the y direction of the plate, then we could assume that h is represented by the expression

$$h = h_0\,[1 + \lambda f\,(y)] \tag{6.29}$$

where h_0 is a constant reference thickness, f(y) describes the variation of the plate thickness in the y direction, and λ is a constant that keeps the plate thin enough to fall within the range of thin plate theory.

If we substitute Eq. (6.29) into Eq. (2.28), we obtain

$$D = D_0\,[1 + \lambda f\,(y)]^3 \tag{6.30}$$

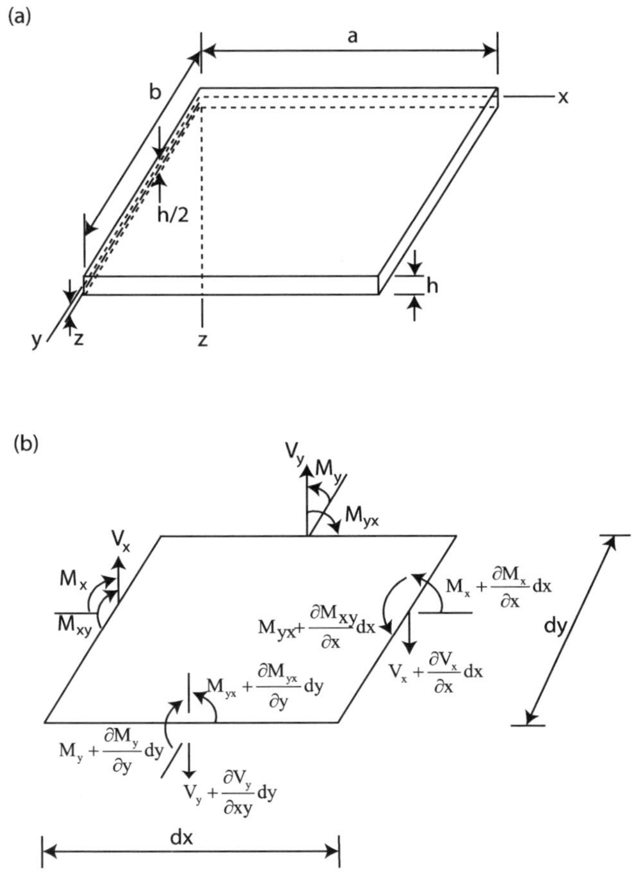

Fig. 6.13. (**a**) Thin rectangular plate of sides a and b and thickness h. (**b**) Free-body diagram of an element of the plate having sides dx and dy

where

$$D_0 = \frac{Eh_0^3}{12\,(1 - \nu^2)} \tag{6.31}$$

When the stresses in the material of the plate are permitted to exceed its elastic limit, then the modulus E in Eq. (6.28) or Eq. (6.31) becomes variable along the x and y directions of the plate. Its variation along these two directions of the plate must be determined in order to obtain a solution for the plate.

The methodology developed in Chap. 4 and Sect. 6.6 of this text (see also [3,6]), regarding the determination of a reduced (or equivalent) modulus for beams, will be extended in this section to apply for the inelastic analysis of thin rectangular plates of uniform and variable thickness that are subjected to deformations guided by the small deflection theory.

In order to use this theory and methodology for rectangular thin plates, we need to know the stress–strain relationships of the material of the plate in the x and y directions. Stress–strain diagrams are usually obtained experimentally for the various existing and new materials and many of them can be found in literature. We assume here that the stress–strain diagrams for the stresses σ_x and σ_y in the x and y directions, respectively, of the plate are as shown in Fig. 6.14a and b, respectively.

For convenience regarding the application of the theory and the equation, we assume that the compressive yield strength of the material is equal in magnitude to its tensile yield strength, i.e., symmetry. This is not a requirement for the application of the method, but it follows typical engineering practices for steels. The curve A0B, as discussed in Chap. 4, would represent the stress

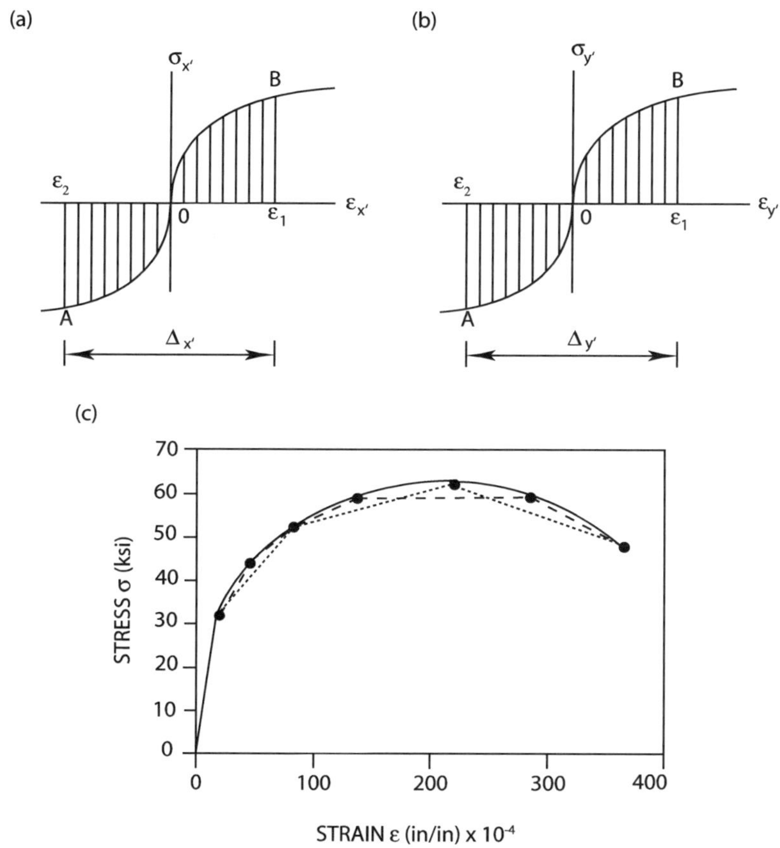

Fig. 6.14. (**a**) Stress–strain curve for σ_x vs. ε_x. (**b**) Stress–strain curve for σ_y vs. ε_y. (**c**) Stress–strain curve of the material of a plate with its shape approximated with four and six straight lines (1 ksi = 6,894.757 Pa, 1 in. = 0.0254 m)

distribution along the depth h of the thin plate if h is substituted for Δ in Figs. 6.14a and b.

Let it be assumed that the curvature of the neutral surface produced by the bending moment M_y is r_y. At a distance z from the neutral surface, see Fig. 6.13a, the unit elongation ε of a fiber can be written as

$$\varepsilon = \frac{z}{r_y} \tag{6.32}$$

Therefore,

$$z = \varepsilon r_y \tag{6.33}$$

$$dz = r_y \, d\varepsilon \tag{6.34}$$

If h_1 and h_2 are the distances from the neutral surface to the lower and upper surfaces of the plate, respectively, we have

$$\varepsilon_1 = \frac{h_1}{r_y} \tag{6.35}$$

$$\varepsilon_2 = \frac{h_2}{r_y} \tag{6.36}$$

Consequently,

$$\Delta_y = \varepsilon_1 + \varepsilon_2 = \frac{h_1}{r_y} + \frac{h_2}{r_y} = \frac{h}{r_y} \tag{6.37}$$

Thus the total strains $\varepsilon_x\prime$ and $\varepsilon_y\prime$ and the stresses σ_x and σ_y can be written as shown below:

$$\varepsilon'_x = \frac{1}{1-\nu^2}(\varepsilon_x + \nu\varepsilon_y) \tag{6.38}$$

$$\varepsilon'_y = \frac{1}{1-\nu^2}(\varepsilon_y + \nu\varepsilon_x) \tag{6.39}$$

$$\sigma_x = \frac{E_x}{1-\nu^2}(\varepsilon_x + \nu\varepsilon_y) \tag{6.40}$$

$$\sigma_y = \frac{E_y}{1-\nu^2}(\varepsilon_y + \nu\varepsilon_x) \tag{6.41}$$

On this basis, we can rewrite Eqs. (6.40) and (6.41) as follows:

$$\sigma_x = E_x \varepsilon'_x \tag{6.42}$$

$$\sigma_y = E_y \varepsilon'_y \tag{6.43}$$

At any cross section which is under consideration, the bending moment M_y of the plate is

$$M_y = \int_A \sigma_y z \, dA = \int_{h_2}^{h_1} \sigma_y z \, dz \tag{6.44}$$

where the sides of dA are assumed to be unity. Thus, by using Eqs. (6.33) and (6.34), we obtain

$$M_y = r_y^2 \int_{\varepsilon_2}^{\varepsilon_1} \sigma_y \varepsilon_y \, d\varepsilon_y \tag{6.45}$$

By using Eq. (6.37) and rearranging, we can write Eq. (6.45) as

$$M_y = \frac{h^3}{\Delta_y^3} \frac{1}{r_y} \int_{\varepsilon_2}^{\varepsilon_1} \sigma_y \varepsilon_y \, d\varepsilon_y \tag{6.46}$$

On this basis, the reduced rigidity D_{ry} along the y coordinate may be written as

$$D_{ry} = \frac{h^3}{\Delta_y^3} \frac{1}{1-\nu^2} \int_{\varepsilon_2}^{\varepsilon_1} \sigma_y \varepsilon_y \, d\varepsilon_y \tag{6.47}$$

For elastic analysis, Eq. (6.46) reduces to $M_y = D_{ry}/r_y$.

In a similar manner, the bending moment M_x and the reduced rigidity D_{rx} along the x coordinate of the plate, may be written as

$$M_x = \frac{h^3}{\Delta_x^3} \frac{1}{r_x} \int_{\varepsilon_2}^{\varepsilon_1} \sigma_x \varepsilon_x \, d\varepsilon_x \tag{6.48}$$

$$D_{rx} = \frac{h^3}{\Delta_x^3} \frac{1}{1-\nu^2} \int_{\varepsilon_2}^{\varepsilon_1} \sigma_x \varepsilon_x \, d\varepsilon_x \tag{6.49}$$

The integrals in Eqs. (6.49) and (6.47) represent the first moment with respect to the vertical axis through the origin 0 of the shaded area in Fig. 6.14a and b, respectively.

By using the well-known curvature–moment expressions for rectangular plates and rearranging, we may write the expressions

$$\frac{\partial^2 w}{\partial x^2} = \frac{1}{1-\nu^2}\left(\frac{M_x}{D_{rx}} - \nu \frac{M_y}{D_{ry}}\right) \tag{6.50}$$

$$\frac{\partial^2 w}{\partial y^2} = \frac{1}{1-\nu^2}\left(\frac{M_y}{D_{ry}} - \nu \frac{M_x}{D_{rx}}\right) \tag{6.51}$$

where w is the vertical deflection along the z axis of the plate.

We can facilitate the solution of the thin plate problem and the related equations by introducing, as we have done it in previous chapters and sections of this text, a trial-and-error procedure. For example, we can start by using Fig. 6.14a and b and assuming values for the quantities Δ_x and Δ_y. This step is equivalent to assuming values for the strains ε_1 and ε_2. Since we have assumed that the numerical values of the compressive and the tensile strength of the material are equal, we have in this case $\varepsilon_1 = \varepsilon_2 = \varepsilon$ and $\Delta = 2\varepsilon$.

For every assumed value of Δ_x and Δ_y, we can compute the required moments $M_{x_{req}}$ and $M_{y_{req}}$ by using Eqs. (6.48) and (6.46), respectively. The values of the reduced plate rigidities D_{rx} and D_{ry} can be determined by utilizing Eqs. (6.49) and (6.47), respectively. We note at this point that the integral in Eqs. (6.49) and (6.47) represent, respectively, the first moment of the shaded area in Figs. 6.14a and b, with respect to the origin 0.

We can repeat the procedure until the required moments $M_{x_{req}}$ and $M_{y_{req}}$, for statically determinate plates, are equal to the actual moments M_x and M_y, respectively, of the plate. The values of the reduced rigidities D_{rx} and D_{ry}, as stated earlier, can be determined by using Eqs. (6.49) and (6.47), respectively.

Since M_x, M_y, D_{rx}, and D_{ry} are now known quantities, the vertical deflection w of the rectangular thin plate along its x and y axes can be determined by integrating Eqs. (6.50) and (6.51), respectively. It should be pointed out here that the actual moments M_x and M_y of the plate can be evaluated by performing the elastic analysis of the plate. The method of the equivalent systems, as discussed in [3,6], can facilitate the solution for thin plates of variable thickness.

As stated above, for statically determinate plates, i.e., simply supported edges, the required moments $M_{x_{req}}$ and $M_{y_{req}}$ which are obtained from the inelastic analysis would have to match the corresponding actual moments M_x and M_y which are obtained from the elastic analysis, so that static equilibrium requirements are satisfied.

When the plate is statically indeterminate, i.e., all edges fixed, the required moments $M_{x_{req}}$ and $M_{y_{req}}$ will not have the same values as the corresponding moments M_x and M_y that are obtained from elastic analysis, because they depend on both the external load and the distribution of the inelastic moments along the dimensions of the plate.

For statically indeterminate plates, a reasonable solution may be obtained by utilizing the well-established procedures for statically indeterminate problems. Some modifications, however, may have to be made. This may be initiated by first defining the redundants. For example, if the one edge of the plate is fixed and the other three are simply supported, the fixed-end moment along the fixed edge may be taken as the redundant.

At this point, we can assume that all edges of the plate are simply supported and apply inelastic analysis the way it is discussed above. This analysis will produce the rotation along the side of the plate that is supposed to be fixed. The purpose of the redundant moment is to make this rotation zero. Thus, by assuming again that all edges of the plate are simply supported, we apply a moment along the side of the plate that is supposed to be fixed. By following inelastic analysis, the value of this moment should be the one required to counterbalance the rotations produced by the plate loading when all sides were assumed simply supported. Since the rotation along the side of the plate may vary, the fixed-end moment along this side may also be variable.

The procedure for statically indeterminate plates is more complicated, but not unrealistic. The following numerical examples illustrate the application of the theory to thin plates of uniform and variable thickness.

Example 6.2 **Linear Thickness Variation Along the y Axis**

The method developed in this section will be used here to carry out the inelastic analysis of a rectangular thin plate with linear thickness variation along the y axis and loaded with a uniformly distributed load as shown in Fig. 6.15. The sides of the plate are $a = b = 120.0$ in. (3.048 m), the thickness $h_0 = 8.0$ in. (0.2032 m), the parameter $\lambda = 0.3$, and the uniformly distributed load $q = 700.0$ psi (4,826,329.9 Pa). The material of the plate is mild steel, with a stress–strain curve as shown in Fig. 6.14c. In the same figure, the approximation of the stress–strain curve with four and six straight lines is also shown. The values of the modulus E and stress σ corresponding to these approximations of the stress–strain curve are shown in Table 4.7 of Sect. 4.3 in this text.

Solution: The inelastic analysis is carried out by using the four-line approximation of the stress–strain curve of the mild steel. The iteration procedure discussed above is initiated by using the stress–strain curve in Fig. 6.14c and assuming values of Δ_x, and Δ_y. This is equivalent to assuming values of strains ε_1 and ε_2 in Fig. 6.14a and b. Note that because of symmetry we have $\varepsilon_1 = \varepsilon_2 = \varepsilon$ and $\Delta = 2\varepsilon$.

Next, for every assumed value of Δ_x and Δ_y, we calculate the required moments $M_{x_{req}}$ and $M_{y_{req}}$ by using Eqs. (6.48) and (6.46), respectively. At the same time we calculate the values of the reduced rigidities D_{rx} and D_{ry} by using Eqs. (6.49) and (6.47), respectively. This procedure can be repeated until the assumed pair of Δ_x, and Δ_y will make $M_{x_{req}}$ and $M_{y_{req}}$ equal to the actual moments M_x and M_y, respectively, of the rectangular plate.

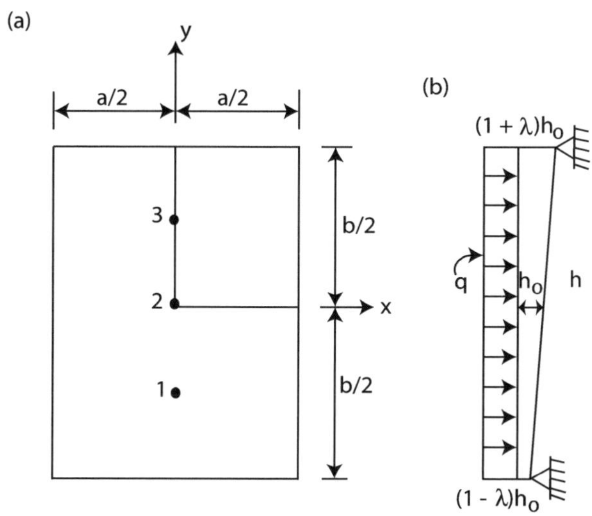

Fig. 6.15. (**a**) Rectangular plate with linear thickness variation in the y direction. (**b**) Uniformly distributed load q acting over the entire upper surface of the plate

The actual moments M_x and M_y produced by the application of the distributed load q, can be determined by using known methods of analysis for thin rectangular plates. In this problem, the convenient method of the equivalent systems for plates of variable thickness, discussed in [3], Chap. 6, is used.

By following this methodology, the values of $\varepsilon_x, \varepsilon_y, M_{x_{req}}, M_{y_{req}}, D_{rx}$, and D_{ry} are determined along the y axis at $x = 0$, and they are shown in Table 6.5. With known M_x, M_y, D_{rx}, and D_{ry}, the deflections and rotations at any point of the plate may be determined from Eqs. (6.50) and (6.51), by integration. The Simpson's rule may be used for this purpose. For points along the y axis with $x = 0$, the deflections and rotations are determined in this manner, and they are shown in Table 6.6. From this table, we observe that the maximum vertical deflection $w = 1.9245$ in. (0.0489 m), and occurs at the point of the plate where $x = 0$ and $y = -12.0$ in. (0.3048 m). The maximum rotation is 0.069279 rad, or 3.9694°, and occurs at $y = -60.0$ in. (1.524 m), as shown in the same table.

In Fig. 6.16, the variation of the vertical deflection w at the quarter points 1, 2, and 3 along the y axis, with distributed load q, is shown. We note that w increases very rapidly for loads $q > 750$ psi (5,171,067.75 Pa), suggesting that at this level of loading the plate is getting very deep into the inelastic range, and with any further increase of the applied load we may have a catastrophic failure. This information is very useful to the practicing design engineer who wants to establish criteria for ultimate or permissible design loads. This information would be particularly useful in the design of engineering structures that are designed to resist the effects of blast and earthquake.

Example 6.3 **Uniform Thickness Variation**

If $\lambda = 0$, the rectangular plate in Fig. 6.15 becomes a flat plate of uniform thickness $h_0 = 8.0$ in. (0.2032 m). The variation of the vertical deflection w at the quarter points 1, 2, and 3 of this plate is plotted in Fig. 6.17 by increasing the vertical load q.

For this case we observe that for values of $q > 800$ psi (5,515,805 Pa), the slope of the curves starts to increase very rapidly, even if we have small increases of the vertical load q beyond this value. This indicates that the ultimate capacity of the plate to resist load and deformation is reached. For the same plate problem, along the y direction of the plate and at $x = 0$, the variation of the rigidity functions

$$g(x) = \frac{D_{rx}}{D_x} \tag{6.52}$$

and

$$g(y) = \frac{D_{ry}}{D_y} \tag{6.53}$$

in the x and y directions, respectively, when $q = 800$ psi (5,515,805 Pa), is shown in Fig. 6.18.

Table 6.5. Reduced rigidities and required bending moments for simply supported rectangular plate with linear thickness variation (1 in. = 0.0254 m, 1 in.lb = 0.113 N m)

y (in.)	h	$\varepsilon_x(10)^{-4}$	$\varepsilon_y(10)^{-4}$	$M_{x_{req}} = M_x\,(10)^4$ (in.lb)	$M_{y_{req}} = M_y\,(10)^4$ (in.lb)	$D_{rx}(10)^8$	$D_{ry}(10)^8$
−60	5.60	0	0	0	0	1.5176	1.5176
−54	5.84	5.229	17.871	−6.242	−11.459	1.7212	1.7212
−48	6.08	10.976	29.012	−12.573	−22.681	1.9422	1.9422
−42	6.32	16.085	37.262	−18.821	−29.613	2.1814	2.0233
−36	6.56	15.775	55.100	−26.403	−36.151	2.4395	1.8034
−30	6.80	12.525	74.275	−30.344	−40.976	2.6972	1.6247
−24	7.04	16.400	76.525	−35.420	−44.223	2.8827	1.7393
−18	7.28	24.625	62.000	−39.809	−45.989	3.0506	2.1954
−12	7.52	31.525	45.825	−43.374	−46.518	3.2781	2.8792
−6	7.76	34.725	34.400	−46.086	−45.985	3.6124	3.6228
0	8.00	34.550	27.400	−47.828	−44.575	4.0705	4.2964
+6	8.24	32.550	23.450	−48.689	−42.786	4.6114	4.8295
+12	8.48	29.100	23.575	−48.610	−44.121	5.1851	5.2696
+18	8.72	26.544	20.720	−47.297	−37.695	5.7282	5.7298
+24	8.96	27.189	17.055	−49.257	−34.981	6.2161	6.2161
+30	9.20	24.137	14.737	−41.777	−32.151	6.7291	6.7291
+36	9.44	20.234	12.788	−37.072	−29.044	7.2696	7.2696
+42	9.68	15.816	10.854	−30.887	−25.261	7.8382	7.8382
+48	9.92	10.932	8.518	−22.940	−20.066	8.4358	8.4358
+54	10.16	5.638	5.178	−12.830	−12.255	9.0630	9.0630
+60	10.40	0	0	0	0	9.7206	9.7206

Table 6.6. Deflections and rotations along the y axis at x = 0 for rectangular plate of linear thickness variation along the y axis (1 in. = 0.0254 m)

y (in.)	Tapered Plate	
	Slope $(10)^{-2}$	w (in.)
−60	6.9279	0
−54	6.7346	0.4113
−48	6.2336	0.8013
−42	5.5326	1.1549
−36	4.5852	1.4596
−30	3.2972	1.6973
−24	1.8353	1.8513
−18	0.5484	1.9216
−12	−0.4111	1.9245
−6	−1.0990	1.8784
0	−1.6128	1.7966
+6	−2.0240	1.6872
+12	−2.3906	1.5547
+18	−2.6971	1.4017
+24	−2.9396	1.2324
+30	−3.1456	1.0498
+36	−3.3228	0.8556
+42	−3.4714	0.6517
+48	−3.5906	0.4397
+54	−3.6736	0.2216
+60	−3.7057	0

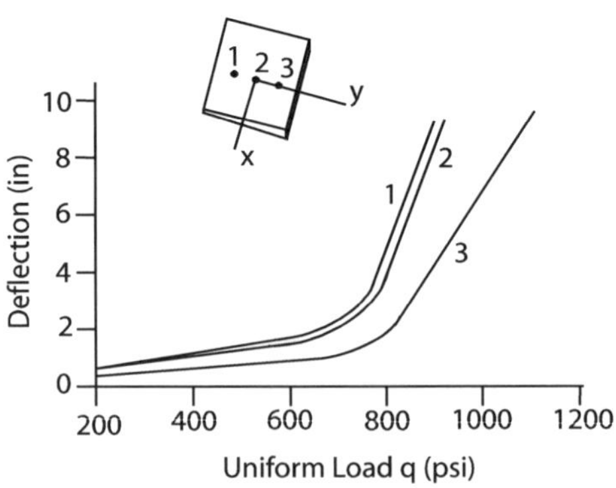

Fig. 6.16. Variation of the deflection w with distributed load q at points 1,2, and 3, for rectangular plate of linear thickness in the y direction (1 in. = 0.0254 m, 1 psi = 6,894.757 Pa)

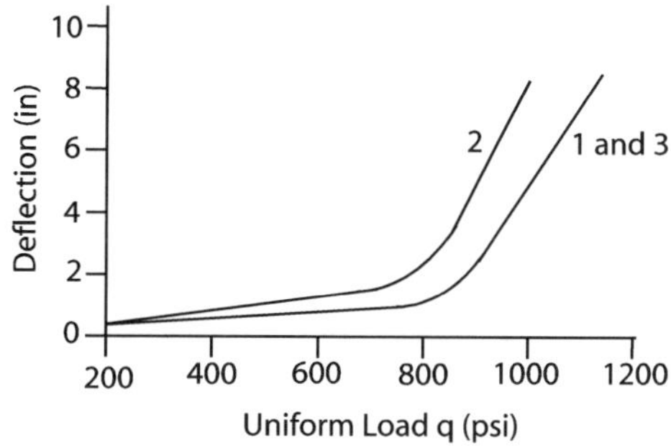

Fig. 6.17. Variation of vertical deflection of the plate at the quarter points 1, 2, and 3, when b/a = 1 and $\lambda = 0$ (1 in. = 0.0254 m, 1 psi = 6,894.757 Pa)

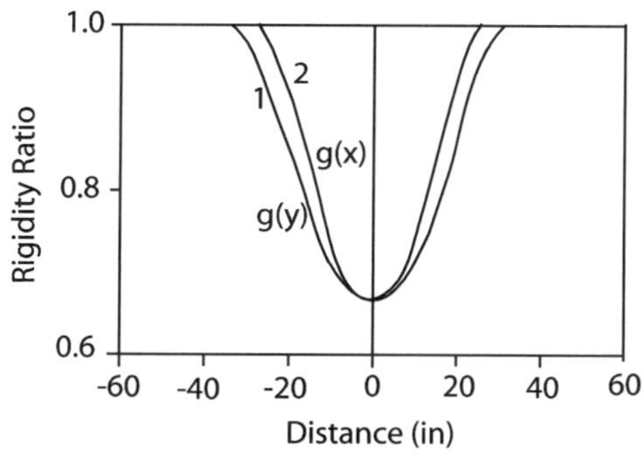

Fig. 6.18. Variation of the plate rigidity functions g(x) and g(y) for b/a = 1 and q = 800 psi (1 in. = 0.0254 m, 1 psi = 6,894.757 Pa)

Note that the smallest value of the rigidity ratios is at the centre of the plate. The plate will remain elastic, that is g(x) = 1.0, for y in the range of about ± 35 in $\leq$ y $\leq \pm 60$ in.

Other types of thin rectangular plates with various loading conditions and thickness variations, may be analyzed in a similar manner. For circular plates you may consult [2, 3].

6.9 Inelastic Earthquake Response of Multistory Buildings

For moderate to low intensity earthquakes, the analysis of structures based on elastic response would be in many cases appropriate. For strong earthquakes, however, an inelastic or elastoplastic analysis would be more appropriate. This reasoning is based on the fact that such earthquakes are experienced by a structure only a few times, if at all, in its life span, and some damage to the structure could be tolerated. For such severe dynamic loads it becomes, also, very uneconomical to have designed a structure entirely within the elastic range, having in mind that such loads could only occur a few times, if any, during the life span of the structure.

Many ductile materials are characterized by a large yielding range that permits them to absorb large amounts of energy before complete failure occurs. For such special cases, the important decision to be made by a designer is the amount of energy that a structure should be permitted to absorb for a still-safe design.

The inelastic analysis of buildings and other types of structures and structural analysis, is based on how we define the resistance R of a structure or a structural element.

6.9.1 Resistance R of a Structure

The resistance R of a structure, or a structural component such as a beam, a plate, and so on, is defined here as the internal force that tends to restore the structure, or a structural component, to its unloaded static position. It may have a variety of forms. For example, if a structural element is made up of brittle material, the resistance R will be represented by curve 1 in Fig. 6.19a. On the other hand, R will have the form given by curve 2 if the element is made up of a ductile material. Curve 3 would represent the resistance R when the material is plain concrete.

The resistance functions are, in most cases, idealized in order to simplify the analysis of the problem at hand. For many practical dynamics problems, it is often reasonable, and convenient, to use the bilinear resistance function form shown in Fig. 6.19b. see also Fig. 4.16a in Chap. 4.

For structures that can be idealized as a one-degree spring-mass system, see [5, 15, 84], and Fig. 6.20a, the resistance R may be assumed to have the bilinear form shown in Fig. 6.20b. In this case, R increases linearly until the elastic limit displacement y_{el} is reached, and the slope is equal to the spring constant k, as shown in the figure. At y_{el}, the resistance function attains its maximum value R_m, and it will remain constant and equal to R_m with increasing y. The extent of this yielding range is dependent on the ductility limit of the material.

If y reaches its maximum value y_m before the ductility limit is attained, while R_m remains constant, the structural system is said to rebound, and during rebound the resistance R is assumed to decrease linearly along a line parallel to its initial elastic line, as shown in Fig. 6.20b, and the decrease will continue until R becomes equal to $-R_m$. The other possibility is for y to continue to increase while R_m remains constant until the ductility limit is attained. In this case, the resistance function can be assumed to be composed of two lines, as shown in Fig. 6.19b.

For the spring-mass system in Fig. 6.20a, the resistance function R is represented by the spring force ky. By using the free-body diagram in Fig. 6.20c

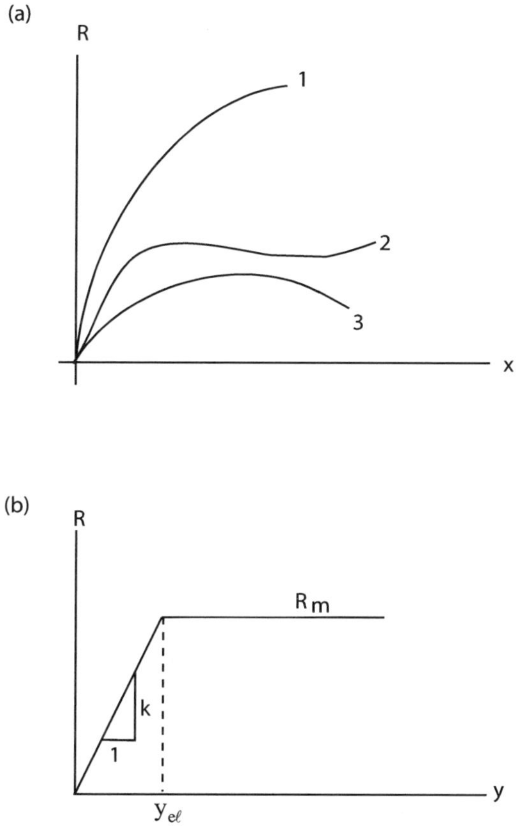

Fig. 6.19. (**a**) Resistance functions for various types of materials. (**b**) Bilinear form of the resistance R

and applying the second law of motion, the following differential equation representing the spring-mass system is obtained:

$$m\ddot{y} + R - F(t) = 0 \tag{6.54}$$

By using the resistance function shape shown in Fig. 6.20b, R is equal to ky for the elastic range, and Eq. (6.54) yields

$$m\ddot{y} + ky - F(t) = 0, \qquad 0 \le y \le y_{el} \tag{6.55}$$

For values of y between y_{el} and y_m, R_m remains constant, and Eq. (6.54) yields

$$m\ddot{y} + R_m - F(t) = 0, \qquad y_{el} \le y \le y_m \tag{6.56}$$

When R_m starts to decrease linearly until it becomes $-R_m$, y will attain values between y_m and $y_m - 2y_{el}$, and Eq. (6.54) takes the form:

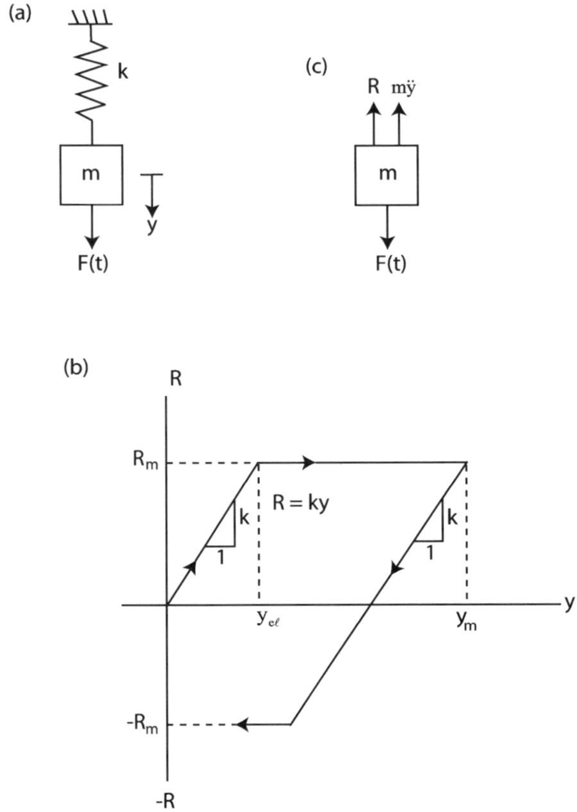

Fig. 6.20. (**a**) Spring-mass system with one degree of freedom. (**b**) Resistance function representation for the one-degree spring-mass system. (**c**) Free-body diagram of mass m

$$m\ddot{y} + R_m - k\,(y_m - y) - F\,(t) = 0, \qquad (y_m - 2y_{el}) \le y \le y_m \qquad (6.57)$$

Additional equations may be obtained for the negative plastic range in a similar manner, but this is not usually required, because this range is not of interest to the practicing engineer.

Equations (6.55)–(6.57) can be solved by using a rigorous solution, or by using an appropriate numerical method. A convenient numerical method to solve these equations is the acceleration impulse extrapolation method (AIEM) which is discussed in [5,84], and also briefly in Appendix A.

If a rigorous solution is used, the initial conditions of Eq. (6.55) are the initial conditions of the problem under consideration. The initial conditions of Eq. (6.56) are the final displacement and final velocity obtained from the solution of Eq. (6.55), and the initial conditions of Eq. (6.57) are the final displacement and final velocity obtained from the solution of Eq. (6.56). This procedure would be sufficient if F(t) is a continuous function of time and it does not reduce to zero before the analysis for y is completed. If F(t) is of finite duration, or has discontinuities, additional stages of the equations, one for each discontinuity, should be included in the solution. See [5, 15, 84].

In practical applications, the ductility ratio η is introduced, which is defined as the ratio of y_m and y_{el}; that is,

$$\eta = \frac{y_m}{y_{el}} \tag{6.58}$$

The ductility ratio η provides a measure that can be used to decide how far into the inelastic range a member or a structure can be permitted to be exposed. In practice, ductility ratios of 5, and even higher, are used for structures that are subjected to blast, earthquake, and dynamic loading conditions that occur only a few times during their life span.

The following simple example illustrates the application of the preceding methodology and equations.

Example 6.4 The uniform simply supported beam in Fig. 6.21a supports the heavy weight W = 20 kips as shown, and also carries the dynamic force F(t) whose time variation is as shown in Fig. 6.21b. The cross section of the beam is a W24 × 76 wide flange with moment of inertia $I = 2,100.0$ in.^{4}, section modulus $S = 196.3$ in.^{3}, modulus of elasticity $E = 30 \times 10^6$ psi, and plastic modulus $Z_p = 200$ in.^{3}. By following elastoplastic analysis, determine the maximum vertical displacement of the beam that is produced by F(t). Use an appropriate idealized one-degree spring-mass system. Assume that the beam is weightless, and neglect damping. The various quantities in this problem can be converted to metric units by using the conversion factors 1 in. = 0.0254m, 1 ksi = 6,895 kPa, and 1 kip = 4,448 N.

Solution: Since the beam and loading conditions are symmetrical, the maximum vertical displacement will occur at its center C. For point C, the idealized one-degree spring-mass system is as shown in Fig.6.21c. The spring constant k represents the stiffness of the beam at its center C, which is the required vertical force at C to produce a vertical deflection equal to unity. Therefore,

$$y_c = \frac{PL^3}{48EI} = 1$$

or

$$P = k = \frac{48EI}{L^3} = \frac{(48)(30)(10)^3(2,100)}{(20)^3(12)^3} = 218.75 \text{ kips in.}^{-1}$$

The plastic moment M_p, based on a yield-point stress of 30 ksi, is

$$M_p = (30)(200) = 6,000 \text{ kip in.}$$

The maximum resistance R_m is equal to the load at the center C of the beam, which produces the ultimate moment M_p. For this beam case and loading,

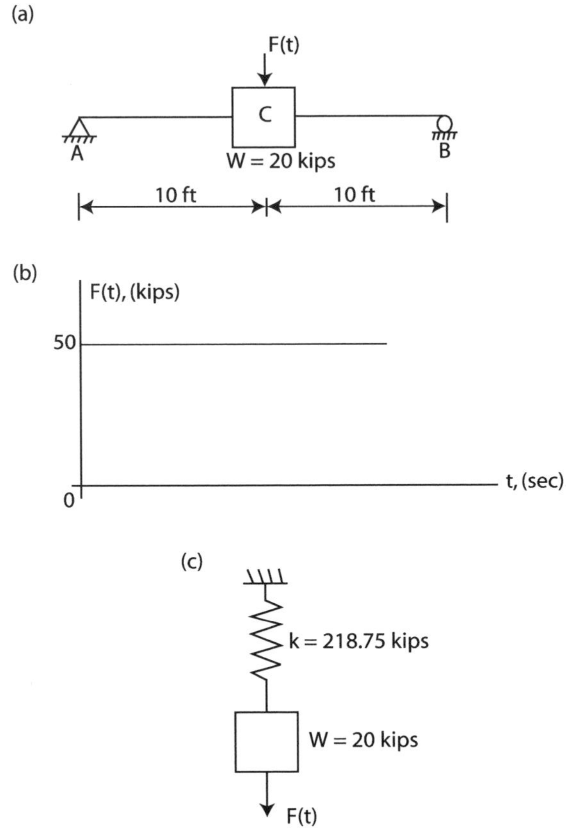

Fig. 6.21. (**a**) Uniform simply supported beam supporting a weight W and a dynamic force F(t). (**b**) Time variation of F(t). (**c**) Idealized one-degree spring-mass system. (1 ft = 0.3048 m, 1 kip = 4,448 N)

the plastic hinge develops at the center C. Thus,

$$M_p = \frac{R_m L}{4}$$

or

$$R_m = \frac{4M_p}{L} = \frac{(4)\,(6,000)}{(20)\,(12)} = 100 \text{ kips}$$

Since the beam carries also the weight W = 20 kips, the maximum resistance that is available to resist the dynamic load is

$$R_m = 100.0 - 20.0 = 80.0 \text{ kips}$$

The resistance function R is assumed to have the form shown in Fig. 6.20b. The elastic deflection y_{el} of the idealized one-degree system in Fig. 6.21c is

$$y_{el} = \frac{R_m}{k} = \frac{80.0}{218.75} = 0.3657 \text{ in.}$$

By using Eq. (6.56) and substituting the appropriate values for each term we find

$$0.05176\,\ddot{y} + R_m - F(t) = 0 \tag{6.59}$$

or

$$\ddot{y} = 19.3199\,F(t) - 19.3199\,R_m \tag{6.60}$$

For the various intervals of the vertical deflection y, we have the following three equations:

$$\ddot{y} = 965.995 - 4,226.228\,y, \qquad 0 \le y \le 0.3657 \tag{6.61}$$

$$\ddot{y} = -579.597, \qquad 0.3657 \le y \le y_m \tag{6.62}$$

$$\ddot{y} = -579.597 + 4,226.228\,(y_m - y)\,, \quad (y_m - 0.7314) \le y \le y_m \tag{6.63}$$

The numerical AIEM is used here to carry out the analysis.

The period of vibration τ of the spring-mass system in Fig. 6.21c is

$$\tau = 2\pi\sqrt{\frac{m}{k}} = 2\pi\sqrt{\frac{0.05176}{218.75}} = 0.0967\,s$$

A time interval $\Delta t = 0.01\,s$, which is about one tenth of τ, would be a reasonable one to use for this numerical method. Application of the AIEM yields the results shown in Table 6.7. The maximum deflection y_m occurs at time $t = 0.05\,s$, and it is equal to 0.4811 in. Note that Eq. (6.61) is used to determine $\ddot{y}$ up to $t = 0.04\,s$, Eq. (6.62) is used for $0.04 \le t \le 0.06$, and Eq. (6.63) is used for the remaining computations shown in the table. The computations, however, could go on indefinitely until other peaks of the displacement y are obtained.

Table 6.7. Variation of the displacement y with time (1 in. = 0.0254 m)

t (s)	$\ddot{y}$ (in.s^{-2})	$\ddot{y}\,(\Delta t)^2$ (in.)	y (in.)
0	965.995	0.0966	0
0.01	761.868	0.0762	0.0483
0.02	235.703	0.0236	0.1728
0.03	−390.202	−0.0390	0.3209
0.04	−579.597	−0.0580	0.4300
0.05	−579.597	−0.0580	0.4811
0.06	−550.436	−0.0550	0.4742
0.07	−288.833	−0.0289	0.4123
0.08	94.909	0.0095	0.3215
0.09	435.543	0.0436	0.2402
0.10	597.830	0.0598	0.2025
0.11	504.430	0.0504	0.2246
0.12			0.2971

The ductility ratio η in this case is

$$\eta = \frac{y_m}{y_{el}} = \frac{0.4811}{0.3657} = 1.316$$

A larger dynamic force F(t) would produce larger η. A larger η would be also obtained if a lighter wide-flange section is used to support W and F(t). The design engineer can decide at this point how far into the inelastic range the beam should be permitted to go. This would largely depend upon the type of F(t), and upon how many times in the lifespan of the beam the dynamic force is applied to the beam structure.

The solution of this problem is based on elastoplastic response, and certain amount of permanent deformation y_p will take place. In this solution, the amount of y_p is

$$y_p = y_m - y_{el} = 0.4811 - 0.3657$$
$$= 0.1154 \text{ in.}$$

When the elastoplastic system attains its maximum deflection y_m in Fig. 6.20b and starts to rebound, the application of Eq. (6.63) shows that it will vibrate harmonically for an indefinite period of time since there is no damping to die out the motion. This residual vibration is elastic and it takes place about an equilibrium position y, which can be determined from the following equation:

$$y = y_m - \frac{R_m - F}{k} \tag{6.64}$$

that is,

$$y = 0.4811 - \frac{80.0 - 50.0}{218.75}$$
$$= 0.344 \text{ in.}$$

The peak amplitude Y of the residual vibration is

$$Y = \frac{R_m - F}{k} = \frac{80.0 - 50.0}{218.75}$$
$$= 0.1371 \text{ in.}$$

It may be also stated that the residual vibration takes place about a position y that is equal to $y_p + y_{st}$, where $y_{st} = F/k$ is the static equilibrium position about which the system would vibrate if its response were considered to be elastic at all times. Figure 6.22 illustrates the motion of the system.

6.9.2 Multistory Buildings Subjected to Strong Earthquakes

The preceding analysis and methodology will be extended here to apply for the dynamic inelastic response of multistory structures that are exposed to

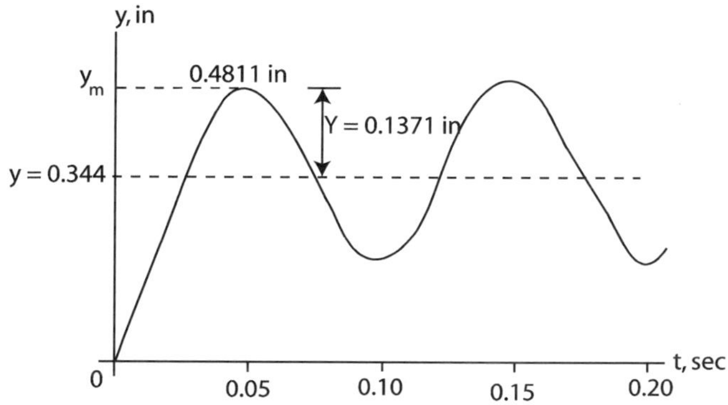

Fig. 6.22. Variation of the displacement y with respect to time t (1 in. = 0.0254 m)

the effects of rather strong earthquakes. For such kind of inelastic response, it is reasonable to neglect the flexibility of the girders and assume that their stiffness is infinite compared to the stiffness of the columns. For such cases, yielding will usually take place in the story which by comparison is the weakest in transmitting the magnitudes of the shear forces.

Practical experience and research have shown that in many cases yielding occurs near the base of the structure. Based on this reasoning, the magnitudes of the shear forces in the upper part of the structure are reduced when they are compared with the values obtained by an elastic analysis for the same base earthquake motion. Thus, if the design of a structure is based on some fraction of the maximum value of the critical shear obtained by an elastic analysis, yielding will occur in the weakest story and the shear forces in the remaining parts of the structure will have appropriately revised values.

We start the development of the required equations by considering the two-story frame in Fig. 6.23a, and assuming that its girders are infinitely stiff in comparison with its columns. The idealized two-degree spring-mass system for the frame is shown in Fig. 6.23b, where u_s is the earthquake support motion, x_1 and x_2 are the first and second story horizontal displacements, respectively, and k_1 and k_2 are the spring constants. The resistance functions R_1 and R_2 are characterized by the total shear forces V_1 and V_2, respectively, and they are shown in Fig. 6.23c. Based on this reasoning we have

$$R_1 = k_1 (x_1 - u_s) \tag{6.65}$$

$$R_2 = k_2 (x_2 - x_1) \tag{6.66}$$

By considering the free-body diagrams of masses m_1 and m_2 in Fig. 6.23c and applying the second law of motion, the following two differential equations are obtained:

$$m_1\ddot{x}_1 + R_1 - R_2 = 0 \tag{6.67}$$

$$m_2\ddot{x}_2 + R_2 = 0 \tag{6.68}$$

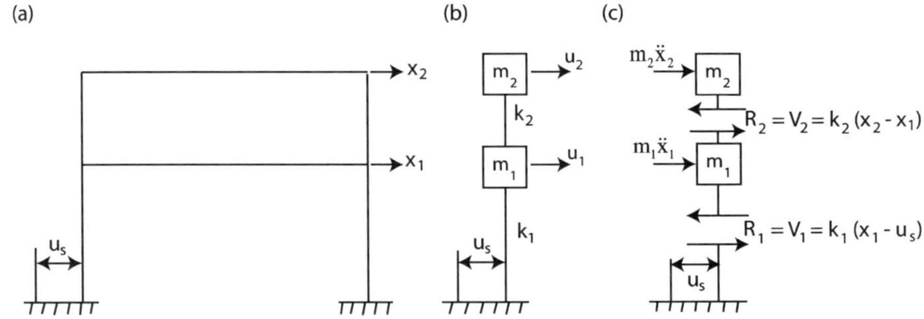

Fig. 6.23. (**a**) Two-story frame subjected to an earthquake support motion u_s. (**b**) Idealized two-degree spring-mass system for the frame. (**c**) Resistance functions R_1 and R_2

The resistance functions R_1 and R_2 in Eqs. (6.67) and (6.68) are given by Eqs. (6.65) and (6.66), respectively.

The dynamic elastoplastic analysis of the frame may be initiated by assuming that R_1 and R_2 are bilinear, and assuming maximum plastic story resistances R_{1m} and R_{2m} for R_1 and R_2, respectively. If an elastic analysis is performed, then the maximum plastic story resistances R_{1m} and R_{2m} may be taken as equal to one-half the maximum shears obtained form the elastic analysis. This is equivalent to selecting elastic limits for the story shears. With this in mind, the dynamic response of the system may be determined easily by using the AIEM to solve Eqs. (6.67) and (6.68). In the solution, R_1 and R_2 should be assumed to remain constant after the maximum values R_{1m} and R_{2m} are attained. The analysis presupposes that some plastic distortions of the system are permitted, and that their extent can be controlled by the ductility ratio η, as stated earlier.

If an elastic analysis is not carried out, a trial-and-error procedure may be used until a satisfactory design is obtained. The following example illustrates the application of the methodology.

Example 6.5 The two-story building in Fig. 6.24a is subjected to an earthquake motion with ground displacement vs. time diagram as shown in Fig. 6.25. By considering the frame in Fig. 6.24b, perform an elastoplastic analysis to arrive at a suitable column size so that the ductility ratio η is somewhere between 2.5 and 3. All units in this problem can be converted to metric units by using the conversion factors 1 in. $= 0.0254$ m, 1kip $= 4,448$ N, and 1 kip $=$ 1,000 lbs.

Solution: The graph in Fig. 6.25 is, in reality, the ground displacement vs. time graph of the Kern County, California, earthquake. In Fig. 6.25, however,

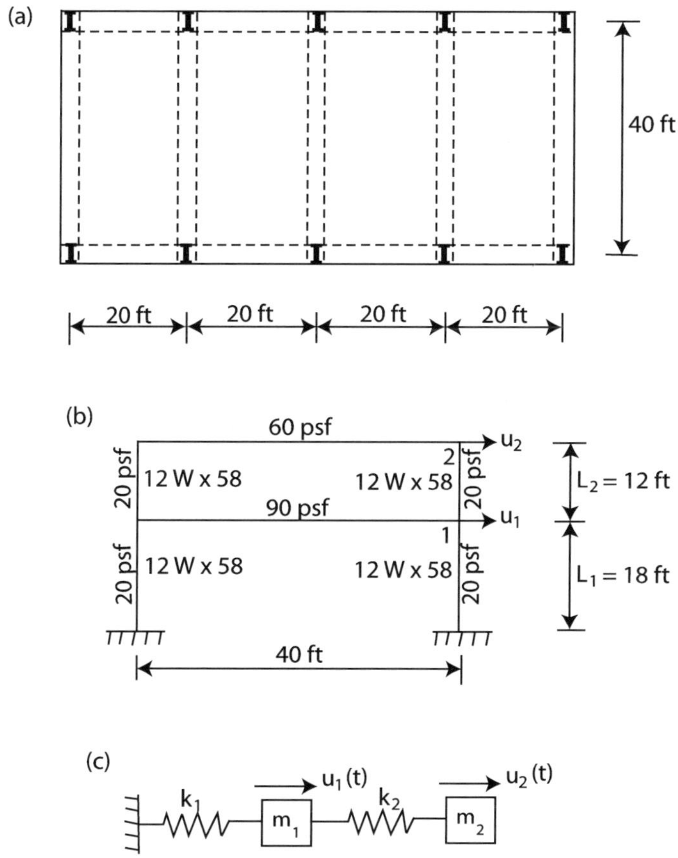

Fig. 6.24. (**a**) Top view of a two-story building. (**b**) An interior frame of the building. (**c**) Idealized two-degree spring-mass system for the frame. (1 ft = 0.3048 m, 1 psi = 6,895 Pa)

the horizontal time scale is taken as one tenth of the actual duration of the earthquake. Table 6.8 shows the values of the ground displacements u_s, in inches, at time increments $\Delta t = 0.02$ s.

We begin the analysis by assuming that all column sizes in Fig. 6.24b are W12 × 58 wide-flange sections. In addition, we assume that the girders are infinitely stiff when they are compared with the stiffness of the columns. On this basis, the idealized two-degree spring-mass system is the one shown in Fig. 6.24c.

Since the elastic properties of the building in Fig. 6.24a are uniform throughout its length, the dynamic analysis for an interior frame, such as the one in Fig. 6.24b, together with associated wall and floor areas, would be

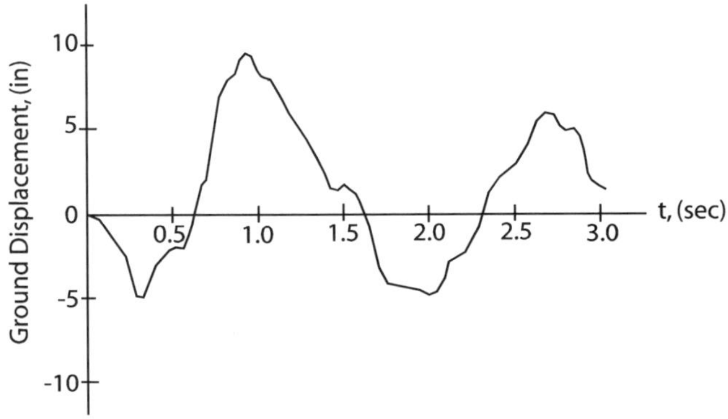

Fig. 6.25. Earthquake ground displacement curve (1 in. = 0.0254 m)

Table 6.8. Values of earthquake ground displacements obtained from the graph in Fig. 6.25 (1 in. = 0.0254 m)

t	u_s	t	u_s	t	u_s	t	u_s	t	u_s
0.02	−0.02	0.56	−0.40	1.10	6.20	1.64	−1.90	2.18	−3.55
0.04	−0.05	0.58	+0.30	1.12	5.80	1.66	−2.40	2.20	−3.40
0.06	−0.15	0.60	0.80	1.14	5.60	1.68	−2.60	2.22	−3.00
0.08	−0.29	0.62	1.40	1.16	4.80	1.70	−2.80	2.24	−2.40
0.10	−0.41	0.64	1.45	1.18	4.60	1.72	−3.10	2.26	−2.00
0.12	−0.70	0.66	2.00	1.20	4.10	1.74	−3.30	2.28	−1.40
0.14	−0.90	0.68	3.40	1.22	3.70	1.76	−3.50	2.30	−0.80
0.16	−1.15	0.70	4.40	1.24	3.40	1.78	−3.60	2.32	0.00
0.18	−1.40	0.72	5.80	1.26	3.00	1.80	−3.80	2.34	+0.60
0.20	−1.70	0.74	6.80	1.28	2.60	1.82	−4.00	2.36	1.00
0.22	−2.00	0.76	7.60	1.30	2.20	1.84	−4.20	2.38	1.20
0.24	−2.50	0.78	7.85	1.32	1.90	1.86	−4.40	2.40	1.40
0.26	−3.20	0.80	7.90	1.34	1.60	1.88	−4.50	2.42	1.60
0.28	−4.00	0.82	8.20	1.36	1.20	1.90	−4.55	2.44	1.80
0.30	−4.20	0.84	8.60	1.38	0.80	1.92	−4.59	2.46	1.95
0.32	−4.70	0.86	9.20	1.40	0.50	1.94	−4.70	2.48	2.05
0.34	−4.00	0.88	9.45	1.42	0.40	1.96	−4.90	2.50	2.15
0.36	−3.30	0.90	9.20	1.44	0.50	1.98	−5.00	2.52	2.40
0.38	−3.00	0.92	9.00	1.46	0.65	2.00	−4.90	2.54	2.60
0.40	−2.60	0.94	8.40	1.48	0.65	2.02	−4.80	2.56	3.00
0.42	−2.20	0.96	7.90	1.50	0.60	2.04	−4.75	2.58	3.20
0.44	−1.90	0.98	7.50	1.52	0.50	2.06	−4.60	2.60	3.40
0.46	−1.88	1.00	7.40	1.54	0.49	2.08	−4.30	2.62	3.60
0.48	−2.00	1.02	7.50	1.56	0.35	2.10	−4.00	2.64	3.90
0.50	−1.80	1.04	7.60	1.58	−0.40	2.12	−3.80	2.66	4.35
0.52	−1.60	1.06	7.40	1.60	−1.00	2.14	−3.70	2.68	4.50
0.54	−1.00	1.08	6.60	1.62	−140	2.16	−3.65	2.70	4.60

sufficient for the analysis of the building. On this basis we have,

$$\begin{aligned} m_1 &= \frac{1}{386.4}\left[(90)(20)(40) + (76)(40) + (2)(20)\left(\frac{18+12}{2}\right)(20)\right] \\ &= 0.22526\ \frac{\text{kip s}^2}{\text{in.}} \end{aligned}$$

$$\begin{aligned} m_2 &= \frac{1}{386.4}\left[(60)(20)(40) + (76)(40) + (2)(20)\left(\frac{12}{2}\right)(20)\right] \\ &= 0.14451\ \frac{\text{kip s}^2}{\text{in.}} \end{aligned}$$

By considering the two columns of the first story, the total stiffness k_1 is

$$\begin{aligned} k_1 = (2)\frac{12EI}{L^3} &= \frac{(2)(12)(30)(10)^3(476)}{(18)^3(12)^3} \\ &= 34.01\ \text{kips in.}^{-1} \end{aligned}$$

For the second-story columns, the total stiffness k_2 is

$$\begin{aligned} k_2 = (2)\frac{12EI}{L^3} &= \frac{(2)(12)(30)(10)^3(476)}{(12)^3(12)^3} \\ &= 114.78\ \text{kips in.}^{-1} \end{aligned}$$

For the W12 $\times$ 58 columns, note that we have moment of inertia $I_x = 476\ \text{in.}^4$, section modulus $S_x = 78.1\ \text{in.}^3$, and plastic modulus $Z_p = 86.5\ \text{in.}^3$. We calculate now the maximum elastic moment M_e in the column by using the equation

$$M_e = S_x f_y = (78.1)(36) = 2,811.6\ \text{kip in.}$$

where $f_y = 36$ ksi is the yield stress.

For a column that is fixed at both ends, and where the one end can be displaced by a horizontal displacement x, we have the following well-known expression which relates the moment M and x:

$$M = \frac{6EIx}{L^2} \tag{6.69}$$

where L is the length of the column. On this basis, by using Eq. (6.69), the maximum elastic displacement x_{1el} for the first-story columns is

$$x_{1_{el}} = \frac{(2,811.6)(18)^2(12)^2}{(6)(30)(10)^3(476)} = 1.531\ \text{in.}$$

For the second-story columns, we have

$$(x_2 - x_1)_{el} = \frac{(2,811.6)\,(12)^2\,(12)^2}{(6)\,(30)\,(10)^3\,(476)} = 0.6805\,\text{in.}$$

The second thing to do here is to establish the maximum resistances R_{1m} and R_{2m} for the first and second stories, respectively, of the frame. The plastic moment M_p of the column is

$$M_p = Z_p f_y = (86.5)\,(36) = 3,114\,\text{kip in.}$$

Thus,

$$\begin{aligned} R_{1m} = V_{1m} = \frac{2M_p}{L} &= \frac{(2)\,(3,114)}{(18)\,(12)} \\ &= 28.83\ \text{kips} \\ R_{2m} = V_{2m} = \frac{2M_p}{L} &= \frac{(2)\,(3,114)}{(12)\,(12)} \\ &= 43.25\,\text{kips} \end{aligned}$$

The preceding values of R_{1m} and R_{2m} are assigned here as the maximum plastic story resistance. For a more conservative design, lower values for R_{1m} and R_{2m} may be assigned. This can be also controlled by the ductility ratio η. The value of 3 for η is rather conservative.

The numerical AIEM is used here for the computation of the maximum story displacements x_{1m} and x_{2m}. We select $\Delta t = 0.02\,\text{s}$. At the first time station, the displacements $x_1^{(1)}$ and $x_2^{(1)}$ are as follows:

$$x_1^{(1)} = \frac{1}{6}\left[2\ddot{x}_1^{(0)} + \ddot{x}_1^{(1)}\right](\Delta t)^2 = \frac{1}{6}\left[\ddot{x}_1^{(1)}\right](\Delta t)^2 \tag{6.70}$$

$$x_2^{(1)} = \frac{1}{6}\left[2\ddot{x}_2^{(0)} + \ddot{x}_2^{(1)}\right](\Delta t)^2 = \frac{1}{6}\left[\ddot{x}_2^{(1)}\right](\Delta t)^2 \tag{6.71}$$

By using Eqs. (6.65)–(6.68), we obtain

$$\ddot{x}_1^{(1)} = \frac{R_2 - R_1}{m_1} = \frac{k_2\left(x_2^{(1)} - x_1^{(1)}\right) - k_1\left(x_1^{(1)} - u_s^{(1)}\right)}{m_1} \tag{6.72}$$

$$\ddot{x}_2^{(1)} = -\frac{R_2}{m_2} = -\frac{k_2}{m_2}\left(x_2^{(1)} - x_1^{(1)}\right) \tag{6.73}$$

Thus,

$$x_1^{(1)} = \left[\frac{k_2}{6m_1}\left(x_2^{(1)} - x_1^{(1)}\right) - \frac{k_1}{6m_1}\left(x_1^{(1)} - u_s^{(1)}\right)\right](\Delta t)^2 \tag{6.74}$$

$$x_2^{(1)} = \frac{k_2}{6m_1}\left(x_2^{(1)} - x_1^{(1)}\right)(\Delta t)^2 \tag{6.75}$$

From Eq. (6.75), we obtain

$$x_2^{(1)} = \frac{(k_2/6m_2)(\Delta t)^2}{\left[1+(k_2/6m_2)(\Delta t)^2\right]} x_1^{(1)} = Cx_1^{(1)} \tag{6.76}$$

where

$$C = \frac{(k_2/6m_2)(\Delta t)^2}{\left[1+(k_2/6m_2)(\Delta t)^2\right]} \tag{6.77}$$

By substituting Eq. (6.76) into Eq. (6.74), we find

$$x_1^{(1)} = \frac{\left[k_1 u_s^1 (\Delta t)^2\right]/6m_1}{1+\left\{\left[k_2(\Delta t)^2\right]/Cm_1\right\}(1+C)+\left\{\left[k_1(\Delta t)^2\right]/6m_1\right\}} \tag{6.78}$$

where C is given by Eq. (6.77). Thus, Eqs. (6.78) and (6.76) may be used to determine the displacements $x_1^{(1)}$ and $x_2^{(1)}$, respectively, at the first time station.

The computer program in Appendix B was written in order to perform the AIEM. For given values of I_x, S_x, and Z_p, it performs the following:

1. It calculates k_1 and k_2 values from the equation

$$k = \frac{(2)(12)EI_x}{L^3}$$

2. It calculates $\omega_1, \omega_2, \tau_1$, and τ_2 from the equations

$$\omega_{1,2}^2 = \frac{1}{2}\left(\frac{k_1+k_2}{m_1}+\frac{k_2}{m_2}\right) \pm \frac{1}{2}\left[\left(\frac{k_1+k_2}{m_1}+\frac{k_2}{m_2}\right)^2 - \frac{4k_1k_2}{m_1m_2}\right]^{1/2}$$

$$\tau_1 = \frac{2\pi}{\omega_1}, \qquad \tau_2 = \frac{2\pi}{\omega_2}$$

3. It calculates maximum elastic deflection values for each story column form the equation

$$y = \frac{M_e L^2}{6EI_x} = \frac{S_x(36)L^2}{6EI_x}$$

4. It calculates maximum plastic shears for each story by using the equation

$$R_m = \frac{2M_p}{L} = \frac{2Z_p(36)}{L}$$

For the given value of the time increment Δt and the displacement $u_s^{(1)}$ at the first time station, the program performs the following calculations:

1. It calculates $x_1^{(1)}$ and $x_2^{(1)}$ using Eqs.(6.78) and (6.76).
2. It calculates $x_1^{(1)} - u_s^{(1)}$.

3. It calculates $R_1 = k_1\,(x_1^{(1)} - u_s^{(1)})$.
4. It calculates $x_2^{(1)} - x_1^{(1)}$.
5. It calculates $R_2 = k_2\,(x_2^{(1)} - x_1^{(1)})$.
6. It calculates $\ddot{x}_1 = (R_2 - R_1)/m_1$ and $\ddot{x}_1\,(\Delta t)^2$.
7. It calculates $\ddot{x}_2 = -R_2/m_2$ and $\ddot{x}_2\,(\Delta t)^2$.
8. It calculates

$$\eta_1 = \left|\frac{x_1}{y_{1_{el}}}\right|$$

$$\eta_2 = \left|\frac{x_2 - x_1}{y_{2_{el}}}\right|$$

9. By using the equation

$$x^{(i+1)} = 2x^{(i)} - x^{(i-1)} + \ddot{x}^{(i)}\,(\Delta t)^2$$

it calculates $x_1^{(2)}$ and $x_2^{(2)}$ at the second time station.

Then, by imputing the ground displacement $u_s^{(2)}$, it performs the calculations in steps 2–9 again, and so on. Note that whenever $|R_1| > R_{1m}$ and/or, $|R_2| > R_{2m}$, the program sets $R_1 = R_{1m}$ if $R_1 > 0$ and/or $R_2 = R_{2m}$ if $R_2 > 0$, $R_1 = -R_{1m}$ if $R_1 < 0$ and $R_2 = -R_{2m}$ if $R_2 < 0$.

The first design, which incorporates W12 × 58 wide flange section for the columns, yields $x_{1m} = -5.9501$ in. at time t = 0.68 s, and

$$\eta_1 = \left|\frac{x_{1_m}}{x_{1_{el}}}\right| = 3.89 \tag{6.79}$$

For the second story we have $(x_2 - x_1)_m = -0.3145$ in. at time t = 0.46 s, and

$$\eta_2 = \left|\frac{(x_2 - x_1)_m}{x_{2_{el}}}\right| = 0.46 \tag{6.80}$$

The preceding results show that for the first peak of maximum response, the first story undergoes plastic response and deformation for the indicated column sizes, and the response of the second-story columns is elastic. Since the structure did not yet experience the larger positive peak of the ground displacement, which comes at t = 0.88 s, the value of η may go even higher. Since in this problem we require $2.5 \leq \eta \leq 3.0$, the procedure will be repeated by using a different column size.

We use for this design trial a W10 × 25 wide-flange section. In this case, the AIEM yields $x_1 = 4.9793$ in. occurring at time t = 2.00 s, and

$$\eta_1 = \left|\frac{x_{1_m}}{x_{1_{el}}}\right| = 2.68 \tag{6.81}$$

For the second story, we have $(x_2 - x_1)_m = 0.2777$ in. at time t = 1.52 s, and

$$\eta_2 = \left| \frac{(x_2 - x_1)_m}{x_{2_{el}}} \right| = 0.34 \tag{6.82}$$

The results of this design trial indicate that the first story undergoes plastic response and deformation for its columns, and that the maximum deflections occur well beyond the large positive peak of the ground earthquake displacement. Since the maximum value of the ductility ratio η is between 2.5 and 3.0, as desired, the W10 × 25 column sizes for the frame are acceptable. Note that the lighter columns produced a more satisfactory design for the given earthquake ground motion, indicating that other structural characteristics, such as the structure's periods of vibration, play a key role in its ability to resist the effects of an earthquake.

If a computer and/or a computer program is not available, the AIEM may be carried out manually. For buildings having three or more stories, utilization of computer software would be advisable.

6.10 Elastic and Inelastic Analysis of Thick-Walled Cylinders Subjected to Uniform External and Internal Pressures

In this section we study the response of thick-walled cylinders of constant wall thickness and subjected to uniform external and internal pressures. Both elastic and inelastic responses are considered.

6.10.1 Elastic Analysis of Thick Cylinders

In Fig. 6.26a, a thick-walled cylinder of constant thickness is subjected to a uniformly distributed external pressure p_0 and a uniform internal pressure p_i as shown. In the figure, a and b denote the internal and external radii, respectively, of the cylinder. Under these conditions of symmetry, the deformation of the cylinder is expected to be symmetrical about its centroidal axis and also uniform along its entire length.

In Fig. 6.26a we consider the infinitesimal element ABCD which is constructed as shown in Fig. 6.26b. In this figure, σ_t represents the tangential or hoop normal stress acting on sides AD and BC of the element, and σ_r is the radial normal stress acting normal to its sides AB and DC. Note here that σ_r is variable along its radial direction and, consequently, if it is σ_r on side DC, then, on side AB, will be $\sigma_r + (d\sigma_r/dr)dr$, as shown in Fig.6.26b.

If we consider the element in Fig. 6.26b and set equal to zero all the components of the forces acting in the direction of the bisector of the angle $d\phi$, we find the following expression:

$$\sigma_r r\, d\phi + \sigma_t\, dr\, d\phi - \left[\sigma_r + \frac{d\sigma_r}{dr}dr\right](r + dr)\, d\phi = 0 \tag{6.83}$$

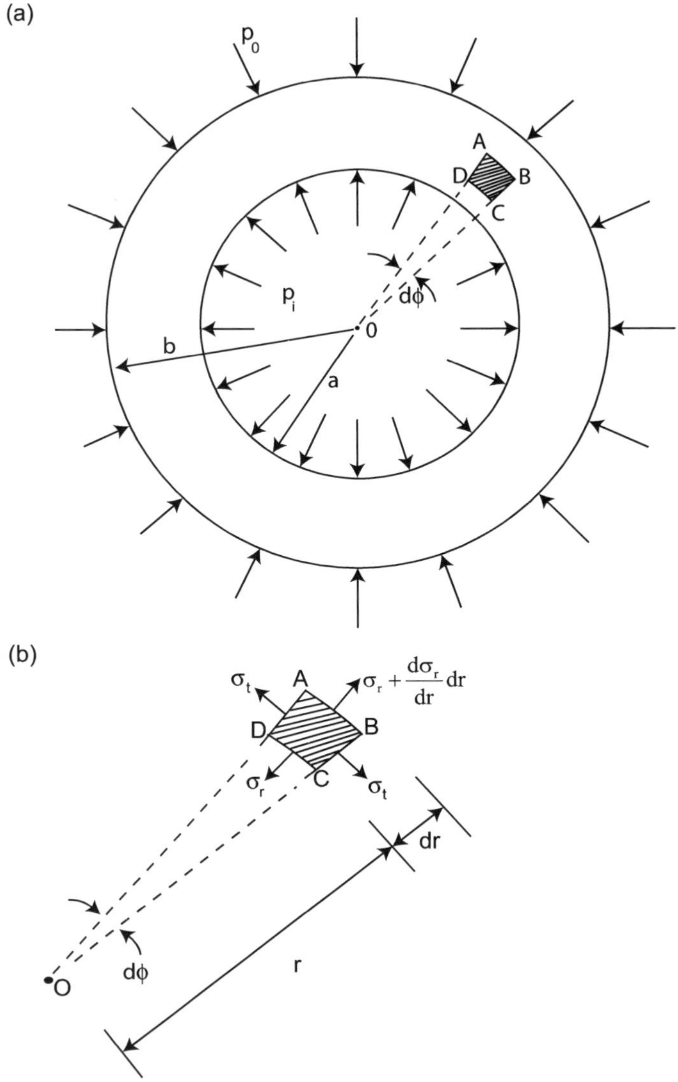

Fig. 6.26. (**a**) Thick-walled cylinder subjected to uniform internal and external pressures. (**b**) Free-body diagram of a radial element ABCD of the cylinder

Equation (6.83) by neglecting small quantities of higher order, gives

$$\sigma_t - \sigma_r - r\frac{d\sigma_r}{dr} = 0 \tag{6.84}$$

The two unknowns in Eq. (6.84) are the stresses σ_r and σ_t which can be determined by utilizing the deformation characteristics of the cylinder. If we denote by the letter u the displacement of the cylinder at a radius r from the center 0 of the cylinder, then at a radius $r + dr$ this displacement will be

$$u + \frac{du}{dr}dr \tag{6.85}$$

Since $(du/dr)dr$ is the total radial deformation of the element ABCD, we conclude that the strain ε_r in the same radial direction is

$$\varepsilon_r = \frac{du}{dr} \tag{6.86}$$

In the circumferential or tangential direction, the strain ε_t is

$$\varepsilon_t = \frac{u}{r} \tag{6.87}$$

From basic mechanics, and by using Eqs. (6.86) and (6.87), the expressions of σ_r and σ_t in terms of the strains ε_r and ε_t are as follows:

$$\sigma_r = \frac{E}{1-\nu^2}\left(\frac{du}{dr} + \nu\frac{u}{r}\right) \tag{6.88}$$

$$\sigma_t = \frac{E}{1-\nu^2}\left(\frac{u}{r} + \nu\frac{du}{dr}\right) \tag{6.89}$$

where ν is the Poisson ratio.

By substituting Eqs. (6.88) and (6.89) into Eq. (6.84), we find

$$\frac{d^2u}{dr^2} + \frac{1}{r}\frac{du}{dr} - \frac{u}{r^2} = 0 \tag{6.90}$$

The general solution of Eq. (6.90) is given by the expression

$$u = Ar + \frac{B}{r} \tag{6.91}$$

where A and B are constants which can be determined by using the stress boundary conditions for σ_r at the inner and outer surfaces of the cylinder. Note that the solution given by Eq. (6.91) can be verified by substitution.

By substituting into Eqs. (6.88) and (6.89) the general solution given by Eq. (6.91), we obtain the following expressions for the stresses σ_r and σ_t:

$$\sigma_r = \frac{E}{1-\nu^2}\left((1+\nu)A - \frac{(1-\nu)}{r^2}B\right) \tag{6.92}$$

$$\sigma_t = \frac{E}{1-\nu^2}\left((1+\nu)A + \frac{(1-\nu)}{r^2}B\right) \tag{6.93}$$

At the inner and outer surfaces of the cylinder the stress boundary conditions for σ_r are as follows:

$$(\sigma_r)_{r=\alpha} = -p_i \tag{6.94}$$

$$(\sigma_r)_{r=b} = -p_0 \tag{6.95}$$

where the tensile stress is considered as positive.

By using Eq. (6.92) and applying the boundary conditions given by Eqs. (6.94) and (6.95), we find the expressions for the constants A and B which are as shown below:

$$A = \left(\frac{1-\nu}{E}\right)\left[\frac{a^2 p_i - b^2 p_0}{b^2 - a^2}\right] \tag{6.96}$$

$$B = \left(\frac{1+\nu}{E}\right)\left[\frac{a^2 b^2 (p_i - p_0)}{b^2 - a^2}\right] \tag{6.97}$$

By substituting Eqs. (6.96) and (6.97) into Eqs. (6.92) and (6.93), the final expressions for the stresses σ_r and σ_t are as follows:

$$\sigma_r = \frac{a^2 p_i - b^2 p_0}{b^2 - a^2} - \frac{a^2 b^2 (p_i - p_0)}{r^2 (b^2 - a^2)} \tag{6.98}$$

$$\sigma_t = \frac{a^2 p_i - b^2 p_0}{b^2 - a^2} + \frac{a^2 b^2 (p_i - p_0)}{r^2 (b^2 - a^2)} \tag{6.99}$$

Note that the sum of σ_r and σ_t is an invariant quantity (constant).

The radial displacement u of the cylinder may be obtained by substituting Eqs. (6.96) and (6.97) into Eq. (6.91). This yields the following equation of u:

$$u = \left(\frac{1-\nu}{E}\right)\left[\frac{a^2 p_i - b^2 p_0}{b^2 - a^2}\right] r + \left(\frac{1+\nu}{E}\right)\left[\frac{a^2 b^2 (p_i - p_0)}{b^2 - a^2}\right]\left(\frac{1}{r}\right) \tag{6.100}$$

If the thick-walled cylinder is subjected only to an internal pressure p_i, then we have $p_0 = 0$, and Eqs. (6.98) and (6.99) yield

$$\sigma_r = \frac{a^2 p_i}{b^2 - a^2}\left(1 - \frac{b^2}{r^2}\right) \tag{6.101}$$

$$\sigma_t = \frac{a^2 p_i}{b^2 - a^2}\left(1 + \frac{b^2}{r^2}\right) \tag{6.102}$$

By examining Eqs. (6.101) and (6.102), we find that σ_r is always a compressive stress while σ_t is always a tensile stress. The maximum value of σ_t is at the inner surface of the cylinder and its value may be obtained from the equation

$$(\sigma_t)_{max} = \frac{(a^2 + b^2)\, p_i}{b^2 - a^2} \tag{6.103}$$

The maximum shear stress τ_{max} occurs at the inner surface of the cylinder and it may be obtained from the equation

$$\begin{aligned} \tau_{max} &= \frac{\sigma_t - \sigma_r}{2} = \frac{1}{2}\left[\frac{p_i (a^2 + b^2)}{b^2 - a^2} + \frac{p_i (a^2 - b^2)}{b^2 - a^2}\right] \\ &= \frac{p_i\, b^2}{b^2 - a^2} \end{aligned} \tag{6.104}$$

Also, for internal pressure only, Eq. (6.100) yields

$$u_{r=\alpha} = \frac{a\,p_t}{E}\left(\frac{a^2+b^2}{b^2-a^2}+\nu\right) \tag{6.105}$$

$$u_{r=b} = -\frac{b\,p_t}{E}\left(\frac{a^2+b^2}{b^2-a^2}-\nu\right) \tag{6.106}$$

If only external pressure acts on the cylinder we have $P_i = 0$, and Eqs. (6.98) and (6.99) yield

$$\sigma_r = -\frac{p_o b^2}{b^2-a^2}\left(1-\frac{a^2}{r^2}\right) \tag{6.107}$$

$$\sigma_t = -\frac{p_o b^2}{b^2-a^2}\left(1+\frac{a^2}{r^2}\right) \tag{6.108}$$

which indicate that both σ_r and σ_t are compressive stresses.

The following example illustrates the application of the above theory.

Example 6.6 For the thick-walled cylinder shown in Fig. 6.26a, determine the tangential and radial stresses at the inner and outer surfaces of the cylinder and at the middle of the wall. Assume that $p_i = 30,000$ psi, $p_o = 0$, $a = 4$ in., and $b = 8$ in. Also determine the radial displacement of its inner surface by assuming that $\nu = 0.3$ and $E = 30 \times 10^6$ psi (1 in. = 0.0254 m, 1 psi = 6,895 Pa).

Solution: Since $p_o = 0$, we use Eqs. (6.101), (6.102), and (6.105), which yield

$$(\sigma_r)_{r=4''} = \frac{(4)^2(30,000)}{(8)^2-(4)^2}\left[1-\frac{(8)^2}{(4)^2}\right] = -30,000\,\text{psi}$$

$$(\sigma_r)_{r=8''} = \frac{(4)^2(30,000)}{(8)^2-(4)^2}\left[1-\frac{(8)^2}{(8)^2}\right] = 0$$

$$(\sigma_r)_{r=6''} = \frac{(4)^2(30,000)}{(8)^2-(4)^2}\left[1-\frac{(8)^2}{(6)^2}\right] = -7,777.78\,\text{psi}$$

$$(\sigma_t)_{r=4''} = \frac{(4)^2(30,000)}{(8)^2-(4)^2}\left[1+\frac{(8)^2}{(4)^2}\right] = 50,000\,\text{psi}$$

$$(\sigma_t)_{r=8''} = \frac{(4)^2(30,000)}{(8)^2-(4)^2}\left[1+\frac{(8)^2}{(8)^2}\right] = 20,000\,\text{psi}$$

$$(\sigma_t)_{r=6''} = \frac{(4)^2(30,000)}{(8)^2-(4)^2}\left[1+\frac{(8)^2}{(6)^2}\right] = 27,778\,\text{psi}$$

$$u_{r=4''} = \left[\frac{(4)^2+(8)^2}{(8)^2-(4)^2}+0.3\right]\frac{(4)(30,000)}{(30)(10)^6} = 7.8667 \times 10^{-3}\,\text{in.}$$

6.10.2 Inelastic Analysis of Thick Cylinders

We consider here the case where the thick-walled cylinder is subjected only to an internal pressure $p_i = p$ and, therefore, its external pressure $p_0 = 0$. In such case, at any radial distance r from the axis of the cylinder, the stresses σ_r and σ_t may be obtained by using Eqs. (6.101) and (6.102), respectively.

At the inner surface of the cylinder, the tangential tensile stress gains its maximum value. At the same surface, we also have the maximum numerical value for the radial compression and also the shear stress will be maximum. The magnitude of the maximum shear stress τ_{max} may be obtained from the equation

$$\tau_{max} = \left(\frac{\sigma_t - \sigma_r}{2}\right)_{r=a} = \frac{pb^2}{b^2 - a^2} \tag{6.109}$$

If the internal pressure p is increased gradually, it will reach in time a value at which the material of the inner surface of the cylinder will start to yield, and this will occur when the maximum shearing stress in Eq. (6.109) becomes equal to the yield point shearing stress τ_{yp}. By substituting τ_{yp} for τ_{max} in Eq. (6.109) and solving for the pressure p, we find that the pressure $p = p_{yp}$ at which yielding is initiated is

$$p = p_{yp} = \tau_{yp}\frac{b^2 - a^2}{b^2} \tag{6.110}$$

If we continue to increase the internal pressure p beyond its p_{yp} value, the plastic deformation will continue to penetrate deeper into the wall of the cylinder. At a certain value of $p = p_{ult}$ the entire wall of the cylinder will reach the state of yielding. If the material is perfectly plastic, the yielding proceeds under the action of a constant shearing stress τ_{yp}, and its value at every point of the plastic region is given by the equation

$$\tau_{yp} = \frac{\sigma_t - \sigma_r}{2} \tag{6.111}$$

If the material of the cylinder is not perfectly plastic, then at an internal pressure $p_{yp} < p_1 < p_{ult}$, part of the cylinder wall will be in the plastic region and the remaining one will be elastic. We designate as c the radius of the cylindrical surface that separates the plastic region of the cylindrical wall from the elastic one. See Fig. 6.27. We designate as σ_{rc} the radial pressure at this surface located at $r = c$ and separating the plastic region of the cylinder wall from the elastic one.

We can determine σ_{rc} by utilizing the outer portion of the cylindrical wall, which is elastic. In this elastic portion, the maximum shear stress τ_{max} can be obtained from Eq. (6.109) by replacing a with c and p with σ_{rc}. This yields

$$\tau_{max} = \frac{\sigma_{rc} b^2}{b^2 - c^2} \tag{6.112}$$

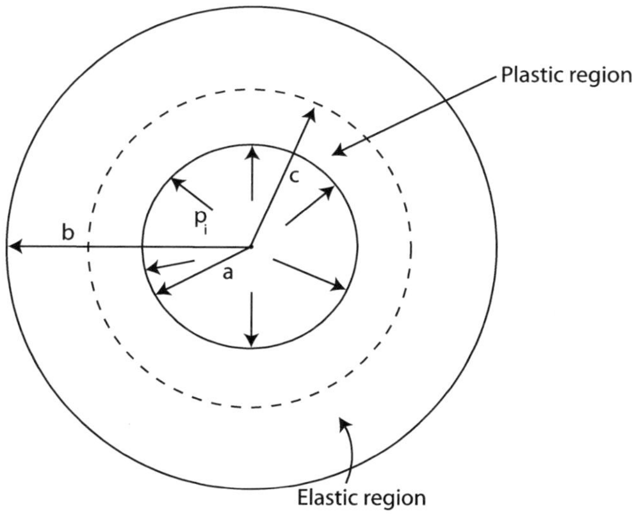

Fig. 6.27. Cross section of the thick-walled cylinder showing the regions of plastic and elastic deformations

At the boundary surface located at $r = c$, the material has just reached the yield point stress and, consequently, $\tau_{max} = \tau_{yp}$. On this basis, Eq.(6.112) may be written as

$$\tau_{yp} = \frac{\sigma_{rc} b^2}{b^2 - c^2} \qquad (6.113)$$

Solving Eq. (6.113) for σ_{rc} we obtain

$$\sigma_{rc} = \frac{\left(b^2 - c^2\right)}{b^2} \tau_{yp} \qquad (6.114)$$

With known σ_{rc}, we can calculate the stresses at any point in the elastic region of the cylindrical wall by using equations similar to Eqs. (6.101) and (6.102). To apply these two equations we have to use c instead of a, σ_{rc} instead of p_i and values of $r \geq c$.

In order to calculate the stresses in the plastic region of the wall of the cylinder, we can use the equilibrium equation of the cylinder given by Eq. (6.84). We rewrite this equation here for convenience.

$$\sigma_t - \sigma_r - r \frac{d\sigma_r}{dr} = 0 \qquad (6.115)$$

From Eq. (6.111) we find

$$\sigma_t - \sigma_r = 2\tau_{yp} \qquad (6.116)$$

By substituting Eq. (6.116) into Eq. (6.115), we find

$$\frac{d\sigma_r}{dr} = \frac{2\tau_{yp}}{r} \tag{6.117}$$

By integrating Eq. (6.117) once, we find

$$\sigma_r = 2\tau_{yp} \log_e r + C \tag{6.118}$$

where C is the constant of integration.

We find C by using the boundary condition that at $r = c$ the stress $\sigma_r = -\sigma_{rc}$, yielding

$$-\sigma_{rc} = 2\tau_{yp} \log_e c + C$$

or, by solving C,

$$C = -\sigma_{rc} - 2\tau_{yp} \log_e c \tag{6.119}$$

By using the value of C obtained in Eq. (6.119) and the value of σ_{rc} obtained in Eq. (6.114) and substituting into Eq. (6.118), we find

$$\sigma_r = 2\tau_{yp} \log_e \left(\frac{r}{c}\right) - \frac{\left(b^2 - c^2\right)}{b^2}\tau_{yp} \tag{6.120}$$

If we take r as equal to the inner radius a of the cylinder, then we can determine the magnitude of the pressure p, which is required to produce plastic flow in the wall of the cylinder up to a depth of radius $r = c$. This pressure can be obtained from the equation

$$p_1 = -2\tau_{yp} \log_e \left(\frac{a}{c}\right) + \frac{\left(b^2 - c^2\right)}{b^2}\tau_{yp} \tag{6.121}$$

If we assume, for example, that $b = 2a$ and $c = 1.5a$, Eq. (6.121) yields $p_1 = 1.248\tau_{yp}$.

The expression of the tangential stress σ_t can be derived from Eq. (6.111), yielding

$$\sigma_t = 2\tau_{yp} + \sigma_r \tag{6.122}$$

or

$$\sigma_t = 2\tau_{yp} \log_e \left(\frac{r}{c}\right) + \frac{\left(b^2 - c^2\right)}{b^2}\tau_{yp} \tag{6.123}$$

On this basis, where the material beyond the yield point yields without an increase in stress, Eqs. (6.121) and (6.123) provide the stresses which are produced in the inner portion of the wall of the cylinder which is undergoing plastic deformation. The outer portion remains elastic and, therefore, equations similar to the ones given by Eqs. (6.101) and (6.102) can be used, as discussed earlier in this section.

By examining Eq. (6.123), we note that for $r = c$, the first term on the right-hand side of the equation becomes equal to zero. In this case, we note that the magnitude of σ_t becomes equal to the magnitude of the tangential stress which is produced by σ_{rc} on the adjacent elastic zone of the cylinder wall. See Eq. (6.114). It should also be noted that the yield-point stress τ_{yp} for various materials, is usually obtained in practice experimentally. See also Sect. 6.12 later in this chapter.

Example 6.7 The thick-walled cylinder in Fig. 6.27 is subjected to an internal pressure p_i only. This pressure is large enough to produce a plastic region on the wall of the cylinder. If the radius a = 5 in. b = 10 in. and c = 7.5 in., determine the pressure p_1 which is required to produce plastic flow in the wall of the cylinder to a depth of radius r = 7.5 in. Also determine σ_t at r = 7.5 in., r = 5 in., and r = 10 in. Assume that $\tau_{yp} = 20,000$ psi (Note that 1 in. = 0.0254 m, 1 psi = 6,895 Pa).

Solution: By applying Eq. (6.121), we find

$$\begin{aligned} p_1 &= -(2)(20,000)\log_e\left(\frac{5}{7.5}\right) + \frac{(100-56.25)}{100}(20,000) \\ &= 16,218.6 + 8,750 \\ &= 24,968 \text{ psi} \end{aligned}$$

From Eq. (6.123), at r = 7.5, we obtain

$$\begin{aligned} (\sigma_t)_{r=7.5} &= (2)(20,000)\log_e\left(\frac{7.5}{7.5}\right) + \frac{(100-56.25)}{100}(20,000) \\ &= 0 + 8750 = 8,750 \text{ psi} \end{aligned}$$

From the same equation, at r = 5 in., we find

$$\begin{aligned} (\sigma_t)_{r=5} &= (2)(20,000)\log_e\left(\frac{5}{7.5}\right) + \frac{(100-56.25)}{10}(20,000) \\ &= -16,218.6 + 8,750 \\ &= -7,468.6 \text{ psi} \end{aligned}$$

At the wall region beyond r = 7.5 in., the material is elastic, and Eq. (6.102) should be used, provided, however, that we replace a by c and p_i by σ_{rc}. Therefore, at r = 10 in., we obtain

$$\begin{aligned} (\sigma_t)_{r=10} &= \frac{c^2\sigma_{rc}}{b^2-c^2}\left(1+\frac{b^2}{b^2}\right) \\ &= \frac{56.25}{100-56.25}(2)\,\sigma_{rc} \\ &= 2.571429\,\sigma_{rc} \end{aligned} \tag{6.124}$$

From Eq. (6.114), we find

$$\begin{aligned} \sigma_{rc} &= \frac{100-56.25}{100}(20,000) \\ &= 8,750 \text{ psi} \end{aligned}$$

Thus, by substituting into Eq. (6.124), we find

$$\begin{aligned}(\sigma_t)_{r=10} &= (2.571429)\,(8,750)\\ &= 22,500\,\text{psi}\end{aligned}$$

6.11 Inelastic Analysis of Members of Non-Rectangular Cross Sections

The inelastic analysis in Chap. 4, was primarily concentrated on beams of rectangular cross section where their depth can vary in any arbitrary manner along their length. On this basis, both the moment of inertia and the modulus of elasticity can vary along their length when their material is stressed beyond its elastic limit. The theory, however, is general, and it can be applied to other types of cross-sectional shapes.

As an illustration, we consider a member who's cross-sectional shape is as shown in Fig. 6.28a. In this case the width of the cross section is not constant along its depth. On this basis, Eqs. (4.4) and (4.5) will have to be written in the following general form which incorporates the possible variation of the width b of the cross section:

$$\int_{-h_2}^{h_1} b\sigma\,dy = r\int_{\varepsilon_2}^{\varepsilon_1} b\sigma\,d\varepsilon = 0 \tag{6.125}$$

$$\int_{-h_2}^{h_1} b\sigma y\,dy = r^2\int_{\varepsilon_2}^{\varepsilon_1} b\sigma\varepsilon\,d\varepsilon = M \tag{6.126}$$

For the cross-sectional shape shown in Fig.6.28a, Eqs. (6.125) and (6.126) are written as follows:

$$\int_{\varepsilon_2}^{\varepsilon'} \sigma\,d\varepsilon + \int_{\varepsilon'}^{\varepsilon_1} \frac{b_1}{b}\sigma\,d\varepsilon = 0 \tag{6.127}$$

$$br^2\left(\int_{\varepsilon_2}^{\varepsilon'} \sigma\varepsilon\,d\varepsilon + \int_{\varepsilon'}^{\varepsilon_1} \frac{b_1}{b}\sigma\varepsilon\,d\varepsilon\right) = M \tag{6.128}$$

In the above two equations, the symbol ε' denotes the longitudinal strain at the junction DE of the web and flange.

By examining Eqs. (6.127) and (6.128), we note that the ordinates of the stress–strain curve A0B in Fig. 6.28b, in the region corresponding to the flange of the cross section between ε' and ε_1 should be increased, or magnified, in the ratio b_1/b. Also note here that the cross section is symmetrical about the y axis.

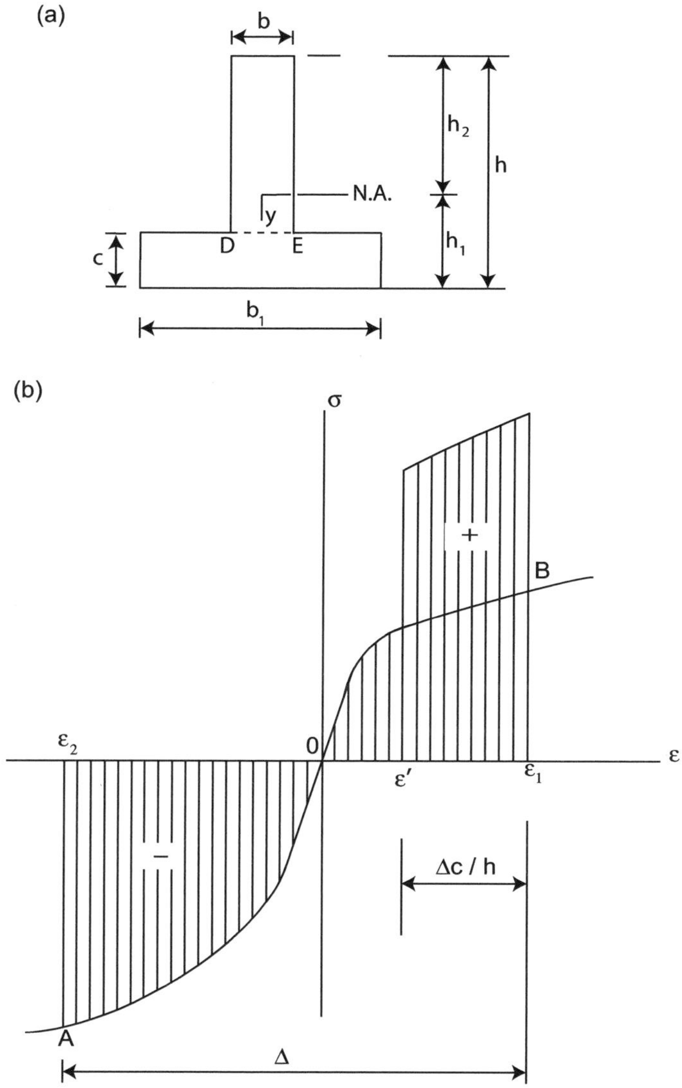

Fig. 6.28. (**a**) Cross-sectional shape symmetrical about the y axis. (**b**) Stress–strain diagram

In order to determine the position of the neutral axis (N.A.), we can follow a procedure similar to the one used in Chap.4. That is, we can use the stress-strain diagram in Fig.6.28b and on the horizontal axis we place the assumed length $\Delta = h/r$ in a way that makes the two shaded areas under the curve A0B numerically equal. This procedure provides the values of the strains ε_1 and ε_2 in the extreme fibers of the cross section located at distances h_1 and h_2, respectively, from the neutral axis.

The strain ε' at the junction of the web and flange of the cross section, can be determined by using the following expression:

$$\frac{c}{h} = \frac{\varepsilon_1 - \varepsilon'}{\Delta} \tag{6.129}$$

or, by solving for ε', we find

$$\varepsilon' = \frac{h\varepsilon_1 - c\Delta}{h} \tag{6.130}$$

Having located the position of the neutral axis and realizing that the two integrals in the parenthesis at the left-hand side of Eq. (6.128), represent the first moment of the shaded area in Fig. 6.28b about a vertical axis through the origin 0, then for any assumed $\Delta = h/r$, we can determine the moment M by using Eq. (6.128).

For statically determinate members that are subjected to transverse loadings, the bending moment obtained by using Eq. (6.128) as discussed above, should be equal to the bending moment at the section that is produced by the static application of the transverse loads. This is easily determined by using simple statics as discussed in Chap. 4. This moment is also related to the bending stiffness by the following equation, as discussed in Chap. 4:

$$M = \frac{E_r I}{r} \tag{6.131}$$

where E_r is the reduced modulus, $E_r I$ is the reduced bending stiffness, I is the moment of inertia of the cross section about its neutral axis, and r is the radius of curvature.

In Eq. (6.131), since M, r, and I, are known or calculated, the reduced modulus E_r, or the reduced bending stiffness at sections along the length of the member, can be determined by using Eq. (6.131). If the elastic limit of the material is not exceeded, then E_r in Eq. (6.131) is equal to the elastic modulus E. If the depth h of the member also varies along the length of the member, the above procedure remains the same, except that now the moment of inertia of the member along its length is also variable. The inelastic analysis in Chap. 4 takes into consideration cross-sectional moment of inertia variations, and it can be used here.

If we neglect the effect of shear on the deflection of a member which is acted upon by transverse loads, then Eq. (6.131), which is derived for pure bending, provides the relation between bending moment and curvature, and the methods discussed in Chap. 4 can be used to determine the deflections of the member when its material is stressed beyond its proportional limit. However, in this case, we have to take into consideration that the bending stiffness $E_r I$ is a variable quantity and it varies with the magnitude of the bending moment. Such cases are amply illustrated in Chap. 4.

If the transversely loaded member is statically indeterminate, the calculation of the redundant forces will require more complicated investigation, because beyond the proportional limit the redundant forces and moments are no longer proportional to the applied loads and the superposition principle does not apply. However, for many problem cases, trial-and-error approaches can be used to calculate such quantities. See also [3] for more information on this subject.

6.12 Torsion Beyond the Elastic Limit of the Material

To illustrate the procedure we consider a solid circular shaft whose external radius is a. The procedure, however, remains the same if the circular shaft is hollow. If we twist the circular shaft beyond the elastic limit of its material, we make the assumption that the cross sections of the shaft, which were plane before deformation, they remain plane after deformation and also their radii remain straight.

Under these conditions, the shearing strain γ at a distance r from the axis of the shaft, may be obtained by using basic mechanics, and it is given by the following equation:

$$\gamma = r\theta \tag{6.132}$$

In Eq. (6.132), θ is used to denote the angular twist of the shaft per unit length.

We can determine the magnitude of the torque that produces the angular twist θ if we know how the shearing strain γ is related to the shearing stress τ if the elastic limit of the material is exceeded. We assume that such a relationship is represented by the shearing stress shearing strain curve shown in Fig. 6.29. In practice, such curves can be obtained experimentally.

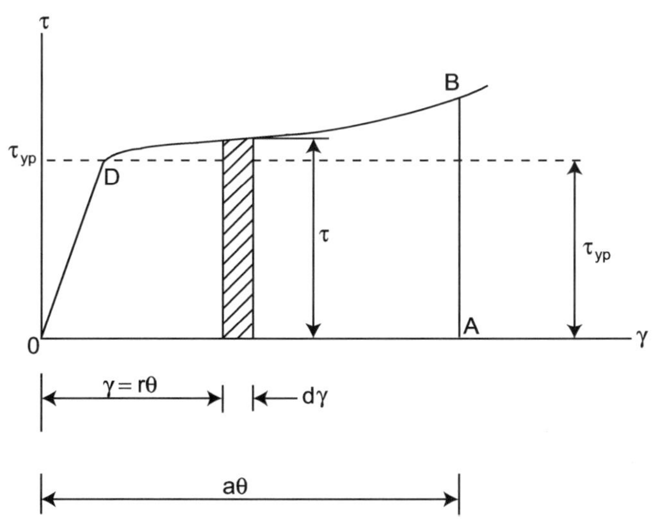

Fig. 6.29. Diagram showing the relation between the shearing strain γ and the shearing stress τ of the material

The maximum shear stress τ_{max} of the shaft occurs at $r = a$, and at this location, for $r = a$, the maximum shearing strain is $a\theta$. From statics, the torque T which is required to produce the angular twist θ, may be obtained from the following expression:

$$\int_0^a 2\pi r^2 \tau \, dr = T \tag{6.133}$$

From Eq. (6.132) we find that

$$r = \frac{\gamma}{\theta} \tag{6.134}$$

and

$$dr = \frac{d\gamma}{\theta} \tag{6.135}$$

By substituting Eqs. (6.134) and (6.135) into Eq. (6.133), we find

$$\frac{2\pi}{\theta^3} = \int_0^{a\theta} \gamma^2 \tau \, d\gamma = T \tag{6.136}$$

By examining Eq. (6.136), we note that the integral on its left side represents the moment of inertia of the area OBA in Fig. 6.29 about the vertical axis 0τ. After having calculated this moment of inertia using the curve in Fig. 6.29, we can return to Eq. (6.136) and determine the torque T. The curve 0A in Fig. 6.29, represents also the shearing stress distribution along the radius of the shaft.

If we assume that the material remains elastic at all times during loading and obeys Hooke's law, we have

$$\tau = \gamma G = r\theta G \tag{6.137}$$

where G in this equation is the constant of proportionality, known in practice as the shear modulus.

By substituting Eq. (6.137) into Eq. (6.133), we obtain

$$2\pi\theta G \int_0^a r^3 \, dr = \theta G I_P = T \tag{6.138}$$

where I_P in this equation is known as the polar moment of inertia of the circular cross section of the shaft, and it is given by the equation

$$I_P = 2\pi \int_0^a r^3 \, dr \tag{6.139}$$

Therefore, Eq. (6.138) is the torsion equation of circular shafts when the elastic limit of the material is not exceeded.

In practice, when the yield point of the material is very noticeable, the curvilinear portion DB of the diagram in Fig. 6.29 is often replaced by the horizontal dashed line shown in the same figure, whose abscissa is equal to the yield point shearing stress τ_{yp}. In this manner, we create a situation where the shearing stress distribution along the radius of the shaft approaches that of a uniform distribution.

Under this condition, the magnitude of the corresponding ultimate torque T_{ult} may be obtained from Eq. (6.133) by substituting τ_{yp} for τ. On this basis, Eq. (6.133) yields

$$T_{ult} = \frac{2\pi a^3}{3}\tau_{yp} \tag{6.140}$$

When the torque reaches the value given by Eq. (6.140), further twist of the shaft takes place without any further increase of the torque, and this condition will continue to exist until the strain hardening of the material becomes significant.

The above condition, in reality, represents a bilinear approximation of the stress–strain curve in Fig. 6.29. However, if desired, more straight lines can be used, as discussed in detail in Chap. 4, to approximate the actual shape of the stress–strain curve of the material and obtain a more accurate value for the ultimate torque T_{ult}.

6.13 Vibration Analysis of Inelastic Structural Members

Prismatic and nonprismatic members are commonly used as structural components for numerous engineering structures and machines, such as highway bridges, various machine types, buildings, space and aircraft structures, and so on, just to name a few. Under severe weight and loading conditions, as discussed in Chap. 4, the material of such members can be stressed well beyond their elastic limit, a situation which causes the modulus of elasticity E_x to vary along the length of the member. The moment of inertia I_x of the member could also be variable along its length.

A static or dynamic analysis of such inelastic members must take into consideration the variation of both E_x and I_x, if a reasonably accurate solution of this problem is required. The static analysis of such types of inelastic members, plates, etc, is carried out in detail in Chap. 4 and in various sections of Chap. 6. See also the work of the author in [2, 3, 5, 6, 49, 84, 103]. In this section, the fundamental frequency of vibration of such inelastic members, and their corresponding mode shapes, will be investigated by using dynamically equivalent pseudolinear systems.

The procedure will be illustrated by using a few numerical examples.

Example 6.8 By using inelastic analysis, determine the fundamental free frequency of vibration and the corresponding mode shape of the inelastic tapered cantilever beam discussed in Examples 4.1 and 4.2. Assume that the distributed uniform load w acting on the member represents its dead weight and possibly other additional weights securely attached to the member. Provide a complete solution to this problem by using a dynamically equivalent system of constant stiffness E_1I_B, where E_1 and I_B are the reference values for E_x and I_x, respectively.

Solution: If we assume that the weight $w = 1,600\,\text{lb in.}^{-1}$ $(280,203\text{N m}^{-1})$, a portion of the member will be stressed well beyond its elastic limit, and the member will become inelastic. Since vibrations are taking place about the static equilibrium position, the vibration will take place from this position with the member being inelastic.

With respect to the static equilibrium position, an equivalent system of constant stiffness E_1I_B, where E_1 and I_B are the reference values for E_x and I_x, respectively, may be derived as discussed in Examples 4.1 and 4.2. Such an equivalent system is already derived for this member and it is shown in Fig. 4.5b. The static equilibrium positions of the equivalent system and of the initial system in Fig. 4.2 are practically identical. The equivalent system, however, has a constant stiffness E_1I_B, and it is loaded with equivalent concentrated weights. The equivalent system will be used here to determine the fundamental frequency of free vibration of the initial member in Fig. 4.2. Stodola's method and the iteration procedure will be used for this purpose. This method is thoroughly discussed in [5].

It should be pointed out, however, that any appropriate method of vibration analysis, by including the finite element method [3, 5], may be used to solve the equivalent system. It should be also pointed out that since the variation, along the length of the member, of both E_x and I_x are known and are determined as shown in Examples 4.1 and 4.2, the original variable stiffness member could be used directly to determine its free frequencies of inelastic vibration by including the fundamental.

For many practical applications we usually need to know only the fundamental frequency, and the equivalent system will provide it in a simpler manner. For higher frequencies of free vibration, the accuracy in using the equivalent system is not yet examined by the author. If higher frequencies of inelastic vibration are needed, they can be determined by using the original system, having first established the variation of E_x and I_x as discussed in Example 4.1 and Chap. 4, in general.

The original system and the equivalent system are again shown in Fig.6.30a and b, respectively. By applying unit loads at points 1, 2, and 3 of the equivalent system in Fig. 6.30b, the α_{ij} deflection coefficients are determined by using formulas, and they are as follows:

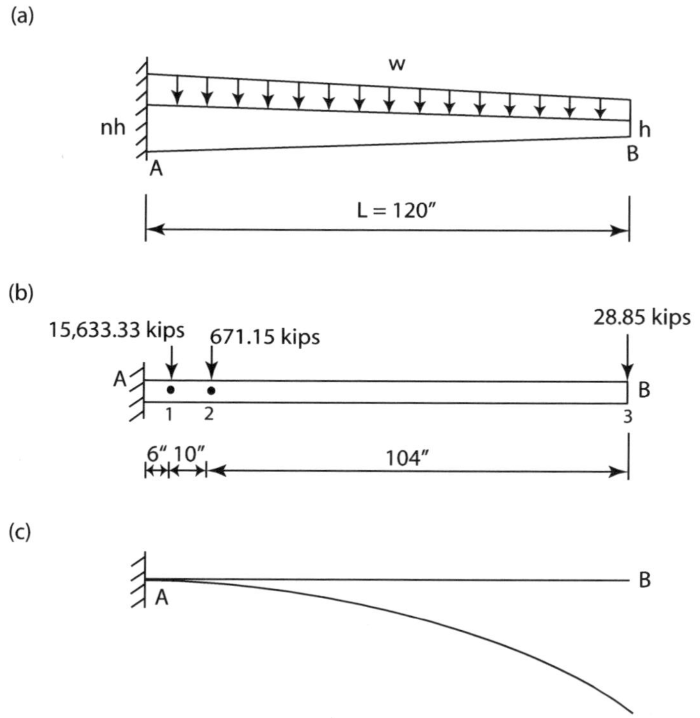

Fig. 6.30. (**a**) Initial tapered cantilever beam. (**b**) Equivalent system of uniform stiffness E_1I_B. (**c**) Shape of the fundamental mode of vibration

$$\alpha_{11} = \frac{72}{E_1I_B}, \quad \alpha_{12} = \alpha_{21} = \frac{252}{E_1I_B}, \quad \alpha_{13} = \alpha_{31} = \frac{2{,}124}{E_1I_B}$$

$$\alpha_{21} = \frac{252}{E_1I_B}, \quad \alpha_{22} = \frac{1{,}365.33}{E_1I_B}, \quad \alpha_{23} = \alpha_{32} = \frac{14{,}677.33}{E_1I_B}$$

$$\alpha_{31} = \frac{2{,}124}{E_1I_B}, \quad \alpha_{32} = \frac{14{,}677.33}{E_1I_B}, \quad \alpha_{33} = \frac{576\times 10^3}{E_1I_B}$$

In applying Stodola's method, the required general matrix equation is as follows (see Eq. (5.40) of [5]):

$$\begin{bmatrix} y_1 \\ y_2 \\ \vdots \\ y_i \\ \vdots \\ y_n \end{bmatrix} = \omega^2 \begin{bmatrix} \alpha_{11}m_1 \ \alpha_{12}m_2 \cdots \alpha_{1i}m_i \cdots \alpha_{1n}m_n \\ \alpha_{21}m_1 \ \alpha_{22}m_2 \cdots \alpha_{2i}m_i \cdots \alpha_{2n}m_n \\ ------------- \\ \alpha_{i1}m_1 \ \alpha_{i2}m_2 \cdots \alpha_{ii}m_i \cdots \alpha_{in}m_n \\ ------------- \\ \alpha_{n1}m_1 \ \alpha_{n2}m_2 .... \alpha_{ni}m_i \cdots \alpha_{nn}m_n \end{bmatrix} \begin{bmatrix} y_1 \\ y_2 \\ \vdots \\ y_i \\ \vdots \\ y_n \end{bmatrix} \tag{6.141}$$

where y_i $(i = 1, 2, 3, \ldots, n)$ are the deflection amplitudes under the mass concentration points $i = 1, 2, 3, \ldots, n$, α_{ij} are the deflection coefficients which are determined by applying unit loads at points $i = 1, 2, 3, \ldots, n$, m_i $(i = 1, 2, 3, \ldots, n)$ are the concentrated masses at points i, and ω are the frequencies of vibration of the member.

By using Eq. (6.141), the matrix equation for the problem at hand is written as follows:

$$\begin{bmatrix} y_1 \\ y_2 \\ y_3 \end{bmatrix} = \frac{\left(61,277 \times 10^3\right)\omega^2}{E_1 I_B g} \begin{bmatrix} 18.369 & 2.760 & 1.000 \\ 64.292 & 14.954 & 6.910 \\ 567.399 & 160.757 & 271.188 \end{bmatrix} \begin{bmatrix} y_1 \\ y_2 \\ y_3 \end{bmatrix} \tag{6.142}$$

The iteration procedure of Eq. (6.142) may be initiated by assuming that $y_1 = y_2 = y_3 = 1.000$. This yields

$$\begin{bmatrix} y_1 \\ y_2 \\ y_3 \end{bmatrix} = \frac{\left(61,277 \times 10^3\right)(22.129)\,\omega^2}{E_1 I_B g} \begin{bmatrix} 1.000 \\ 3.893 \\ 45.160 \end{bmatrix} \tag{6.143}$$

The new amplitudes $y_1 = 1.000$, $y_2 = 3.893$, and $y_3 = 45.160$, shown in Eq. (6.143), may be used in Eq. (6.142) to repeat the procedure and find new and more correct values for y_1, y_2, and y_3. The procedure is repeated until the values of y_1, y_2, and y_3 of the last repetition are closely identical, for practical purposes, to the ones found in the preceding trial. For this problem, after six iterations (or repetitious), the matrix equation (6.142) converges as follows:

$$\begin{bmatrix} y_1 \\ y_2 \\ y_3 \end{bmatrix} = \frac{\left(61,277 \times 10^3\right)(278)\,\omega^2}{E_1 I_B g} \begin{bmatrix} 1.000 \\ 6.587 \\ 241.382 \end{bmatrix} \tag{6.144}$$

By using the first row of the above matrix, we find

$$\frac{\left(61,277 \times 10^3\right)(278)\,\omega^2}{E_1 I_B g} = 1$$

or

$$\omega^2 = \frac{E_1 I_B g}{(61,277 \times 10^3)(278)} \tag{6.145}$$

The reference value E_1 that is used to derive the equivalent system in Fig. 6.30b is equal to 26×10^6 psi $(17.927 \times 10^{-6}\,\text{kPa})$, because the bilinear stress–strain curve is used to derive E_r and, consequently, $g(x)$. The moment of inertia I_B at the free end B is $256\,\text{in.}^4$ $(106.555 \times 10^{-6}\,\text{m}^4)$, and the acceleration of gravity $g = 386.4\,\text{in}\,\text{s}^{-2}$ $(9.81\text{m}\,\text{s}^{-2})$. On this basis, Eq. (6.145) yields

$$\omega^2 = \frac{\left(26 \times 10^6\right)(256)(386.4)}{(61,277 \times 10^3)(278)} = 150.976\,(\text{rps})^2$$

$$\omega = 12.287\,\text{rps}$$

$$f = \frac{\omega}{2\pi} = 1.956\,\text{Hz}$$

If elastic analysis had been used the fundamental frequency of vibration would be equal to 2.397 Hz, which is 22.55% higher than the one obtained by using inelastic analysis.

The shape of the fundamental mode is characterized by the amplitudes $y_1 = 1.000$, $y_2 = 6.587$, and $y_3 = 241.382$, which are shown in the column on the right-hand side of Eq. (6.144). The mode shape is shown plotted in Fig. 6.30c.

This problem was also solved by using the equivalent system and applying Rayleigh's energy method. The results obtained, for all practical purposes, are identical. The actual values obtained are $\omega = 12.292\,\mathrm{rps}$, and the mode shape is characterized by the amplitudes $y_1 = 1.000$, $y_2 = 6.505$, and $y_3 = 234.358$. These results were obtained by repeating the procedure three times in the utilization of Rayleigh's method. For practical experience, the reader can verify these results.

Example 6.9 By using inelastic analysis, determine the fundamental frequency of vibration and the corresponding mode shape of the uniform simply supported beam shown in Fig. 6.31a. The weight $w = 2,500\,\mathrm{lb\,in.}^{-1}$ ($437,817.5\,\mathrm{N\,m}^{-1}$) acting on the member is assumed to be composed of the weight at the beam and other possible weights that are securely attached to the member. The width b of the member is 6 in. (0.1524 m) and its depth $h = 8\,\mathrm{in.}$ (0.2032m).

Solution: In this case, the weight $w = 2,500\,\mathrm{lb\,in.}^{-1}$ ($437,817.5\,\mathrm{N\,m}^{-1}$) acting on the member is large enough to cause the member to become inelastic at the static equilibrium position. Therefore, the modulus of elasticity at sections along the length of the member will be variable. In our analysis here we will assume that the material of the member is monel, and Tables 4.1, 4.2 provide the required data for a two-line, three-line, and a six-line approximation of the stress–strain curve of monel.

The modulus function g(x) and, consequently, the constant stiffness equivalent system, are derived by using the two-line approximation of the stress–strain curve of monel. By proceeding as in Examples 4.1 and 4.2 of Chap. 4, the values of Δ, E_r, $g(x)$, $M_{req} = M_x$, and $M_e = M_x/f(x)g(x)$ are determined, and they are shown in Table 6.9. The moment diagram M_e of the equivalent system is plotted as shown by the solid line in Fig. 6.31b. Its shape approximation with six straight-line segments, as shown by the dashed lines in Fig. 6.31b, leads to the constant stiffness equivalent system shown in Fig. 6.31c.

The equivalent system in Fig. 6.31c is loaded with equivalent concentrated weights acting at points 1, 2, 3, 4, and 5. The static deflections y_1, y_2, y_3, y_4, and y_5 at points 1, 2, 3, 4, and 5, respectively, caused by the equivalent weights, establish the static equilibrium position of the member, and they can be determined by using known handbook formulas. They are as follows:

$$y_1 = 0.7390\,\mathrm{in.}, \qquad y_1^2 = 0.5461\,\mathrm{in.}^2,$$
$$y_2 = 1.0701\,\mathrm{in.}, \qquad y_2^2 = 1.1451\,\mathrm{in.}^2,$$

Table 6.9. Values of f(x), E_r, g(x), M_{req}, and M_e for inelastic analysis of the simply supported beam

(1) x (in.)	(2) h_2 (in.)	(3) f(x)	(4) Δ (10^{-3} in. in.$^{-1}$)	(5) E_r (10^6 psi)	(6) g(x)	(7) $M_{req} = M_r$ (10^6 in. lb)	(8) $M_e = M_x/f(x)g(x)$ (10^6 in. lb)
0	8	1	0	26.000	1.0000	0	0
6	8	1	1.0276	26.000	1.0000	0.855	0.855
12	8	1	1.9421	26.000	1.0000	1.620	1.620
18	8	1	2.7584	26.000	1.0000	2.295	2.295
24	8	1	3.4615	26.000	1.0000	2.880	2.880
30	8	1	4.0962	25.858	0.9945	3.389	3.408
36	8	1	4.7962	24.571	0.9451	3.771	3.990
42	8	1	5.7961	22.082	0.8493	4.096	4.822
48	8	1	6.9962	19.282	0.7416	4.317	5.821
54	8	1	8.2961	16.787	0.6457	4.457	6.902
60	8	1	8.8961	15.813	0.6082	4.502	7.402
66	8	1	8.2961	16.787	0.6457	4.457	6.902
72	8	1	6.9962	19.282	0.7416	4.317	5.821
78	8	1	5.7961	22.082	0.8493	4.096	4.822
84	8	1	4.7962	24.571	0.9451	3.771	3.990
90	8	1	4.0962	25.858	0.9945	3.389	3.408
96	8	1	3.4615	26.000	1.0000	2.880	2.880
102	8	1	2.7584	26.000	1.0000	2.295	2.295
108	8	1	1.9471	26.000	1.0000	1.620	1.620
114	8	1	1.0276	26.000	1.0000	0.855	0.855
120	8	1	0	26.000	1.0000	0	0

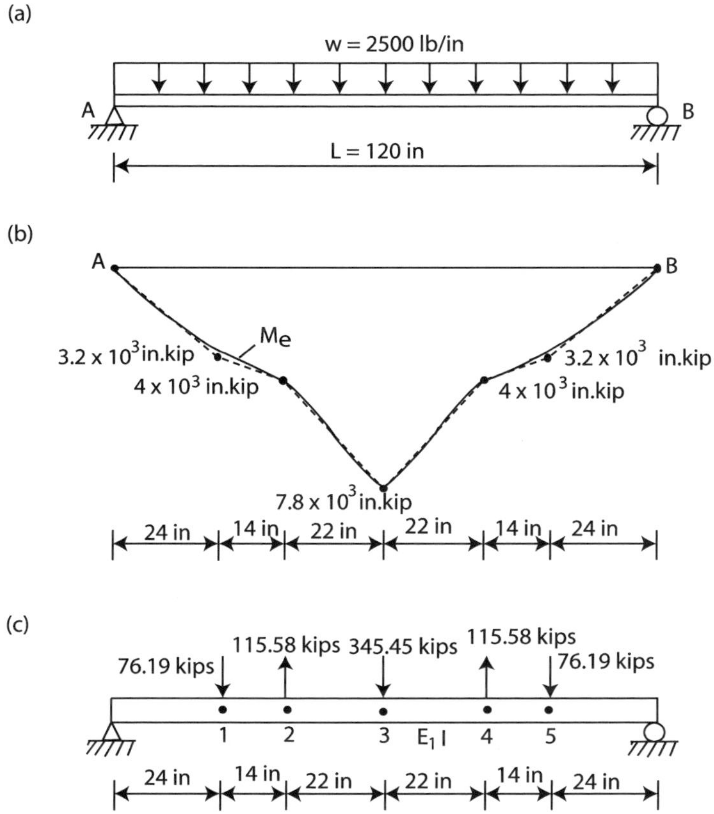

Fig. 6.31. (**a**) Initial uniform simply supported beam. (**b**) Moment diagram M_e of the equivalent system with its shape approximated with six straight-line segments. (**c**) Equivalent system of constant stiffness E_1I

$$y_3 = 1.3076\,\text{in.}, \qquad y_3^2 = 1.7098\,\text{in.}^2,$$
$$y_4 = y_2 = 1.0701\,\text{in.}, \quad y_4^2 = 1.1451\,\text{in.}^2,$$
$$y_5 = y_1 = 0.7390\,\text{in.}, \quad y_5^2 = 0.5461\,\text{in.}^2,$$

The method of Lord Rayleigh, represented by Eqs. (5.28) and (5.30) in Chap. 5 of [5], will be used here to determine the fundamental frequency of vibration. For the first approximation, we use Eq. (5.28), which is as follows:

$$\omega^2 = g\frac{\sum_{i=1}^{5} W_i y_i}{\sum_{i=1}^{5} W_i {y_i}^2} \tag{6.146}$$

On this basis, we have

$$\sum_{i=1}^{5} W_i y_i = (76.19)(0.7390) + (115.58)(1.0701) + (345.45)(1.3076) + (115.58)(1.0701) + (76.19)(0.7390) = 811.6836$$

$$\sum_{i=1}^{5} W_i {y_i}^2 = (76.19)(0.5461) + (115.58)(1.1451) + (345.45)(1.7098) + (115.58)(1.1451) + (76.19)(0.5461) = 918.5667$$

Thus, Eq. (6.146) yields

$$\omega^2 = (386.4)\,\frac{811.6836}{938.5667} = 334.1633$$

or

$$\omega = 18.28\,\text{rps}$$

The procedure may be repeated for better accuracy by using Eq. (5.30) of [5], which is written again below:

$$\omega^2 = g\frac{\sum_{i=1}^{5} W_i' y_i'}{\sum_{i=1}^{r} W_i\,(y_i')^2} \tag{6.147}$$

In order to use Eq. (6.147), we need to know the new loads $W_i{}'$ and the new deflections $y_i{}'$ produced by these loads. The $W_i{}'$ loads are as follows [5]:

$$\begin{aligned}
W_1' &= W_1 y_1 = (76.19)(0.739) = 56.30\,\text{kips}\\
W_2' &= W_2 y_2 = -(115.58)(1.0701) = -123.68\,\text{kips}\\
W_3' &= W_3 y_3 = (345.45)(1.3076) = 451.71\,\text{kips}\\
W_4' &= W_2' = -123.68\,\text{kips}\\
W_5' &= W_1' = 56.30\,\text{kips}
\end{aligned}$$

By applying statically the new $W_i{}'$ loads at points 1, 2, 3, 4, and 5 of the equivalent system in Fig. 6.31c, we obtain

$$\begin{aligned}
y_1' &= 0.9501\,\text{in.}, & (y_1')^2 &= 0.9027\,\text{in.}^2,\\
y_2' &= 1.3773\,\text{in.}, & (y_2')^2 &= 1.8970\,\text{in.}^2,\\
y_3' &= 1.6880\,\text{in.}, & (y_3')^2 &= 2.8493\,\text{in.}^2,\\
y_4' &= 1.3773\,\text{in.}, & (y_4')^2 &= 1.8970\,\text{in.}^2,\\
y_5' &= 0.9501\,\text{in.}, & (y_5')^2 &= 0.9027\,\text{in.}^2,
\end{aligned}$$

On this basis, we have

$$\sum_{i=1}^{5} W_i' y_i' = 1,210.1567$$

$$\sum_{i=1}^{5} W_i (y_i')^2 = 1,560.3547$$

Thus, Eq. (6.147) yields,

$$\begin{aligned}
\omega^2 &= (386.4)\,\frac{1,210.1567}{1,560.3547} = 299.6784 \\
\omega &= 17.3112\,\mathrm{rps} \\
f &= \frac{\omega}{2\pi} = 2.7552\,\mathrm{Hz}
\end{aligned}$$

No further repetition is required for practical purposes, because the next repetition yields about the same results.

If elastic analysis is used to solve the original member in Fig. 6.31a, we find that

$$\begin{aligned}
\omega &= \frac{\pi^2}{L^2}\sqrt{\frac{EI}{m}} \\
&= \frac{\pi^2}{(120)^2}\sqrt{\frac{(26 \times 10^6\,(256)\,(386.4))}{2500}} \\
&= 21.9833\,\mathrm{rps}
\end{aligned}$$

which is 26.99% higher than the one obtained by using inelastic analysis.

If we assume $y_1 = 1.000$, the normalized shape of the fundamental mode is characterized by the amplitudes

$$\begin{aligned}
y_1 &= 1.0000, & y_4 &= 1.4496, \\
y_2 &= 1.4496, & y_5 &= 1.0000, \\
y_3 &= 1.7767,
\end{aligned}$$

Other types of inelastic beam problems may be treated in a similar manner. We should remember, however, that only an introduction to this important problem and the required methodologies, are presented in this section.

6.14 Inelastic Analysis of Flexible Members

In this section, the flexible member will be stressed well beyond the elastic limit of its material, and therefore the modulus E will vary along its length. Consequently, the problems to be solved here include flexible members subjected to large deformations and acted upon by loading conditions that stress the flexible member heavily into the inelastic range. The flexible member is permitted to have a uniform or a variable thickness along its length, and it can be subjected to complicated loading conditions. See also [2, 3] for additional information on this subject.

The inelastic analysis of flexible members will be carried out by using the concept of pseudolinear equivalent systems discussed in Chap. 4 of this text. If the flexible member is subjected to complicated combined loading conditions, a simplified nonlinear equivalent system of constant stiffness may be first obtained as discussed in Chap. 4. The simplified nonlinear system is then used to solve it by utilizing pseudolinear analysis. This approach can be used to solve very complicated problems in a relatively easy and convenient way.

The general nonlinear differential equation of the elastic line of a flexible member with variable moment of inertia I_x and variable modulus of elasticity E_x, is written again here as

$$\frac{y''}{\left[1+(y')^2\right]^{3/2}} = -\frac{M_x}{E_x I_x} \tag{6.148}$$

where M_x is the bending moment at a cross section along the length of the member.

If both E_x and I_x are permitted to vary along the length of the member, the variable stiffness $E_x I_x$ is expressed as

$$E_x I_x = E_1 I_1 g(x) f(x) \tag{6.149}$$

where g(x) represents the variation of E_x with respect to a reference value E_1 and f(x) represents the variation of I_x with respect to a reference value I_1. If E and I are both constant, then f(x) = g(x) = 1.00.

By substituting Eq. (6.149) into Eq. (6.148), we find

$$\frac{y''}{\left[1+(y')^2\right]^{3/2}} = -\frac{M_e}{E_1 I_1} \tag{6.150}$$

where

$$M_e = \frac{M_x}{g(x) f(x)} \tag{6.151}$$

For convenience, Eq. (6.148) may be rewritten as

$$y'' = -\frac{M_e z_e}{E_1 I_1} \tag{6.152}$$

where

$$z_e = \left[1+(y')^2\right]^{3/2} \tag{6.153}$$

We may also write it as

$$y'' = -\frac{M'_e}{E_1 I_1} \tag{6.154}$$

where

$$M'_e = M_e z_e = \frac{M_x z_e}{g(x) f(x)} \tag{6.155}$$

The quality M_e' in Eq. (6.155) represents the variation of the bending moment along the x axis of an equivalent pseudolinear system of constant stiffness $E_1 I_1$, that has the same elastic line and boundary conditions as the initial variable stiffness member represented by Eq. (6.148). The elastic line and boundary conditions are also identical with those of the equivalent nonlinear system represented by Eq. (6.150). These concepts are discussed in detail in Sect. 1.8 of the first chapter.

The purpose in this section is to develop the required methodology which is suitable to carry out the inelastic analysis of flexible members when both E and I can vary in any manner along their length.

The procedure here is illustrated by considering the tapered cantilever bar in Fig. 6.32a, which is loaded as shown. In Fig. 6.32b we note that for

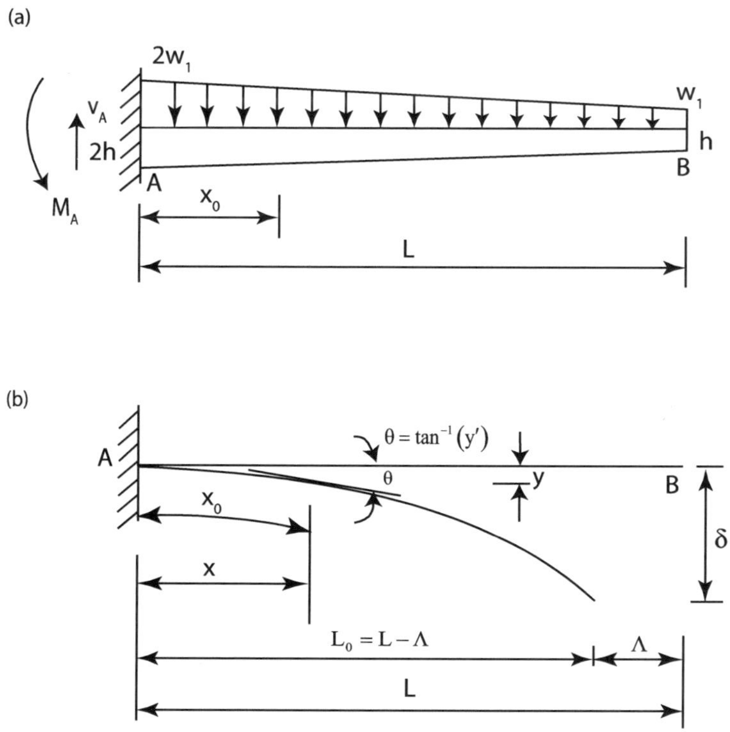

Fig. 6.32. (**a**) Flexible cantilever bar loaded with a trapezoidal loading. (**b**) Large deformation configuration of the flexible bar

every length x_o there corresponds a length x, where $0 \leq x \leq L_o = L–\Lambda$, and $L_o = L - \Lambda$. Thus, we can express x_o as

$$x_o = x + \Lambda\,(x) \tag{6.156}$$

where $\Lambda(x)$ represents the horizontal displacements of x_o. We also know that

$$x_o = \int_o^x \left[1 + (y')^2\right]^{1/2} dx \tag{6.157}$$

and

$$L = \int_o^{L-\Lambda} \left[1 + (y')^2\right]^{1/2} dx \tag{6.158}$$

which establish the correspondence between x and x_o and $L_o = L - \Lambda$ with L.

For the flexible member in Fig. 6.32a, the expressions for V_A and M_A can be determined by using statics and they are as follows:

$$V_A = \frac{3w_1L}{2} \tag{6.159}$$

$$M_A = \frac{2w_1LL_o}{3} \tag{6.160}$$

Again, by using statics, the bending moment M_x at any section of distance $0 \leq x \leq L_o$, is determined and it is given by the expression

$$M_x = \frac{2w_1LL_o}{3} - \frac{3w_1L}{2} + \frac{w_1\,(2L - x_0)\,x_ox}{2L} + w_1\left[2 - \frac{(2L - x_o)}{L}\right]\frac{x_ox}{3} \tag{6.161}$$

In Eq. (6.161) the expression for the coordinate x_o is given by Eq. (6.156), and the distributed vertical load w at any $0 \leq x_o \leq L$ is given by the equation

$$w = \frac{w_1\,(2L - x_o)}{L} \tag{6.162}$$

The moment of inertia I_x at any $0 \leq x_o \leq L$ is given by the expression

$$I_x = \frac{bh^3}{12}\left[\frac{2L - x_o}{L}\right]^3 \tag{6.163}$$

where b is the width of the rectangular cross section of the member. If the moment of Inertia I_B at the free end B of the member is taken as the reference value for I_x, then we have

$$f\,(x) = \left[\frac{2L - x_o}{L}\right]^3 \tag{6.164}$$

and

$$I_B = \frac{bh^3}{12} \tag{6.165}$$

In order to determine a pseudolinear system of constant stiffness E_1I_1, the rotation y'must be known. This rotation can be determined from Eq. (6.154) by integration, i.e.,

$$y' = \frac{1}{E_1I_1}\int_o^x M_e'\,dx + C_1 \tag{6.166}$$

where C_1 is the constant of integration. In order to carry out the integration of Eq. (6.166), the horizontal displacement Λ at the free end of the beam and the function g(x) must be known. Thus, inelastic analysis should be initiated for the computation of g(x), and nonlinear analysis for flexible members must be utilized for the computation of Λ, as discussed in the first four chapters of the text.

The methodology regarding the computation of the large deflections and rotations using inelastic analysis, may be carried out by employing trial and error procedures and the equivalent pseudolinear system represented by Eq. (6.154). The required integrations may be carried out with sufficient accuracy by using Simpson's rule. The following step-by-step procedure is suggested here for convenience:

Step 1. The first approximation of $g(x) = E_r/E_1$, where E_r is the reduced modulus, may be obtained by assuming that $g(x) = 1$. Proceed with the integration of Eq. (6.154), as shown in Eq. (6.166), to determine y'. The boundary condition of zero rotation at the fixed end of the member may be used to determine the constant of integration C_1. Since M_e' in Eq. (6.166) depends on $L_o = (L\text{–}\Lambda)$, the value of the horizontal displacement Λ of the free end of the bar must first be determined. The value of Λ can be determined from Eq. (6.158) by applying a trial and error procedure, i.e., assume a value of Λ and determine L from Eq. (6.158) in conjunction with Eq. (6.166). The procedure may be repeated until the correct length L of the bar is obtained. With known Λ, the values of y' can be determined from Eq. (6.166) and the values of M_x at any $0 \leq x \leq L_o$ can be determined from Eq. (6.161). The procedure converges rather fast.

Step 2. With known M_x, we can start the procedure to determine E_r, and consequently $g(x) = E_r/E_1$, by using Timoshenko's method as explained in Chap. 4, Sect. 4.2. Since the problem is statically determinate, we have $M_{req} = M_x$.

Step 3. By using the values of y' from Step 1 we determine z_e from Eq. (6.153). With known z_e and using the values of g(x) from Step 2, we determine M_e' from Eq. (6.155).

Step 4. By using M_e' from Step 1 and g(x) from Step 2, we determine a new y' from Eq. (6.166) which incorporates the inelastic behavior of the member. The boundary condition of zero rotation at the fixed end of the beam may be used to determine C_1.

Step 5. By utilizing Eq. (6.158) and the new y' obtained from Step 4, a new horizontal displacement Λ and a new $L_o = L\text{–}\Lambda$ may be obtained. Thus, a new $M_{req} = M_x$ may be determined from Eq. (6.161), and a new g(x) may be calculated as discussed in Step 2.

Step 6. By using the new y' from Step 4, a new M_e' may be obtained from Eq. (6.155) and a new y' may be calculated by using Eq. (6.166) and the $g(x)$ from Step 5.

The procedure may be repeated until the last y' is almost identical to the one obtained from the preceding trial. Usually four to seven repetitions are sufficient. The utilization of pseudolinear analysis, i.e., M_e', greatly facilitates the solution and convergence of the trial and error procedure. With known y', the large deflections may be obtained by either (1) integrating Eq. (6.166) once and satisfying the boundary condition of zero deflection at the fixed end or (2) using the moment area method, since M_e' is known. The second procedure is the most convenient (see Chaps. 1–4).

The following numerical example illustrates the procedure.

Example 6.10 By applying the methodology discussed in this section, perform an elastoplastic analysis and determine a pseudolinear equivalent system of constant stiffness E_1I_1 for the cantilever beam in Fig. 6.32a. Use this equivalent system to determine deflections and rotations at the free end of the member. Assume that $w_1 = 3.0\,\text{lb in.}^{-1}$ ($525.38\,\text{N m}^{-1}$); $h = 3.0$ in. (0.0762 m), and the constant width $b = 3.0$ in. (0.0762 m). The material of the member is monel, with a stress–strain curve as shown in Fig. 4.3a. The values of E and σ for a two-, three-, and a six-line approximation of this stress–strain curve, are shown, respectively, in Tables 4.1 and 4.2. For this problem, use the three-line approximation of the stress-strain curve.

Solution: By applying the step-by-step procedure stated above and repeating it about four times, the final values of y', $g(x)$, z_e, M_e', and strain Δ are obtained, and they are shown in Table 6.10. The moment diagram M_e' of the equivalent pseudolinear system of constant stiffness E_1I_1, where $E_1 = 22 \times 10^6$ psi (151.68×10^6 kPa) and $I_1 = 6.75\,\text{in.}^4$ ($2.8096 \times 10^{-6}\,\text{m}^4$), is shown in Fig. 6.33a. The approximation of its shape with three straight line segments as shown in the same figure, leads to the pseudolinear equivalent system of constant stiffness E_1I_1 shown in Fig. 6.33b.

By using the pseudolinear system and applying the well-known moment–area method, the rotation y' and deflections y along the length L_o of the member may be easily obtained. Since the pseudolinear system yields y', the actual rotation $\theta = \tan^{-1}(y')$. Table 6.11 shows the values of the vertical displacement δ, horizontal displacement Λ, and rotation θ at the free end of the beam for various values of the distributed load w_1. The column notation "Exact PES," which means exact pseudolinear equivalent system, is used to denote the results obtained by using the pseudolinear equivalent system. For example, for $w_1 = 3.0\,\text{lb in}^{-1}$ (525.38 N m), we find $\delta = 646.57$ in. (16.42 m), $\Lambda = 263.90$ in. (6.70 m), and $\theta = 52.03°$.

For the results obtained as discussed above, the required integrations of Eq. (6.154) during the application of the step-by-step procedure, were carried out by subdividing each time the length L_o into 40 equal segments and carrying out the required integration by using Simpson's rule. This procedure yields practically exact results.

Table 6.10. Final values of y', $g(x)$, z_e, Δ, and M'_e

x (in.)	f(x)	Δ	g(x)	y'	z_e	$M_{req} = M_x$ (in. kip)	$M'_e = \frac{M_x z_e}{f(x)g(x)}$ (in. kip)
0.0	8.0	4.6214×10^{-2}	0.1609	0.0	1.0	1,472.20	1,143.80
36.8	7.56	2.8364×10^{-2}	0.2466	2.3658×10^{-1}	1.0848	1,334.00	776.01
73.6	7.12	1.4914×10^{-3}	0.4395	3.8983×10^{-1}	1.2356	1,201.20	474.04
110.4	6.69	9.1136×10^{-3}	0.6710	4.8703×10^{-1}	1.3749	1,073.90	329.23
147.2	6.25	6.8136×10^{-3}	0.8331	5.6109×10^{-1}	1.5061	952.90	275.72
184.0	5.83	5.6136×10^{-3}	0.9327	6.2626×10^{-1}	1.6409	834.49	253.38
220.8	5.41	4.8636×10^{-3}	0.9850	6.8793×10^{-1}	1.7862	730.90	244.96
257.6	5.00	4.3538×10^{-3}	1.0	7.4875×10^{-1}	1.9468	630.28	245.33
294.4	4.60	3.9185×10^{-3}	1.0	8.0992×10^{-1}	2.1279	536.73	248.12
331.2	4.22	3.4866×10^{-3}	1.0	8.7156×10^{-1}	2.3310	450.35	249.03
368.0	3.84	3.0592×10^{-3}	1.0	9.3310×10^{-1}	2.5552	371.27	247.12
404.8	3.48	2.6381×10^{-3}	1.0	9.9368×10^{-1}	2.7983	299.60	241.25
441.6	3.13	2.2256×10^{-3}	1.0	1.0522	3.0551	235.49	230.20
478.4	2.79	1.8249×10^{-3}	1.0	1.1072	3.3171	179.05	212.83
515.3	2.47	1.4404×10^{-3}	1.0	1.1570	3.5725	130.38	188.37
552.1	2.17	1.0782×10^{-3}	1.0	1.1999	3.8064	89.55	156.84
588.9	1.89	7.4935×10^{-3}	1.0	1.2343	4.0035	56.57	119.53
625.7	1.64	4.5567×10^{-3}	1.0	1.2590	4.1506	31.33	79.43
662.5	1.40	2.2060×10^{-3}	1.0	1.2739	4.2418	13.67	41.39
669.3	1.19	6.0220×10^{-3}	1.0	1.2803	4.2820	3.35	12.05
736.1	1.0	0.0	1.0	1.2815	4.2880	0.00	0.00

Note: 1 in. = 0.0254 m, 1 in. kip = 112.9848 N m

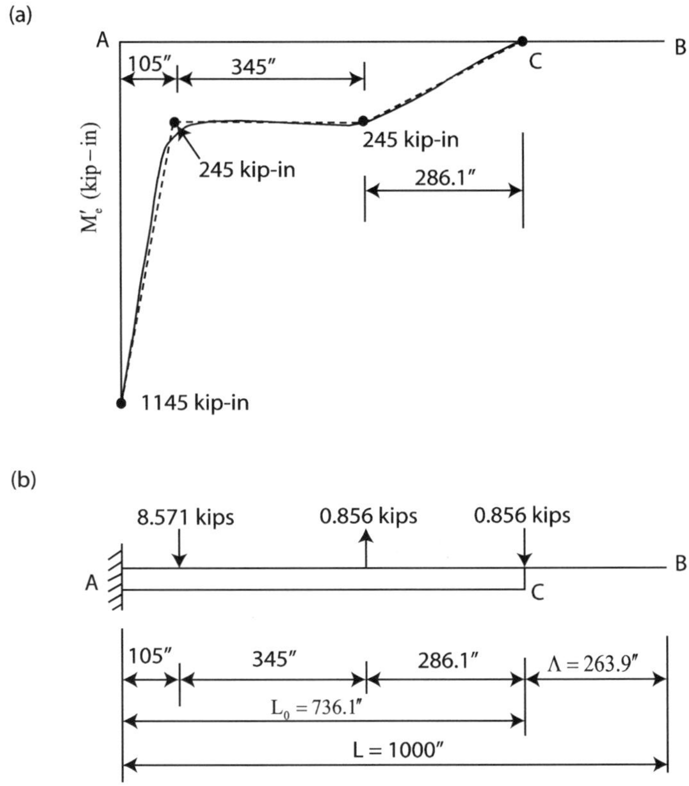

Fig. 6.33. (**a**) Moment diagram M'_e of the pseudolinear equivalent system with its shape approximated with three straight lines. (**b**) Equivalent pseudolinear system of constant stiffness E_1I_1 (1 in. = 0.0254 m, 1 kip in. = 112.9848 Nm, 1 kip=4.448 kN)

Inelastic Analysis Using Simplified Nonlinear Equivalent Systems

This method and procedure is particularly useful for flexible beam problems where the moment of inertia of the member and its loading conditions are complicated, and often arbitrary. The application of this methodology is illustrated by using the variable stiffness cantilever beam loaded as shown in Fig. 6.32a. In order to compare results, the geometric dimensions and loading values for the problem in Fig. 6.32a are considered to be the same as the ones used to carry out the inelastic analysis in Example 6.10.

Along the undeformed length L of the member, at any $0 \leq x \leq L$, the moment M_x is obtained by using simple statics, and it is as follows (note that $x = x_o$ in this case):

$$M_x = \frac{2w_1L^2}{3} - \frac{3w_1Lx}{2} + \frac{w_1x^2}{2L}(2L - x) + \frac{w_1x^3}{3L} \tag{6.167}$$

Table 6.11. Values of δ, Λ, and θ at the free end of the member vs. values of distributed load w_1 using the exact PES and the simplified NES

w_1 (lb in.$^{-1}$)	δ Exact PES (in.)	δ Simplified NES (in.)	Difference (%)	Λ Exact PES (in.)	Λ Simplified NES (in.)	Difference (%)	θ Exact PES (°)	θ Simplified NES (°)	Difference (%)
1	184.66	184.6	−0.03	20.9	20.80	−0.48	15.63	15.67	0.26
2	361.41	361.99	0.16	80.1	81.08	1.22	30.52	31.02	1.63
3	646.57	639.53	−1.09	263.9	257.20	−2.54	52.03	51.95	−0.15
4	779.18	778.25	−0.12	414.3	413.98	−0.08	64.01	63.98	−0.05
5	841.07	850.74	1.15	516.1	519.90	0.74	70.84	71.59	1.06
6	881.7	881.52	−0.02	587.1	584.10	−0.51	75.26	75.45	0.25

Note: 1 in. = 0.0254 m, 1 lb in.$^{-1}$ = 175.1268 N m^{-1}

The straight undeformed configuration of the member, where $x = x_o$, is used here to derive M_x. Along the same x axis, we have

$$I_x = \frac{bh^3}{12}\left[\frac{2L - x}{L}\right]^3 \tag{6.168}$$

and

$$f(x) = \left[\frac{2L - x}{L}\right]^3 \tag{6.169}$$

The reference value $I_1 = bh^3/12$ is the moment of inertia at the free end of the beam.

The moment M_e at any $0 \leq x \leq L$ of the nonlinear equivalent system of constant stiffness EI_1, may be obtained from Eq. (6.151). Thus, by substituting Eq. (6.167) and (6.169) into Eq. (6.151), we obtain

$$\begin{aligned} M_e &= \frac{M_x}{f(x)} \\ &= \frac{L^2}{6(2L - x)^3}\left[4w_1L^3 - 9w_1L^2x + 3w_1x^2(2L - x) + 2w_1x^3\right] \end{aligned} \tag{6.170}$$

Note that g(x) in Eq. (6.151) is taken to be equal to one, because the inelastic analysis is not yet initiated.

By assuming the same values for w_1 and L as for the pseudolinear analysis in Example 6.10, the plot of M_e is shown by the solid line in Fig. 6.34a. The approximation of its shape with one straight line as shown by the dashed line, leads to the simplified nonlinear equivalent system of constant stiffness EI_1, shown in Fig. 6.34b. Accurate large deflections and rotations may be obtained here by using the simplified nonlinear system in Fig. 6.34b, which replaces the original variable stiffness nonlinear system in Fig. 6.32a. The solution of the simplified nonlinear problem in Fig. 6.34b may now be obtained by either (1) using simple nonlinear existing methods of analysis, if such methods are readily available or (2) by applying pseudolinear analysis and the step-by-step procedure stated earlier.

The inelastic analysis here was carried out by using the simplified nonlinear equivalent system in Fig. 6.34b and applying pseudolinear analysis using the step-by-step procedure discussed in this section. This procedure produces the final values for g(x), Δ, y', z_e, and $M_e{}'$ shown in Table 6.12. Thus, with known $M_e{}'$, the values of the large displacement of the member may be determined as stated earlier by either integrating Eq. (6.154) once, or by using the pseudolinear system of constant stiffness E_1I_1 and applying the moment area method. The second approach was used here. Table 6.11, under column heading "Simplified NES" shows values of the vertical displacement δ, horizontal displacement Λ, and rotation θ at the free end of the member, for various values of the distributed load w_1. In the same table, the results obtained using the exact PES are shown for comparison purposes. For practical purposes the

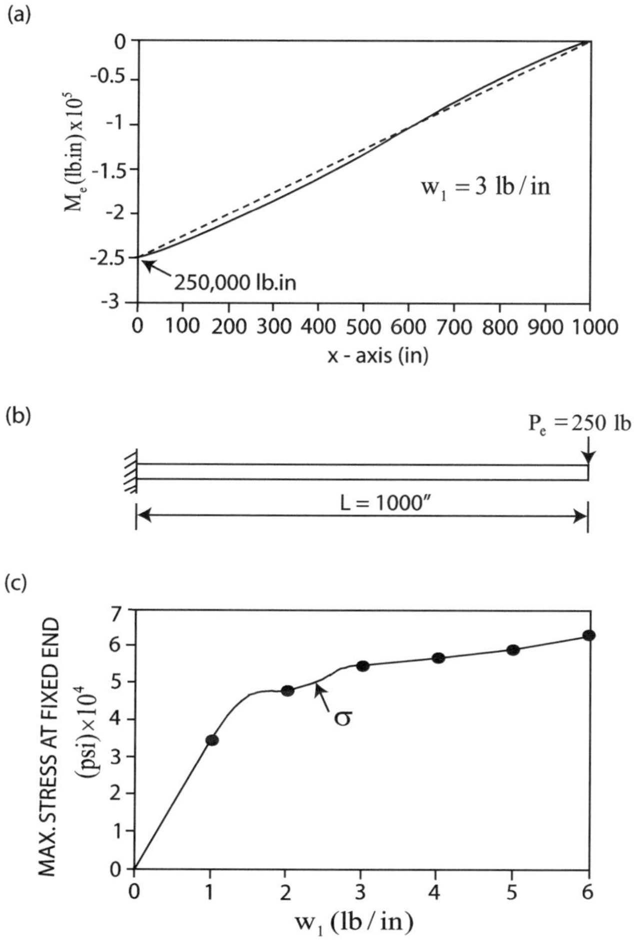

Fig. 6.34. (**a**) Moment diagram M_e of the simplified NES with its shape approximated with one straight line. (**b**) Simplified NES of constant stiffness EI_1. (**c**) Maximum normal stress at the fixed end of the bar with increasing w_1 (1 in. = 0.0254 m, 1 lb in.$^{-1}$ = 175.1268 N m^{-1}, 1 psi = 6,894.757 Pa)

difference is considered to be negligible, but if greater accuracy is required, more straight lines may be used to approximate the M_e diagram. The inelastic levels of maximum stress σ at the fixed end of the beam vs. given values of w_1, are shown plotted in Fig. 6.34c.

In practice, utilization of equivalent simplified nonlinear systems becomes very beneficial, because the practicing engineer prefers the easy and reasonable solution of his practical problem, which gives him/her reasonably accurate answers. For extensive treatment of this subject, the reader may refer to the work of the author in [2, 3].

Table 6.12. Final values of Δ, y', $g(x)$, z_e, and M_e

x (in.)	f(x)	Δ	g(x)	y'	z_e	M_e (in. kip)	$M'_e = \frac{M_e z_e}{f(x)g(x)}$ (in. kip)
0.0	1.0	4.9114×10^{-2}	1.5278×10^{-1}	0.0	1.0	185.71	1,215.60
37.1	1.0	2.8314×10^{-2}	2.4701×10^{-1}	0.2481	1.0929	176.42	780.50
74.3	1.0	1.3714×10^{-2}	4.7354×10^{-1}	0.3978	1.2441	167.14	439.11
111.4	1.0	8.3136×10^{-3}	7.2155×10^{-1}	0.4879	1.3755	157.85	300.68
148.6	1.0	6.3136×10^{-3}	8.7459×10^{-1}	0.5568	1.4976	148.57	254.27
185.7	1.0	5.2636×10^{-3}	9.5959×10^{-1}	0.6177	1.6224	139.28	235.16
222.9	1.0	4.6136×10^{-3}	9.9577×10^{-1}	0.6756	1.7568	130.00	228.83
260.0	1.0	4.1650×10^{-3}	1.0	0.7329	1.9056	120.71	230.03
297.1	1.0	3.7395×10^{-3}	1.0	0.7905	2.0723	111.43	230.90
334.3	1.0	1.6644×10^{-3}	1.0	0.8483	2.2568	102.14	230.51
371.4	1.0	2.9335×10^{-3}	1.0	0.9057	2.4587	92.86	228.30
408.6	1.0	2.5543×10^{-3}	1.0	0.9622	2.6764	83.57	223.67
445.7	1.0	2.1921×10^{-3}	1.0	1.0173	2.9068	74.28	215.93
482.8	1.0	1.8476×10^{-3}	1.0	1.0699	3.1448	64.99	204.41
520.0	1.0	1.5216×10^{-3}	1.0	1.1191	3.3836	55.71	188.51
557.1	1.0	1.2152×10^{-3}	1.0	1.1637	3.6142	46.43	167.80
594.3	1.0	9.2894×10^{-4}	1.0	1.2026	3.8257	37.14	142.10
631.4	1.0	6.6375×10^{-4}	1.0	1.2344	4.0068	27.86	111.61
668.6	1.0	4.2023×10^{-4}	1.0	1.2580	4.1460	18.57	76.99
705.7	1.0	1.9888×10^{-4}	1.0	1.2726	4.2338	9.29	39.31
742.8	1.0	0.0	1.0	1.2775	4.2639	0.0	0.00

Note: 1 in. = 0.0254 m, 1 in. kip = 112.9848 N m

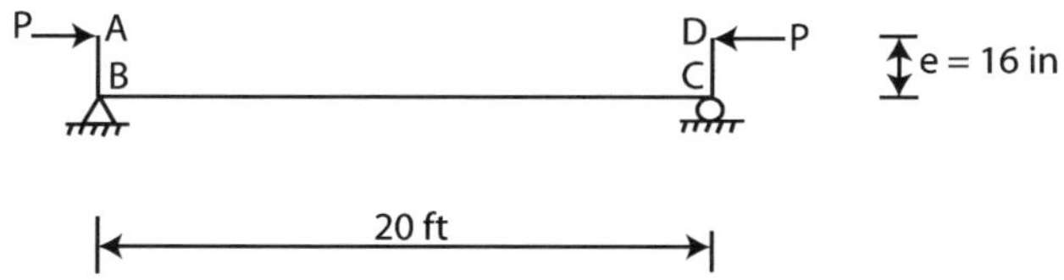

Fig. P(6.1). (1 in. = 0.0254 m, 1 ft = 0.3048 m)

Problems

6.1 The steel frame ABCD in Fig. P(6.1) has a square cross section of sides equal to 6 in. (0.1524 m), and it is acted upon by a load P as shown in the same figure. Determine the value of the load P so that the maximum deflection of member BC does not exceed 1.5 in. (0.0381 m). The modulus of elasticity $E = 30 \times 10^6$ psi (206.85 kPa).

6.2 Repeat Problem 6.1 by assuming that the eccentricity e = 24 in. (0.6096 m), and compare the results.

6.3 The tapered simply supported beam in Fig. 6.5c, is subjected to a uniformly distributed load w = 2.0 kip in.$^{-1}$(350, 254 N m^{-1}). In addition, the beam is acted upon by an axial compressive load $P = 0.1\,P_{cr}$, where $P_{cr} = 386.112$ kip (17,177 kN). At the right support the depth h = 8.0 in. (0.2032 m), and it is 1.5h = 12.0 in. (0.3048 m) at the left support. The width b is constant and equal to 6.0 in. (0.1524 m), and the material of the member is monel. By using the three-line approximation of the stress–strain curve of monel, reproduce the quantities shown in Table 6.4 and compare the results.

6.4 By using the values of M_e obtained in Problem 6.3, determine an equivalent system of constant stiffness EI, where E and I are, respectively, the elastic modulus and moment of inertia of the member at the right support.

6.5 By using the equivalent system determined in Problem 6.4 and applying the moment–area method, determine the vertical deflection at midspan and the rotations at the end supports. Compare the results with the corresponding values shown in Table 6.3.

6.6 Repeat Problem 6.3 by applying the bilinear approximation of the stress–strain curve of monel and compare the results.

6.7 Repeat Problem 6.3 by assuming that $P/P_{cr} = 0.2$ and compare the results.

6.8 By following the procedure for inelastic analysis of plates discussed in Sect. 6.8, reproduce the results for ε_x, ε_y, M_{xreq}, M_{yreq}, D_{rx}, and D_{ry} shown in Table 6.5. Discuss the results.

6.9 By following the methodology for inelastic analysis discussed in Sect. 6.8, verify the results for the vertical deflections and rotations included in Table 6.6.

6.10 By following the methodology for inelastic analysis discussed in Sect. 6.8, reproduce the results for ε_x, ε_y, M_{xreq}, M_{yreq}, D_{rx}, and D_{ry} shown in Table 6.5 by using the six-line approximation of the stress–strain curve shown in Fig. 6.14c, and compare the results.

6.11 Repeat Problem 6.8 by assuming that b/a = 1.5.

6.12 Repeat Problem 6.9 by assuming that b/a = 1.5.

6.13 The fixed-fixed wide-flange steel beam in Fig. P6.13a supports a heavy operating machinery of weight W = 5 kip (22,240 N) located as shown in the figure. During the operation, the machine transmit to the member a dynamic force F(t) that varies as shown in Fig. P6.13b. By using a wide-flange beam cross section, design the member so that its vertical deflection due to F(t) under the machine does not exceed 4 mils (0.0001 m). The idealized one-degree spring-mass system for the beam is shown in Fig. 6.13c. The modulus of elasticity $E = 30 \times 10^6$ psi $(206,850 \times 10^6$ Pa). Use the AIEM in the analysis and neglect the weight of the fixed-fixed beam. The idealized spring-mass system in Fig. P6.13c represents the stiffness at the center C of the beam in Fig. P6.13a having a W24 × 68 wide-flange section. The size of the section can be changed, if deflection criteria are not satisfied.

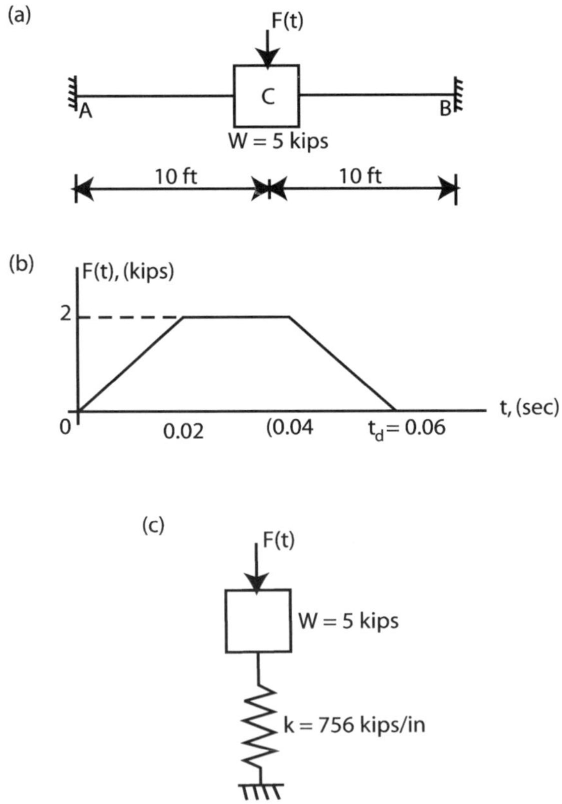

Fig. P6.13. (**a**) Fixed–fixed beam supporting an operating machinery. (**b**) Time variation of F(t). (**c**) Idealized one-degree spring-mass system of the beam based on a W24 × 68 wide-flange steel section (1 kip = 4,448 N, 1 ft = 0.3048 m)

6.14 Repeat problem 6.13 by assuming that the deflection of the member under the machine should be between 10 and 15 mils.

6.15 Repeat the inelastic analysis in Example 6.5 for a ductility ratio η somewhere between 10 and 15 and compare the results.

6.16 Rework the inelastic analysis in Example 6.5 for the ductility ratios $\eta \leq 1$, $1 < \eta \leq 1.5$, and $6 \leq \eta \leq 8$, and compare the results.

6.17 Rework the problem in Example 6.4 and determine the ductility ratio η when both ends of the member are fixed and when F(t) in Fig. 6.21b takes the values of 50, 60, and 70 kip and compare the results.

6.18 Rework the thick cylinder problem in Example 6.6 by assuming that $p_i = 20,000$ psi and $p_0 = 10,000$ psi and compare the results.

6.19 Rework the thick cylinder problem in Example 6.6 by assuming that $p_i = 0$ and $p_0 = 30,000$ psi and compare the results.

6.20 Rework the thick cylinder problem in Example 6.6 by assuming that $p_0 = 0$ and for values of $p_i = 10,000$, 15,000, 20,000, and 25,000 psi and compare the results.

6.21 Rework the thick-walled cylinder problem in Example 6.7 by assuming that $a = 6$ in., $b = 12$ in., and $c = 9$ in., and compare the results.

6.22 Rework Problem 6.21 by assuming that $c = 8$ in. and $c = 10$ in., and compare the results.

6.23 Repeat Example 6.8 by assuming that the weight w acting on the member is equal to 1,550 lb in.$^{-1}$(271,452 N m^{-1}) and compare the results.

6.24 Repeat Example 6.8 by assuming that the weight $w = 1,550$ lb in.$^{-1}$(271, 452 N m^{-1}) and taper $n = 1$. Compare the results.

6.25 Repeat Example 6.9 by assuming that the weight $w = 2,800$ lb in.$^{-1}$ (490, 364 N m^{-1}) and compare the results.

6.26 Having determined the variations of both E_r and I_x along the length of the original member in Fig. 6.30a of Example 6.8, use the original system in Fig. 6.30a and determine the first three frequencies of its free vibration using the finite element method. Compare the results with the ones obtained in Example 6.8.

6.27 Repeat Example 6.10 by assuming that the weight $w_1 = 4.0$ lb in.$^{-1}$ (700.52 N m^{-1}) and compare the results with the ones obtained in Table 6.11.

6.28 Repeat Example 6.10 by assuming that the member is uniform throughout its length with thickness $h = 3.0$ in. (0.0762 m) and compare the results.

Appendix A

Acceleration Impulse Extrapolation Method (AIEM)

In many practical cases, such as forces produced by conventional and nuclear explosions, for example, the force function F(t) is rather arbitrary, and numerical methods of analysis are used to obtain dynamic responses.

An interesting numerical method, known as the acceleration impulse extrapolation method (AIEM), was developed at the Massachusetts Institute of Technology (MIT) during the late 1950s, in order to carry out the dynamic analysis that was required for a successful moon landing [5, 15, 103]. Simple dynamic principles are used to formulate this method, but a very accurate and simple solution may be obtained for very complicated dynamics problems. In this text the AIEM is used to solve idealized one-degree spring-mass systems. The method, however, is simple and general, and it can be used to solve much more complicated problems, as discussed in Sect. 6.9 of this text and other various references, see, for example, [5, 15, 84, 103].

The general discussion is initiated by considering an idealized one-degree spring-mass system, which is under the influence of viscous damping of damping coefficient c. The differential equation of its motion may be derived in the usual way, and it is as follows:

$$\mathrm{m}\ddot{\mathrm{x}} + \mathrm{c}\dot{\mathrm{x}} + \mathrm{kx} = \mathrm{F}(\mathrm{t}) \qquad \text{(A.1)}$$

Solving for the acceleration $\ddot{x}$, we obtain

$$\ddot{\mathrm{x}} = \frac{\mathrm{F}(\mathrm{t}) - \mathrm{c}\dot{\mathrm{x}} - \mathrm{kx}}{\mathrm{m}} \qquad \text{(A.2)}$$

In the AIEM, the acceleration curve in Fig. A.1 is replaced by a series of equally spaced impulses that occur at times $t_0, t_1, \ldots, t_{i-1}, t_i, t_{i+1}, \ldots t_n$. The time interval of each impulse is Δt, and at a time t_i, the magnitude of the acceleration impulse is $\ddot{x}_i(\Delta t)$. Between time stations the velocity $\dot{x}$ is assumed constant and the displacement x is linear.

The solution of Eq. (A.1) involves a step-by-step procedure that starts at time $t = 0$, where the displacement and velocity are know. From there, the displacement is extrapolated from one time station to the next.

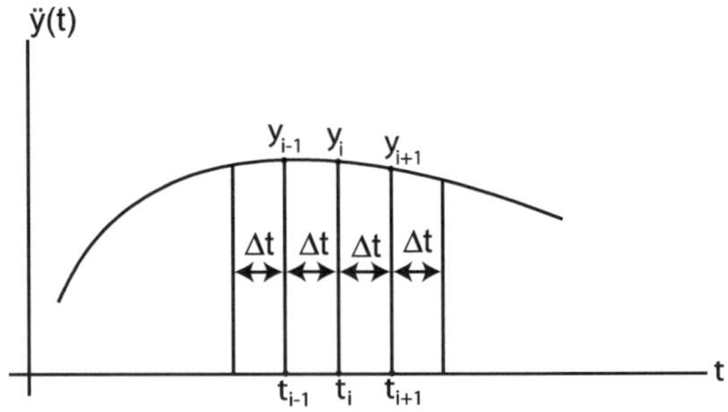

Fig. A.1. Acceleration curve divided into a series of equally spaced impulses

If the displacements $x^{(i-1)}$ and $x^{(i)}$ at time stations $i-1$ and i, respectively, are known, we can determine the acceleration $\ddot{x}^{(i)}$ at station i by using Eq. (A.2). We note, however, that the acceleration $\ddot{x}^{(i)}$ at station i depends on the velocity $\dot{x}^{(i)}$, which is not known. The original version of the method suggests that a reasonable approximate expression for the velocity $\dot{x}^{(i)}$ at station i is the following:

$$\dot{x}^{(i)} = \frac{x^{(i)} - x^{(i-1)}}{\Delta t} + \ddot{x}^{(i)} \left(\frac{\Delta t}{2} \right) \tag{A.3}$$

The second term on the right-hand side of Eq. (A.3) gives an estimate of the amount by which the average velocity of the preceding interval should be increased, in order to obtain the velocity at the next station.

By substituting Eq. (A.3) into Eq. (A.2) and simplifying, we obtain

$$\ddot{x}^{(i)} = \frac{2F(t)}{2m + c(\Delta t)} - \frac{2c\left[x^{(i)} - x^{(i-1)}\right]}{(\Delta t)\left[2m + c(\Delta t)\right]} - \frac{2kx^{(i)}}{2m + c(\Delta t)} \tag{A.4}$$

Equation (A.4) combines Eqs. (A.2) and (A.3), and it may be used here to determine the acceleration $\ddot{x}^{(i)}$ of the idealized one-degree spring-mass system at any station i. If the damping is zero, Eq. (A.4) yields

$$\ddot{x}^{(i)} = \frac{F(t) - kx^{(i)}}{m} \tag{A.5}$$

as it should be expected.

Between time stations i and $i+1$, the average velocity $\dot{x}_{av}$ may be written as follows:

$$\dot{x}_{av} = \frac{x^{(i)} - x^{(i-1)}}{\Delta t} + \ddot{x}^{(i)} (\Delta t) \tag{A.6}$$

Equation (A.6) is also a reasonable expression to use in place of Eq. (A.3) if consistency is desired. If we substitute Eq. (A.6) into Eq. (A.2) and rearrange, we obtain

$$\ddot{x}^{(i)} = \frac{F(t)}{m + c(\Delta t)} - \frac{c\left[x^{(i)} - x^{(i-1)}\right]}{(\Delta t)\left[m + c(\Delta t)\right]} - \frac{kx^{(i)}}{m + c(\Delta t)} \tag{A.7}$$

Equation (A.7) may be also used to determine the acceleration $\ddot{x}^{(i)}$ at any station i.

At any time station i + 1, the displacement $x^{(i+1)}$ may be determined from the equation

$$x^{(i+1)} = x^{(i)} + \dot{x}_{av}\,(\Delta t) \tag{A.8}$$

By substituting Eq. (A.6) into Eq. (A.8), we obtain

$$x^{(i+1)} = 2x^{(i)} - x^{(i-1)} + \ddot{x}^{(i)}\,(\Delta t)^2 \tag{A.9}$$

Equation (A.9) may be used to determine the displacement $x^{(i+1)}$ at time station i + 1, when the preceding two displacements $x^{(i)}$ and $x^{(i-1)}$ at time stations i and i − 1, respectively, are known. Equation (A.4), or Eq. (A.7) may be used to determine the acceleration $\ddot{x}^{(i)}$ shown in Eq. (A.9).

Equation (A.9), however, cannot be used for the evaluation of the displacement $x^{(1)}$ at station 1, because at time t = 0, the displacement $x^{(0)}$ exists, but the displacement $x^{(0-1)}$ at time station 0 − 1 does not exist. We can eliminate this difficulty if for the first time station the displacement $x^{(1)}$ is computed by using one of the following two equations:

$$x^{(1)} = \frac{1}{6}\left[2\ddot{x}^{(0)} + \ddot{x}^{(1)}\right](\Delta t)^2 \tag{A.10}$$

$$x^{(1)} = \frac{\ddot{x}^{(0)}}{2}\,(\Delta t)^2 \tag{A.11}$$

In Eq. (A.10) the acceleration during the first time interval is assumed to vary linearly, and in Eq. (A.11) the acceleration during the first time interval is assumed to be constant and equal to the initial value. Either one of these two equations is reasonable to use, but if the force F(t) is zero at t = 0, then Eq. (A.10) must be used because $\ddot{x}^{(0)} = 0$, and Eq. (A.11) does not apply.

For a great deal of information regarding the application of the AIEM to many types of structural problems, consult [5, 15, 84, 103].

Appendix B

Computer Program Using the AIEM for the Elastoplastic Analysis in Example 6.5

This short computer program was used to carry out the computations in Example 6.5 of Sect. 6.9.2 of Chap. 6, which involves the elastoplastic analysis of a two-story building subjected to the earthquake ground displacements shown in Fig. 6.25.

```
 10      PRINT
 20 100 FORMAT( 'INPUT IX, SX, AND ZX FOR CHOSEN COLUMN')
 30      READ *, XI, SX, ZX
 40      EM1=0.22526
 50      EM2=0.14451
 60      EK1=0.07144*XI
 70      EK2=27.*EK1/8.
 71      EKM=(EK1+EK2)/(EM1)+EK2/EM2
 72      EKK=4.*EK1*EK2/(EM1*EM2)
 73      W12=0.50*EKM+0.50*SQRT(EKM**2-EKK)
 74      W22=0.50*EKM-0.50*SQRT(EKM**2-EKK)
 75      TAU1=2.*3.141593/SQRT(W12)
 76      TAU2=2.*3.141593/SQRT(W22)
 80      Y1EL=9.331*SX/XI
 90      Y2EL=4.*Y1EL/9
100      R1M=0.3333*ZX
110      R2M=0.5000*ZX
120      PRINT 110,EK1,EK2,Y1EL,Y2EL,R1M,R2M
130 110 FORMAT('K1=',F7.3, 'K2=',F7.3, 'Y1EL=',F7.4, 'Y2EL=',F7.4,
140      $' R1M=', F7.3,' R2M=',F7.3)
141      PRINT 115, TAU1, TAU2
142 115 FORMAT ('TAU1=',F7.3,' TAU2==',F7.3)
150      J=0
160      XO1=0.0
170      XO2=0.0
180      PRINT 120
190 120 FORMAT ('ENTER TIME INCREMENT TO BE USED')
```

```
    READ *,DT
    DT2=DT**2
    PRINT 130
130 FORMAT ( ‘ENTER 1ST GROUND DISPLACEMENT’)
 10 READ *,U
    IF(J.GT.1) GO TO 50
    CN=EK2*DT2/(6.*EM2)
    C1A=CN/(1.+CN)
    EK1M=EK1*DT2/(6.*EM1)
    EK2M=EK2*DT2/(6.*EM1)
    X1=K1M*U/(1.+EK2M*(1.-C1A)+EK1M)
    X2=C1A*X1
 50 X1U=1-U
    R1=X1U*EK1
    IF(ABS)R1).LE.R1M) GO TO 80
    IF(R1.LT.0.0) R1 = -R1M
    IF(R1.GT.R1M) R1 = -R1M
 80 X21 =X2-X1
    R2=X21*EK2
    IF(ABS(R2).LE.R2M) GO TO 90
    IF(R2.LT.0.0)R2=-R2M
    IF(R2.GT.R2M) R2 = R2M
 90 X1A=(R2-R1)/EM1
    X1AT2=X1A*DT2
    X2A=-R2/EM2
    X2AT2=X2A*DT2
    ETA1=ABS(X1/Y1EL)
    ETA2=ABS(X21/Y2EL)
    PRINT 150, X1U, R1, X21, R2, X1A, X1AT2, X1,
    ETA1, X2A, X2AT2, X2, ETA2
150 FORMAT( ‘X1-U=’,F8.4, ‘R1=’,F8.2, ‘X2-X1=’,F8.4,‘ R2=,’
    $F8.2/’ X1A=’,F8.2,‘ X1AT2=’,F8.3,‘ X1=’,F8.4,‘ETA1=’,F5.2/
    $‘ X2A=’,F8.2,‘ X2AT2=, ‘ F8.3,‘ X2=’,F8.4,‘ETA2=’,F5.2)
    CX1=X1
    CX2=X2
    X1=2.*X1-XO1+X1AT2
    X2=2.*X2-XO2+X2AT2
    XO1=OX1
    XO2=OX2
    J=2
    GO TO 10
    END
```

References

1. Fertis DG, Afonta AO (1991) Equivalent systems for large deformation of beams of any stiffness variation. *Eur. J. Mech. A/Solids* 10(3), p. 265
2. Fertis DG (1993) Nonlinear Mechanics. First edn, CRC, Boca Raton
3. Fertis DG (1999) Nonlinear Mechanics. Second edn, CRC, Boca Raton
4. Fertis DG, Lee CT (1991) Inelastic analysis of flexible bars using simplified nonlinear equivalent systems. *Int. J. Comput. Struct.* (41)5, p. 947
5. Fertis DG (1995) Mechanical and Structural Vibrations. Wiley, New York
6. Fertis DG (1996) Advanced Mechanics of Structures. Dekker, New York
7. Euler L (1744) Methodus Inveniendi Lineas Curvas
8. Timoshenko SP (1953) History of Strength of Materials. McGraw-Hill, New York
9. Love AEH (1944) A Treatise on the Mathematical Theory of Elasticity. Fourth edn, Dover, New York
10. Todhunter I, Pearson K (1960) A History of the Theory of Elasticity and of the Strength of Materials. Vols. I and II, Dover, New York
11. Lagrange JL (1971) Sur la Force des Ressorts Plies, Mem. Acad. Berlin. 25
12. Dravas GA, McLeau L (August 1966) Historical development of energetical principles in elastomechanics. *Appl. Mech. Rev.*, Part I, 9(8), p. 647, and Part II 19(11), p. 919
13. Plana GAA (1809–1810) Equation de la Courbe Formée Par Une Lame Elastique. Mem. R. Sec. Turin Vol. 18, p. 123
14. Born M (1906) Untersuchungen über der stabilität der elastichen Linie in Ebene und Raum unter venschiedenen Grenzbedingungen. Dissertation, Göttingen
15. Fertis DG (1973) Dynamics and Vibration of Structures. Wiley, New York (This work was republished in English in Taiwan by permission and translated into Chinese and published in Taiwan by permission)
16. Fertis DG (1984) Dynamics and Vibration of Structures. Revised edn, Robert E. Krieger, Malabar, FL
17. Fertis DG (1971) Dynamics of Structural Systems. Vol. I, The University of Akron
18. Fertis DG, Zobel EC (1961) Transverse Vibration Theory. Ronald, New York
19. Fertis DG (1972) Dynamics of Structural Systems. Vol. II, The University of Akron

20. Fertis DG (1964) Deflection and Vibration of Beam-like Structures of Variable and Uniform Stiffness. National Technical University, Athens, Greece
21. Fertis DG, Zobel EC (1958) Equivalent systems for the deflection of variable stiffness members. *J. Struct. Div., Proc. ASCE*, Pap. 1820, 1–13
22. Fertis DG (1966) Dynamic hinge concept for beam vibrations. *J. Struct. Div., Proc. ASCE*, Pap. 4689, pp. 365–380
23. Kosma A, Fertis DG (1962) Solution of the deflection of variable thickness plates by the method of equivalent systems. In: Industrial Mathematics. Vol. 12, Part I, The Industrial Mathematics Society, Detroit, MI, pp. 1–11
24. Fertis DG, Cuningham H (1962) Equivalent systems for shear deflections of variable thickness members. In: Industrial Mathematics. Vol. 12, Part I, The Industrial Mathematics Society, Detroit, MI, pp. 13–29
25. Fertis DG (1995) Vibration of beams and frames by using dynamically Equivalent systems. *J. Struct. Eng. Rev.*, Special Issue SER 65
26. Fertis DG (1964) Contribution to the deflection and free vibration of uniform and variable stiffness members. Doctoral Dissertation, The National Technical University of Athens, Greece
27. Fertis DG, Chaarani W (28 May–2 June, 1989) Vibration of variable stiffness members by using equivalent systems. *Proc. 12th Canadian Cong. Appl. Mech.*, Ontario, Canada
28. Fertis DG (September 1989) Vibration of beams and frames using the dynamic hinge concept. *Proc. 26th Annu. Meet. Soc. Eng. Sci.*, Ann Arbor, MI
29. Fertis DG, Chidiac GE (1989) Dynamically equivalent systems for beams and frames. In: *Proc. 21st Midwestern Mech. Conf.*, Vol. 15, p. 390
30. Fertis DG (1994) Vibration of beams and frames by using dynamically equivalent systems. In: *Proc. Struct. Dyn. Vib. Symp.* Energy Sources Technology Conference and Exhibition, ASME, PD-Vol. 63, p. 53
31. Fertis DG, Pallaki S (1989) Pseudolinear equivalent systems for large deflections of members. *J. Eng. Mech.*, ASCE 115(11), p. 2440
32. Fertis DG, Keene ME (1990) Elastic and inelastic analysis of nonprismatic members. *J. Struct. Eng.*, ASCE 116(2), p. 475
33. Fertis DG, Afonta AO (1990) Large deflection of determinate and indeterminate bars of variable stiffness. *J. Eng. Mech.*, ASCE 116(7), p. 1543
34. Fertis DG, Taneja R (1990) Inelastic analysis of variable stiffness members. *Math. Comput. Model.* 14, p. 942
35. Fertis DG, Taneja R (1991) Equivalent systems for inelastic analysis of prismatic and nonprismatic members. *J. Struct. Eng.*, ASCE 117(2), p. 473
36. Fertis DG, Lee CT (1991) Inelastic analysis of prismatic and nonprismatic members with axial restraints. *Int. J. Mech. Strut. Machines* 19(3), p. 357
37. Fertis DG, Taneja R, Lee CT (1991) Equivalent systems for inelastic analysis of nonprismatic members. *Int. J. Comput. Struct.* 38(1), p. 31
38. Fertis DG, Lee CT (1992) Inelastic response of variable stiffness members under cyclic loading, *J. Eng. Mech.*, ASCE 118(10), p. 1406
39. Fertis DG, Afonta AO (1992) Free vibration of variable stiffness flexible bars. *Int. J. Comput. Struct.* 43(3), p. 445
40. Fertis DG, Afonta AO (1993) Small vibrations of flexible bars by using the finite element method with equivalent uniform stiffness and mass methodology. *Int. J. Sound and Vib.* 163(2), p. 343
41. Fertis DG, Lee CT (1993) Nonlinear vibration and instabilities of elastically supported beams with axial restraints. *Int. J. Eng. Sound Vib.* 165(1), p. 123

42. Fertis DG, Lee CT (1993) Elastic and inelastic analysis of variable thickness plates by using equivalent systems. *Int. J. Eng. Mech. Struct. Mech.* 21(2), p. 201
43. Fertis DG, Schubert FR (1994) Inelastic analysis of prismatic and nonprismatic aluminum members. *Int. J. Comput. Struct.* 52(2), p. 287
44. Fertis DG, Lee CT (November 1990) Equivalent systems for inelastic analysis of nonprismatic members with axial restraints. In: *Proc. F O* Conf., IITT-International, Los Angeles, CA, p. 207
45. Fertis DG, Lee CT (1992) Inelastic analysis of variable thickness rectangular plates. In: *Proc. Euro. Jt. Conf. Eng. Syst., Design and Analysis*, ASME, PD-Vol. 47(5), p. 61, Structural Dynamics and Vibrations, Book No. G068SE
46. Fertis DG, Lee CT (1993) Inelastic analysis of nonprismatic flexible bars with axial restraints. In: *Proc. 16th Annu. Energy-Sources Technol.* Conf. and Exhibition, Structural Dynamics and Vibrations, ASME, PD-Vol. 52, p. 127
47. Fertis DG, Lee CT (1991) Equivalent systems for inelastic analysis of members subjected to large deformations. In: *Proc. Plasticity '91 Symp. Int. Symp. Plast.*, Grenoble, France
48. Fertis DG, Lee CT (1992) Nonlinear vibration of axially restrained elastically supported beams. In: *Proc. Workshop on Dynamics*, ASME, PD-Vol. 44, Dynamic and Vibration, Book No. GO0652
49. Fertis DG (4–7 July 1994) Inelastic vibration of prismatic and nonprismatic members. In: *Proc. 2nd Eur. Jt. Conf. Eng. Syst. Design Anal.*, London, England
50. Fertis DG, Lu Q (1–4 July 1996) Inelastic analysis of members of different material and various cross-sectional shapes. In: *Proc. ESDA 96 Eng. Syst. Design Anal. Nonlinear Syst. Symp.*, Montpelier, France
51. Fertis DG (11–15 August 1997) Free vib. of flexible members. *Int. Symp. Vib. Continuous Syst.*, Estes Park, Colorado
52. Bisshoppe KE, Drucker DC (1945) Large deflection of cantilever beams. *Q. Appl. Math.* 3, p. 272
53. Lau JH (February 1981) Large deflection of cantilever beams. *J. Eng. Mech.*, ASME 107(EM1), p. 259
54. Seide P (September 1984) Large deflections of simply supported beams subjected to moment at one end. *J. Appl. Mech.*, Trans. ASME 51(3), p. 519
55. Goto Y, et al. (1990) Elliptic integral solutions of plane elastica with axial and shear deformations. *Int. J. Solids Struct.* 26(4), p. 375
56. Denman HH, Schmidt R (1968) *Ind. Math.* 18(2), p. 63
57. Wang TM, Lee SL, Zienkiewicz DC (1961) A numerical analysis of large deflections of beams. *J. Mech. Sci.* 3, p. 219
58. Ohtsuki A (July 1986) An analysis of large deflections in a symmetrical three point bending of beams. *Bull. JSME* 29, p. 253
59. Rao BN, Rao GV (1988) Large deflection of a uniform cantilever beam with end rotational load. *Forsh. Ingenieurwes* 54(1)
60. Sundara Raja Iyengar KT, et al. (1955) Large deflection of simply supported beams. *Inst. J.* 259, p. 523
61. Lewis G, Monasa F (1981) Large deflection of cantilever beams of nonlinear material, Comput. Struct. 14(5–6), p. 357
62. Truesdall C (April 1953) *Proc. 1st Midwestern Conf. Solid Mech.* University of Illinois, p. 52

63. Frisch-Fay R (1962) Flexible Bars. Butterworths, Washington, DC
64. Boyd JE (1924) Strength of Materials. Third edn, McGraw-Hill, New York, Appendix D
65. Barton HJ (1944, 1945) On the deflection of a cantilever beam. *Q. Appl. Mech.* 2, p. 168; 3, p. 275
66. Hammel FH, Morton WB (1927) On the large deflection of thin flexible strips and the measurement of their elasticity. Phil. Mag. Ser. 7(4), p. 348
67. Seames AE, Conway HD (1957) A numerical procedure for calculating the large deflection of straight and curved beams. *J. Appl. Mech.* 24, p. 289
68. Leibold R (1948) Die Durchbiegung Einer Beidseiting Fest Eingenspaunten Blatteder, *Z. Angew. Math. Mech.* 28, p. 247
69. Parnes R (1971) Bending of an axially constrained beam. *Int. J. Mech. Sci.* 13, p. 285
70. Hsiao KM, Hou FY (1987) Nonlinear finite element analysis of elastic frames. *Comput. Struct.* 26(4), p. 693
71. Tada Y, Lee G (1970) Finite element solution to an elastic problem of beams. *Int. J. Numer. Meth. Eng.* 2, p. 229
72. Yang TY (1973) Matrix displacement solution of elastica problems of beams and frames. *Int. J. Solids Struct.* 9, p. 829
73. Chajes A (June 1987) Nonlinear frame analysis by finite element methods. *J. Struct. Eng.*, ASME 113(6)
74. Wood RD, Zienkiewicz OC (1977) Geometrically nonlinear finite element analysis of beams, frames, arches, and axisymmetric shells. *Compt. Struct.* 7, p. 725.
75. Wionowsky-Krieger S (1950) The effect of an axial force on the vibration of hinged bars. *J. Appl. Mech.* 17, 35
76. Prathap G (1978) Nonlinear vibration of beams with variable axial restraints. *AIAA J.* 16(6), p. 622
77. Prathap G, Varadan TK (May 1978) Large amplitude vibration of tapered clamped beams. *J. Sound Vib.* 58(1), p. 87
78. Mei C, Decha-Umphai KA (1985) A finite element method for nonlinear forced vibrations of beams. *J. Sound Vib.* p. 369
79. Mei C (March 1972) Nonlinear vibration of beams by matrix displacement method. *AIAA J.* 10(3), p 355.
80. Evensen DA (February 1968) Nonlinear vibrations of beam with various boundary conditions. *AIAA J.* 6(2), p. 370
81. Prathap G, Varadan TK (1976) The inelastic large deformation of beams. *J. Appl. Mech.*, ASME, 43, p. 689
82. Lo CC, Gupta SD (1978) Bending of a nonlinear rectangular beam in large deflection. *J. Appl. Mech.*, ASME, 45, p. 213
83. Monasa F (1979) The effect of material nonlinearity on the bending of the elastica. In: *Proc. 3rd Eng. Mech. Specially Conf.* ASCE, p. 638
84. Fertis DG (1997) Infrastructure Systems. Wiley Series in Infrastructure Management and Design, Wiley, New York
85. Gire JM, Timoshenko SP (1990) Mechanics of Materials. Third edn, PWS-Kent, Boston, p. 514
86. Bleich HH (1955) Response of elastic–plastic structures to transient loads. In: *Proc. NY Acad. Sci.* Vol. 8, (2nd series), p. 135
87. Bleich HH, Salvadory MG (1955) Impulsive motion of elastic–plastic beams. *Trans. ASCE* 120, p. 499

88. Baron ML, Bleich HH, Widlinger P (1961) Dynamic elastic–plastic analysis of structures. *J. Eng. Mech.*, ASCE 89, p. 23
89. Berg GV, da Deppo DA (1960) Dynamic analysis of elastic–plastic structures. *J. Eng. Mech.*, ASCE 86, p. 35
90. Lee EH, Symonds PS (1952) Large plastic deformations of beams under transverse impact. *J. Appl. Mech.*, ASCE, 19, p. 308
91. Toridis T, Wen RK (1966) Inelastic response of beams to moving loads. *J. Eng. Mech.*, ASCE 92, p. 43
92. Bonc R (1967) Forced vibration of hysteretic systems. In: *Proc. 4th Conf. Nonlinear Oscillation*, Prague, Czechoslovakia
93. Wen RK (1975) Lumped elasto-plastic flexibility models. *J. Eng. Mech.*, ASCE 91, p. 289
94. Iyender RN, Dash RK (1976) Study of the random vibration of nonlinear systems by the gaussian closure technique. *J. Eng. Mech.*, ASCE 102, p. 249
95. Sues RH, Mau ST, Wen YK (1988) System identification of degrading hysteretic restoric forces. *J. Eng. Mech.*, ASCE 114(5), p. 833
96. Selby SM (1973) Standard Mathematical Tables. 23rd edn, CRC, Boca Raton, FL
97. Timoshenko *S* (1956) Strength of Materials: Part II, Advanced Theory and Problems. Van Nostrand, New York
98. Simiu E, Scanlan RH (1986) Wind Effects on Structures. Wiley, New York
99. Fuller RC, Zollman DA, Campbell TC (1982) The Puzzle of the Tacoma Narrows Bridge Collapse. The American Association of Physics Teachers. Also videodisc, available from Wiley, New York
100. Petroski H (1994) Design Paradigms: Case Histories of Errors and Judgment in Engineering. Cambridge University Press
101. Gere JM, Timoshenko SP (1990) Mechanics of Materials. Third edn, PWS-Kent, Boston, p. 514
102. Teyssandier J-P (October 2002) Corinthian Crossing. Civil Engineering, *ASCE* 72(10), pp. 42–49
103. Fertis DG (17–20 August 1987) Safety on Long Span Highway Bridges Based on Dynamic Response. *Bridges and Transmission Line Structures, Proceedings Structures Congress* 1987/ST Div/ASCE

Index

A

Acceleration impulse extrapolation method, 323-328
  Analysis of beams using the, 327, 328
  Analysis of building using the, 269-285
  Analysis of frames using the, 269-285
Archimedes, 9
Axial restraints, 240-252

B

Beams, vii, 1-224
  cantilever, 20-22, 25-29, 31,
    combined loading, 90-96
    tapered flexible, vii, 20-22, 25-27, 31, 40, 42, 105-119
    uniform flexible, vii, 20-22, 25-29, 33-38, 64-70
  derivation of equivalent systems for, 29-63
  elastically supported, 7, 50-53
  elastically supported variable stiffness, 50-53,
  inelastic analysis of, 143-184, 240-252, 294-296, 307-322
    large deformations, 307-322
    small deformations, 143-172
    tapered candilever, 155-165
    tapered simply supported, 165-171
    ultimate design loads using, 172-184
    uniform cantilever, 143-154
    uniform simply supported, 143-154
  inelastic analysis of flexible, 307-322
    without axial restraints, 307-322
  inelastic vibration of, 299-306
  nonlinear vibration of flexible, 185-228
  simply supported,
    tapered flexible, 121-128, 129-138
    uniform flexible, 71-86
  solution methodologies for uniform flexible, 63-105,
  solution methodologies for variable flexible, 105-142
  statically indeterminate, 53-58, 82-89, 148
    inelastic analysis of, 144, 148
    large deformation of, 82-89
    uniform flexible, 86-90
  statically indeterminate uniform flexible, 82-90,
  statically ideterminate variable stiffness, 53-58,
  stepped variable stiffness, 48-50,
  variable stiffness, 105-133
  variable stiffness flexible, 105-133
  uniform stiffness flexible, 63-104,
Bending stiffness, 1-228, 11, 17, 22, 30
  uniform, 63-104
  variable, 105-142
Blast, vii
Bernoulli, Daniel, 2
Bernoulli, Jacob, 2, 9
Bernoulli, Johanson, 2, 9
Born, Max, 2

Brahistochrone problem, 9
Bridges, 229-236, 253-258
  suspension, 229-236, 253-258
    failure of, 232-236
Brief history, 1-8
Buildings, vii, 269-284
  dynamic response of multistory, 269-284
  earthquake inelastic response of, 269-284

C

Calculus of variations, 9
Centroidal axis, 144-155
Chebyshev approximation method, 5
Columns, 237-239
  eccentrically loaded, 237-239
Concluding remarks, 224-228
Cylinders, 285-293
  elastic analysis of, 285-289
  inelastic analysis of, 290-293
  thick, 285-293

D

Deflections at any x between zero and L_0, 38-40
Deformations, 1-62
  linear, 44-62
  nonlinear, 1-43
Dynamically equivalent system, ix, 185-228, 195, 199, 202, 207, 209, 223, 299-306

E

Earthquake, 269-284
  analysis of structures subjected to an, 269-284
  inelastic analysis of buildings subjected to an, 269-284
Eccentrically loaded columns, 237-239
Elastic axis, 143-154
Elastica, 1, 2, 16-22
  consept of, 16-22
  methods using, 1-7
  theory of, 1-10, 16-22
Elastic limit, vii, 142-184
Elastic similarity principle, 6
Elasticity,
  modulus of, 3
  reduced modulus of, ix, 144-154
Elliptic integrals, 3, 5, 18, 22
  concept of, 1-7
  complete, 3
  first and second kind, 5
  incomplete, 3
Equivalent systems, 3, 4, 5, 7, 29-58
  alternate approach of, 128-143
  derivation of linear, 44-58
  derivation of pseudolinear 3, 30-40,
  derivation of simplified nonlinear, 3, 4, 7, 40-44, 42, 119-121
  inelastic analysis using, 155-184, 307-322
  inelastic vibrations using, 299-306
  large deformations by using, 29-44
  linear theory of, 44-58
  nonlinear theory of, 29-44
  method of the, 28-58
Euler-Bernoulli law of deformation, viii, 8-10, 17, 25, 37
  nonlinear differential equation of, viii, 8-10
    integration of the, 10-13
Euler, 2, 8, 9
  critical load, 237-239

F

Finite difference method, 5, 213, 224
Finite element method, 2, 6, 213-224
  Galerkin's, 213-224
    using the, 213-224
Flexible beams with axial restraints, 7,
Flexible members,
  nonlinear differential equation for, 8-10, 17, 25
  vibration of, 184-228
Functions of horizontal displacement Δ, 24

G

Galerkin, 201, 213-223
  finite element method, 213-223
    vibration of flexible beams using, 213-223

Gauss, 9
Geodesies, 6

H

Historical development in nonlinear analysis, 1-8
  methods used for nonlinear analysis, 1-8
Hysteretic models, 8

I

Inelastic analysis, 143-184, 240-269, 270-284, 290-295, 322
  beams, 143-184
  beams with axial restraints, 240-252
  cylindrical shells, 285-293
  equivalent systems for, 155-172
  flexible members, 307-322
  inelastic torsion, 297-298
  inelastic vibration of members, x, 299-307
  members with non-rectangular cross section, 294-296
  multistory building, 269-284
  plates, 259-269
    variable stiffness, 259-269
Inelastic torsion, 297, 298
Inelastic vibration, 299-306
Inelastic vibration of members, 8, 299-306
Integral nonlinear differential equation, 26, 28
Instability, 229-239
Isochrone problem, 9
Isoperimetric figures, 9

K

Kirchhoff's dynamic analogy, 6

L

Lagrange, 2
Large deformations, 1-142, 185-228, 307-322
  beams subjected to, 1-142, 307-322
  effect of large mass position change during, 211-213
  geometry of, 22-29
  horizontal displacement functions for, 22-29
  moment and stiffness dependence on the geometry of, 22-29
  variable stiffness members subjected to, 105-142
Leibniz, 9
Lessons to learn, 235, 236
Longest cable-stayed suspension bridge in the world, ix, 253-258
Lundwich type materials, 6

M

Maclaurin's series, 4
Method,
  elastica, 1, 2, 16-22
  elliptic integrals, 4
  equivalent systems, 29-58
  finite difference, 5, 213, 224
  finite element, 2, 6, 213
  forth order Runge-Kutta, 3-5, 76, 112, 119, 125
  Galerkin's finite element, 213-220, 202, 224
  moment-area, 35-38
  Newton-Raaphson, 7
  power series, 3, 4
  Raylleigh's energy, 186, 197-211
Minimum weight, 1
Modulus,
  elastic, 3
  evaluation of reduced, 144-154
  function, 3, 11, 17, 22, 30
  reduced, ix, 144-154
Moment and stiffness dependence on the geometry of deformation, 22-29
Moment of inertia function, 3, 11, 17, 22, 30

N

Neutral axis, 240-252
  position change of, 240-252
Newton, 9
Newton-Raphson method, 7
Nonlinear beam deformations, 1-44, 307-322, 155-184, 294-296

Nonlinear differential equations for beams, 3, 9-11, 17, 25, 30
integral, 26, 28
integration of, 10-13
solution of, 1-43
Nonlinear vibration of flexible beams, 185-228

P

Pedagogical aspects, vii-x,
Plana, 2
Plates, 259-268
equivalent systems for, 259-268
inelastic analysis of thin, 259-268
rectangular thin, 259-268
rigidity functions for, 263
simply supported thin, 259-268
statically indeterminate thin, 264, 265
thin, 259-268
uniform thickness, 266-269
variable thickness, 265, 266
Power series method, 3, 4
Principle of elastic similarity, 6
Pseudolinear differential equation, 32
Pseudolinear systems, vii-x, 30-40

R

Rayleigh's energy method, 186, 197-211
application of, 197-204, 204-211
Reduced modulus, ix, 144-155
Resistance R of a structure, 270-276
Rion-Antirion cable-stayed suspension bridge, ix, 253-258
Rotations at any x between zero and L_0, 38-40
Runge-Kutta and Gill method, 3-5

S

Simplified nonlinear systems, 40-44
Simpson's one-third rule, 13, 14-16
Step-by-step procedure, 307-322
Statically indeterminate flexible members, 82-90
Stiffness,
beam, 8-10, 17
bending, 17
plate, 259, 263
torsional, 298
variable, 17, 22, 25, 30
Stress-strain curves, 145, 150, 161, 166, 175, 179
approximation of, 166
straight-line approximation of, 166, 169, 174, 180
Superposition, 10
flexible members, 8-14
nonlinear structures, 8-14
principle of, 10
Suspension bridges, ix, 229-236, 253-258
cable-stayed, ix, 253-258
failures of, ix, 232-236
fundamental aspects of, 229-231

T

Tacoma Narrows Suspension bridge, ix, 232-234
failure of the, 232-234
other failures, 235, 236
Three-dimensional structures, vii, 269-284
Torsion beyond the elastic limit, 297, 298
Trial-and-error procedure, 16-22
application of the, 16-22
Two-dimensional structures, vii, 259-268

U

Ultimate design loads, 172-184
evaluation of, 172-184
inelastic analysis for, 172-184

V

Variable cross-section, 16-62, 105-142, 155-172, 221-224
members of, 16-62, 105-142, 155-172, 193-224, 240-252
flexible members of, 16-62, 105-142, 193-224, 240-252
plates of, 259-269
Variable stiffness members, 16-62, 105-142, 155-172, 204-224, 240-252
Variational methods, 2

Vibration of flexible members, 7, 186-228
  effect of mass position change, 211-213
  equivalent pseudovariable straight beam for, 185-224
  free, 7, 186-228
  mode shapes, 200, 201, 203, 210, 212, 224
  nonlinear differential equation for, 186-193
    derivation of, 186-193, 188, 192

W

Weight control, ix
  Miminum weight criteria, 1

Printed by Printforce, the Netherlands